Reviews

MATH 1

Fifth Edition

bju press
Greenville, South Carolina

Note: The fact that materials produced by other publishers may be referred to in this volume does not constitute an endorsement of the content or theological position of materials produced by such publishers. Any references and ancillary materials are listed as an aid to the student or the teacher and in an attempt to maintain the accepted academic standards of the publishing industry.

MATH 1 Reviews
Fifth Edition

Writers
Charlene McCall
Lindsey Dickinson, MEd
Rita Lovely

Writer Consultant
L. Michelle Rosier

Biblical Worldview
Ben Adams, MEd
Tyler Trometer, MDiv

Academic Integrity
Jeff Heath, EdD

Instructional Design
Rachel Santopietro, MEd
Danny Wright, DMin

Editors
Krystal Allweil
Carol Myers
Abigail Sivyer

Designers
Michael Asire
Kathryn Ratje

Design Assistant
David Ryan Lompe

Production Designers
Sarah Centers
Olivia Long
Lydia Thompson

Illustration, Design & Production
QBS Learning

Permissions
Ruth Bartholomew
Stacy Stone

Project Coordinators
Abby Ray
Kyla J. Smith

Postproduction Liaison
Peggy Hargis

Cover: FatCamera/E+ via Getty Images
All coin images in this textbook: Fat Jackey/Shutterstock.com

The text for this book is set in Acumin Variable Concept by Robert Slimbach, Adobe Minion Pro, Adobe Myriad Pro, Chalkboard, Free 3 of 9 by Matthew Welch, Grenadine MVB by Akemi Aoki, Helvetica, PreCursive, and Report by Ray Larabie.

All trademarks are the registered and unregistered marks of their respective owners. BJU Press is in no way affiliated with these companies. No rights are granted by BJU Press to use such marks, whether by implication, estoppel, or otherwise.

© 2024 BJU Press
Greenville, South Carolina 29609
First Edition © 1978 BJU Press
Fourth Edition © 2015 BJU Press

Printed in the United States of America
All rights reserved

ISBN 978-1-64626-399-8

15 14 13 12 11 10 9 8 7 6 5 4 3

To the Teacher

The MATH 1 Reviews book provides review activities for the students. There are two pages for each lesson. The first page includes the concepts taught in the lesson, using a format similar to the corresponding Worktext pages. The second page is divided into two parts: review of concepts from an earlier chapter and a fact review.

The Chapter Review pages correspond with the Worktext Chapter Review pages. These pages make an excellent study guide for the Chapter Test. The first page of the Cumulative Review reviews concepts from a previous chapter. The second page is a fact review.

The answer key for the MATH 1 Reviews book can be found in the Teacher Edition on the third and fourth pages of each lesson.

Chapter 1 Review

Write the number in each set.
Circle the set with more.

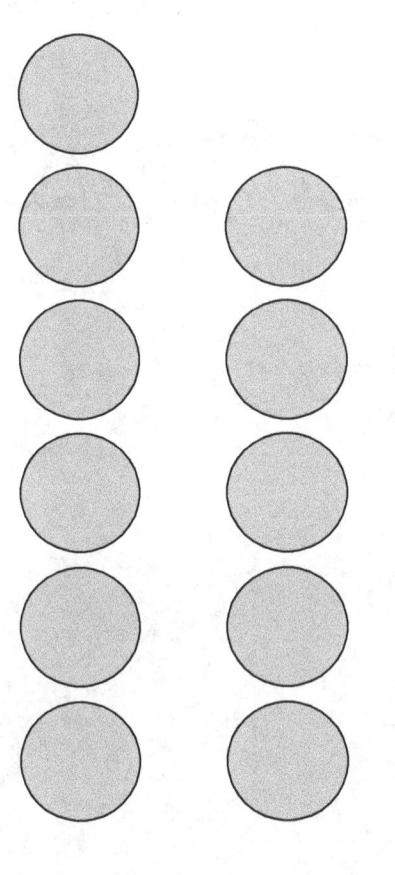

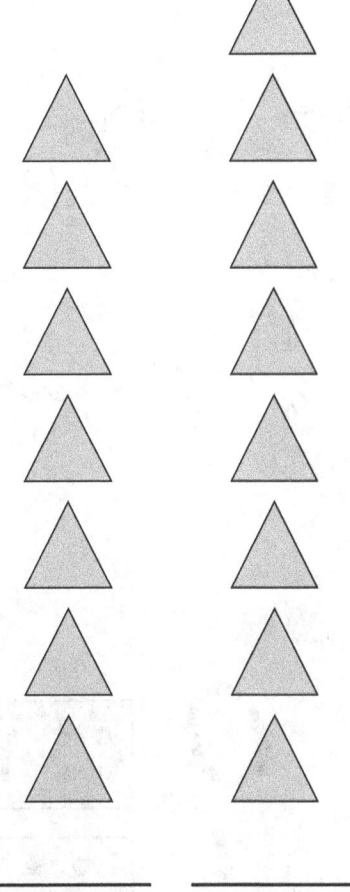

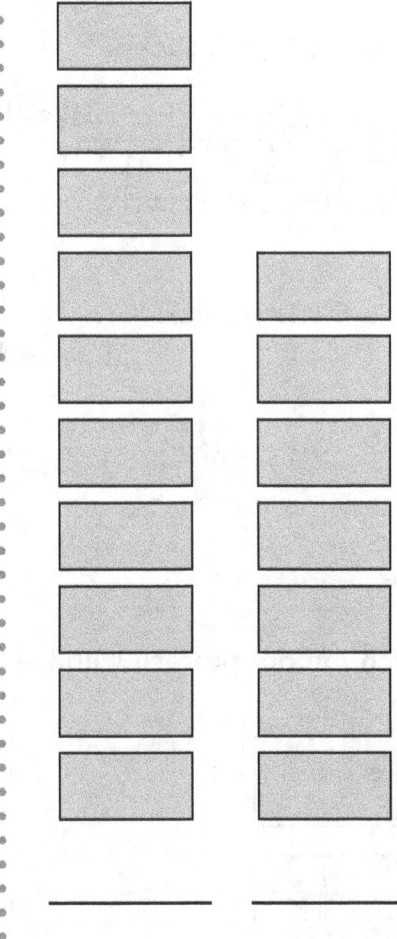

Write the numbers 0–10.

Chapter 1 Review

nine 9

Circle the second horse.

Circle the fourth sheep.

Circle the first duck.

Match the dot pattern with the number.

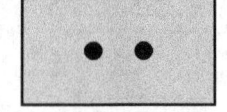

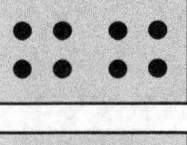

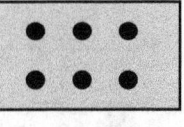

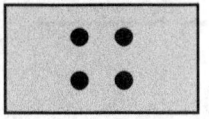

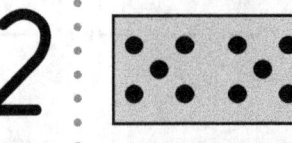

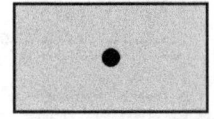

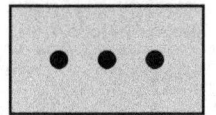

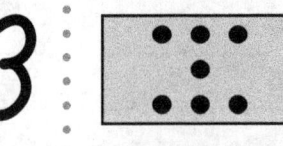

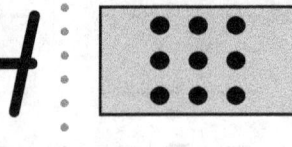

0
1
2
3
4
5
6
7
8
9
10

10 ten

Math 1 Reviews

Kindergarten Review

Circle the shape that completes each pattern.

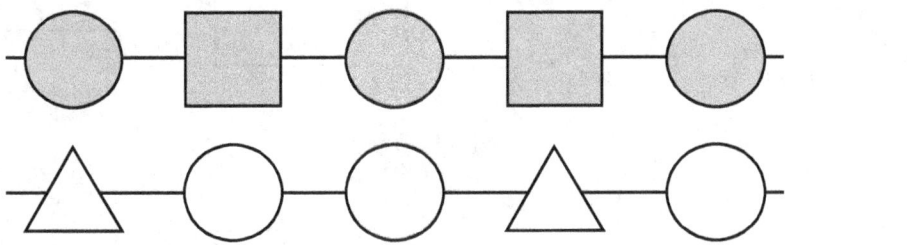

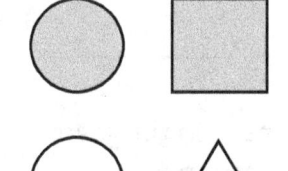

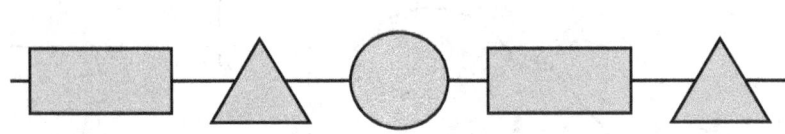

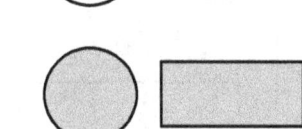

Color each triangle yellow.
Color each circle red.
Color each square green.

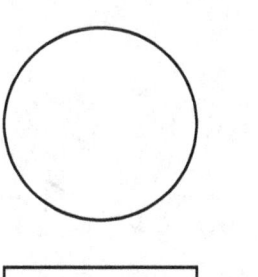

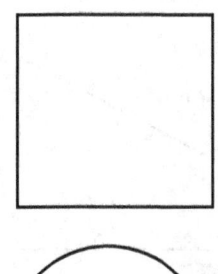

Color each shape that is the same shape as the first one.

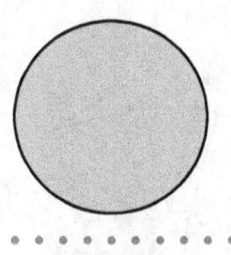

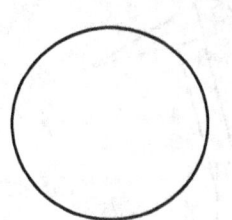

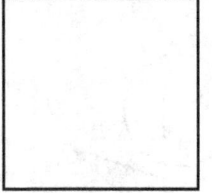

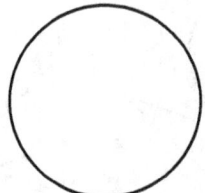

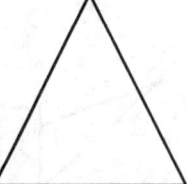

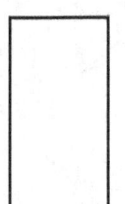

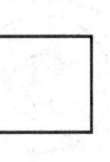

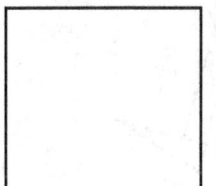

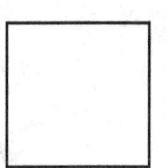

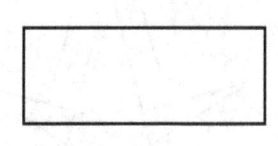

Chapter 1 Kindergarten Review

eleven 11

Color each triangle yellow.
Count the triangles.
Write the number.

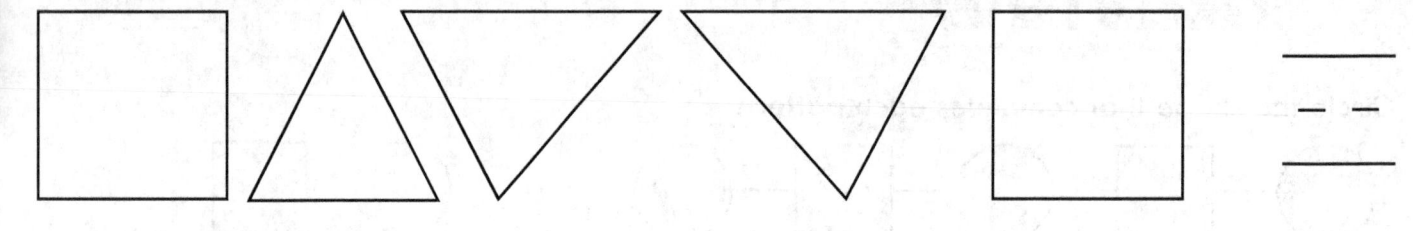

Color each rectangle green.
Count the rectangles.
Write the number.

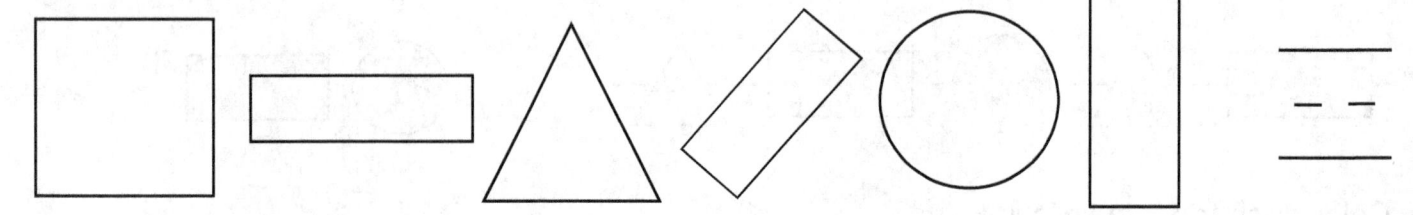

Color the triangle that is on the circle.
Color the square that is off the circle.

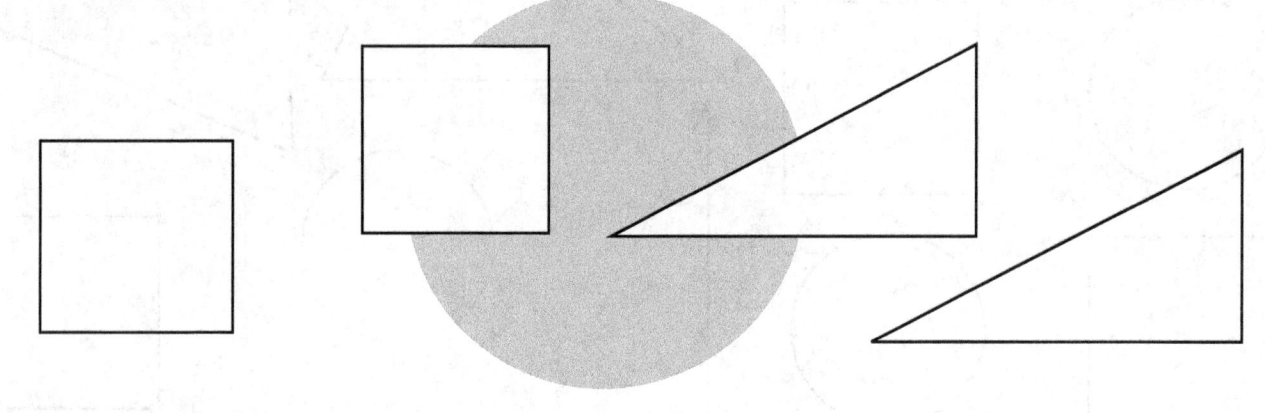

Circle the smaller figure.
Draw a line under the larger figure.

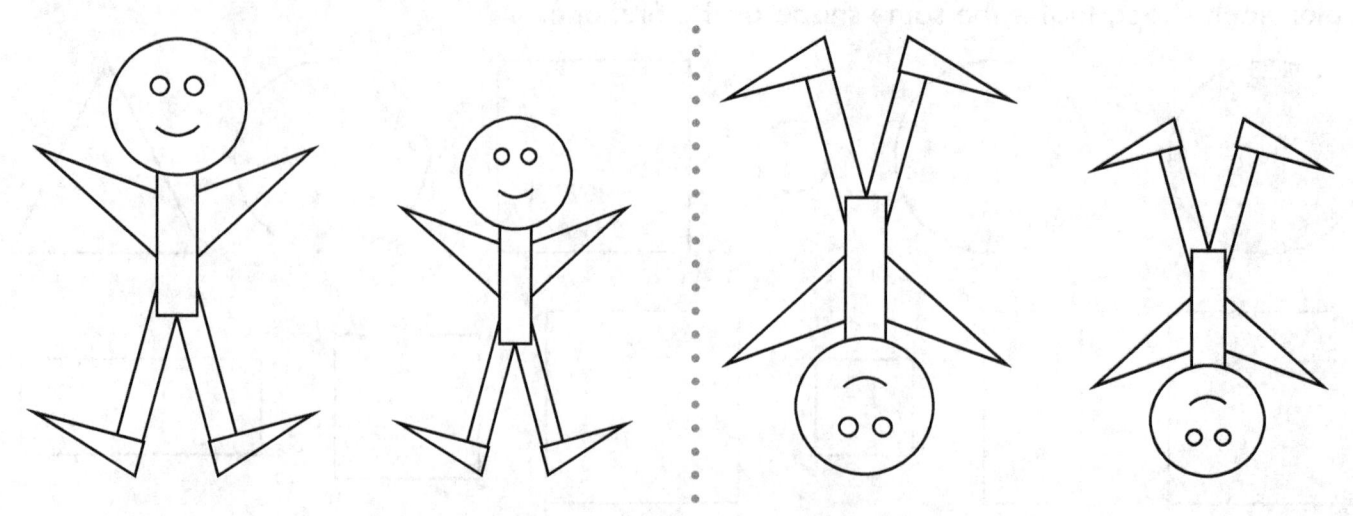

12   twelve

Math 1 Reviews

Using Eleven & Twelve

Count the eggs.
Write the numbers 1–12 in order.

Write the number in each set.

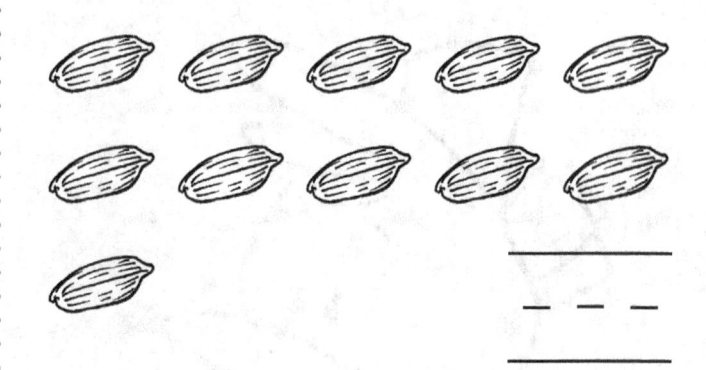

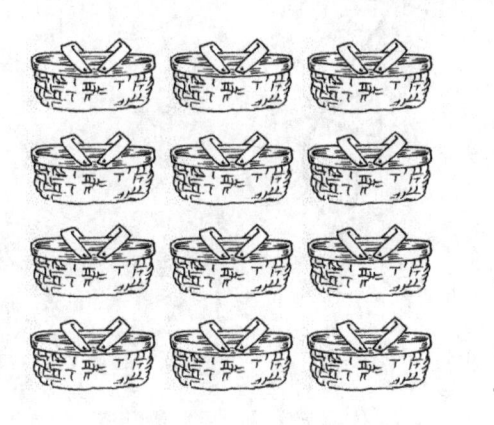

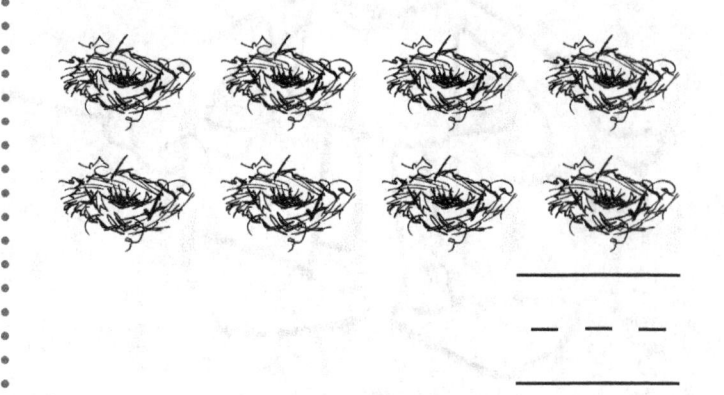

Chapter 2 • Lesson 7

thirteen 13

Number Writing

Write each number.

11

12

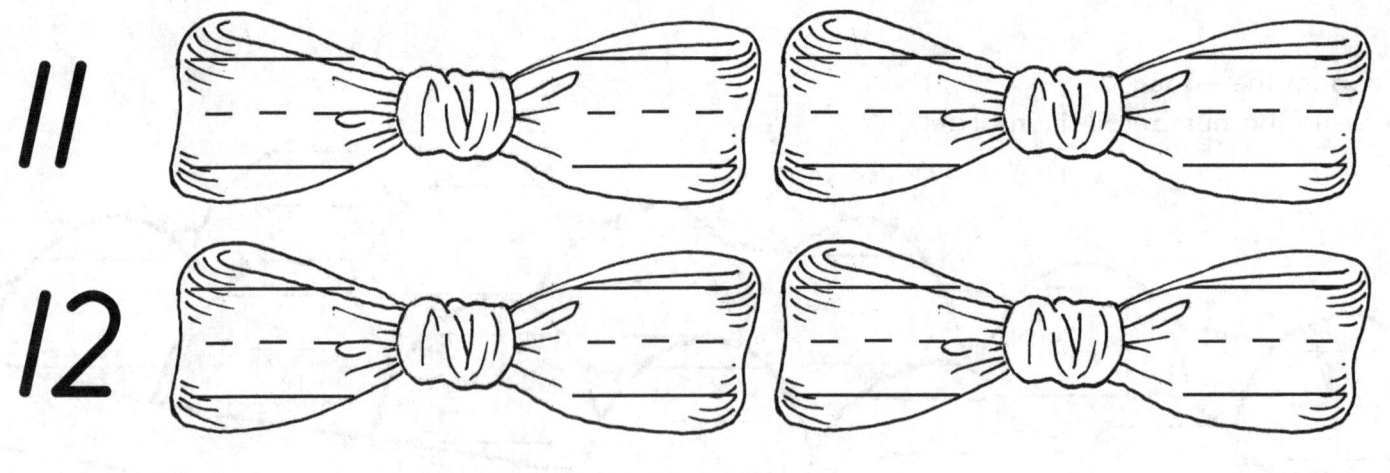

Help DQ fly a kite.
Connect the dots for 1–12.

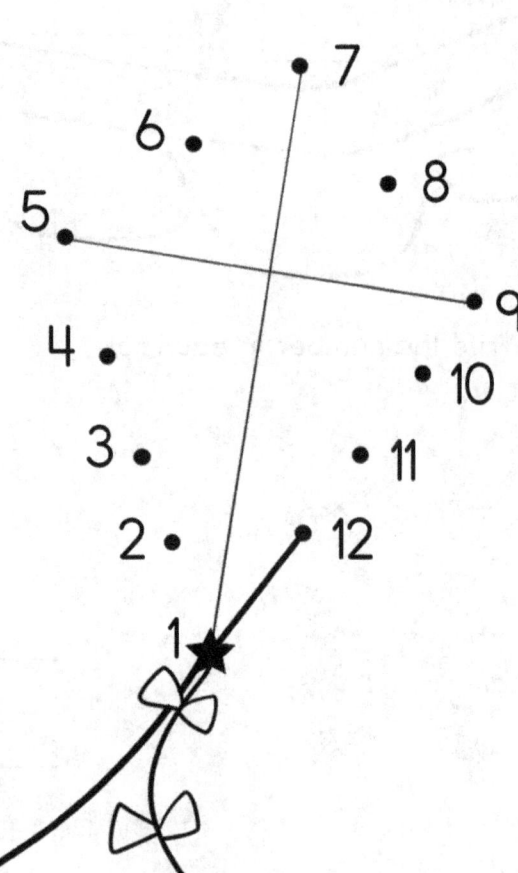

14 fourteen

Math 1 Reviews

© 2024 BJU Press. Reproduction prohibited.

Using Thirteen, Fourteen; Ordinal Numbers

Match the correct number with each set.

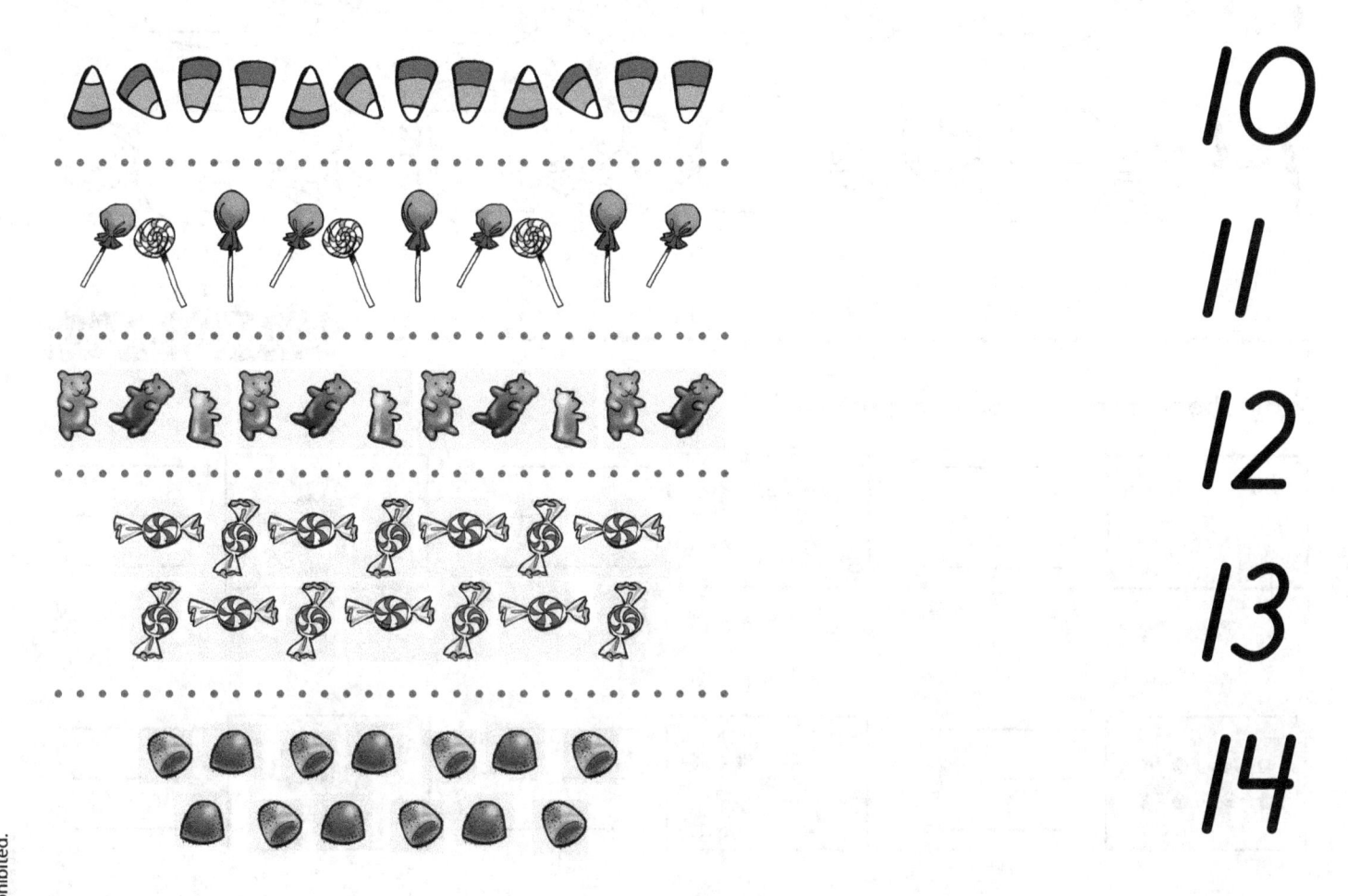

10

11

12

13

14

Circle the correct student.

ninth

sixth

Chapter 2 • Lesson 8

fifteen 15

Number Writing

Write each number.

13

14

Chapter 1 Review

Write the number for each dot pattern.

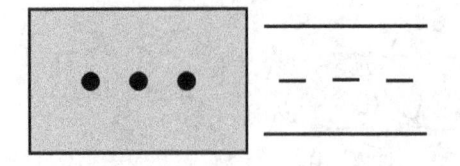

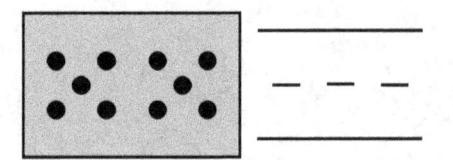

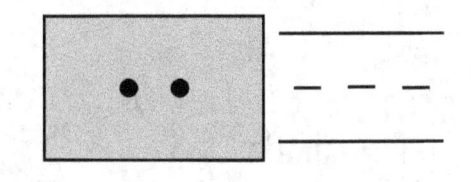

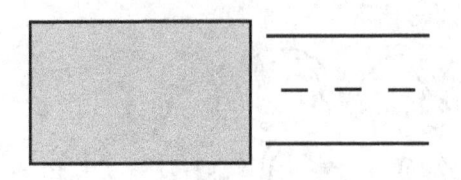

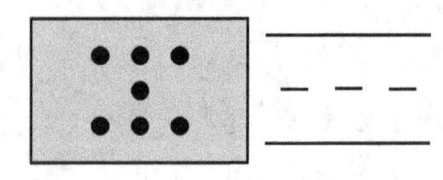

16 sixteen   Math 1 Reviews

Using Fifteen, Sixteen; Pictograph

Name	Number of Flowers in the Garden
Jenny	
Bethany	
Levi	

Look at the pictograph.
Write the number of flowers for each person.
Circle the name of the person who has the least.

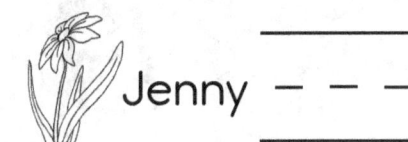

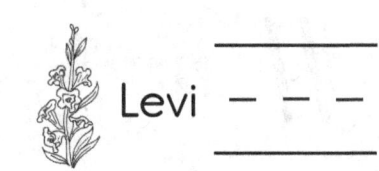

Jenny _____   Bethany _____   Levi _____

Write the missing number.

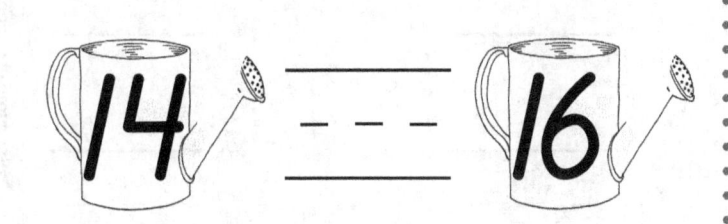

14 _____ 16   _____ 11 12

12 13 _____   14 15 _____

Chapter 2 • Lesson 9   seventeen 17

Number Writing

Write each number.

15

16

Write the numbers 11–14.

11 12 13 14

18 eighteen

Math 1 Reviews

© 2024 BJU Press. Reproduction prohibited.

Using Seventeen, Eighteen; Bar Graph

Mark an *X* on the correct number of ants.

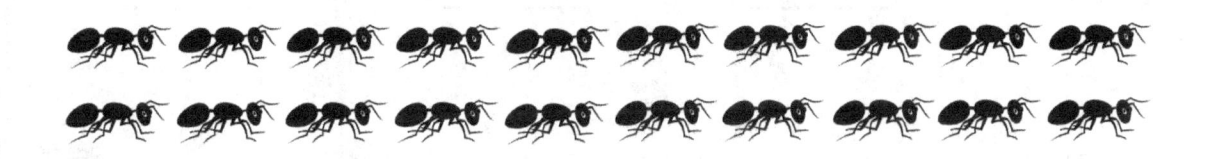

Picnic Supplies

Look at the bar graph.

Circle the item with the most.

Circle the item with the least.

Circle the items with the same.

Circle how many more 🍞 than 🍽.   1   2   3

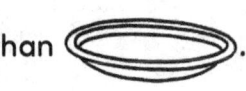

Chapter 2 • Lesson 10   nineteen **19**

Number Writing

Write each number.

17

18

Chapter 1 Review

Write the number in each set.
Circle the set with more.

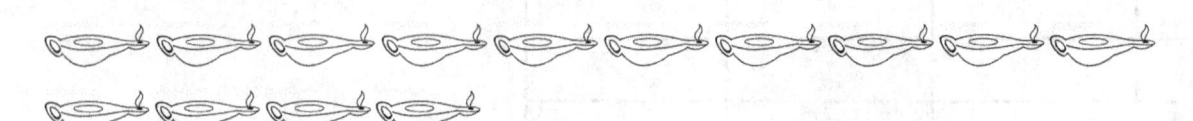

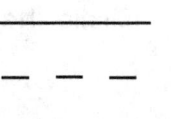

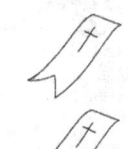

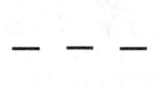

20 twenty   Math 1 Reviews

Using Nineteen, Twenty; Even & Odd Numbers

Write the number. Circle to make pairs.
Circle *yes* if the number is even.
Circle *no* if the number is not even.

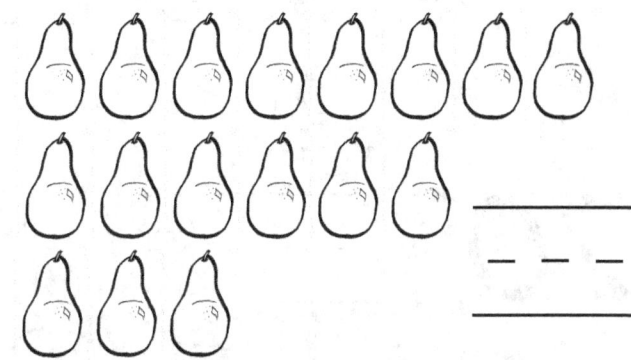

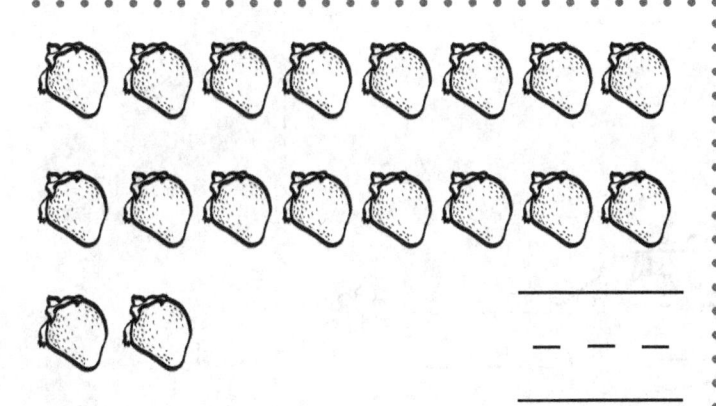

Is the number even? yes   no

Is the number even? yes   no

Is the number even? yes   no

Is the number even? yes   no

Mark an *X* on the correct number of objects.

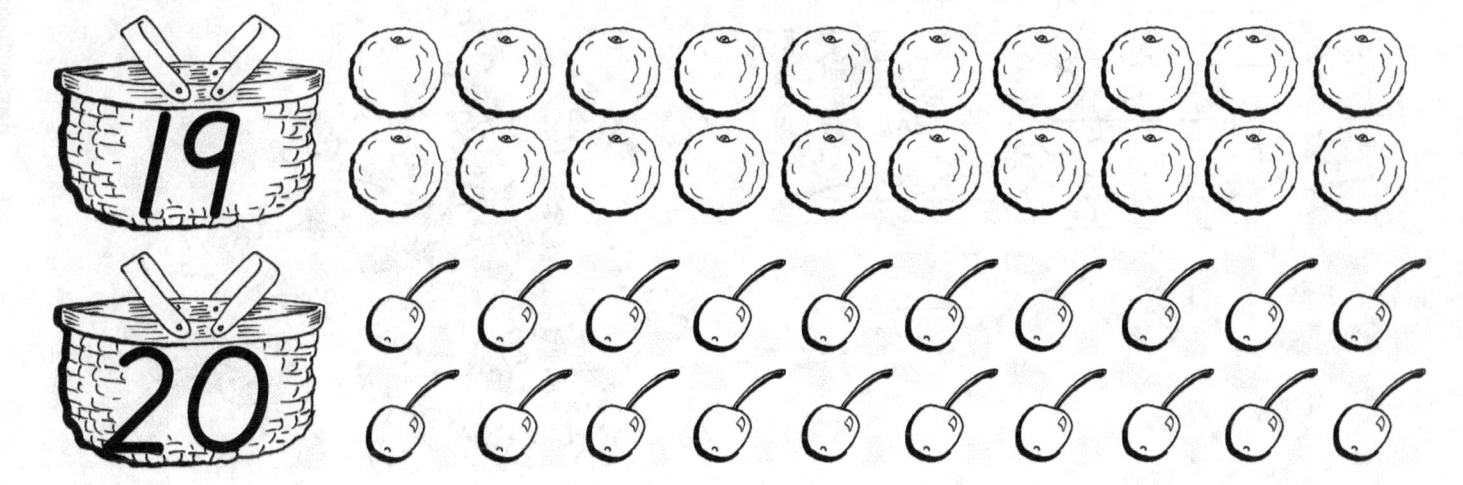

Chapter 2 • Lesson 11

twenty-one **21**

Number Writing

Write each number.

19

20

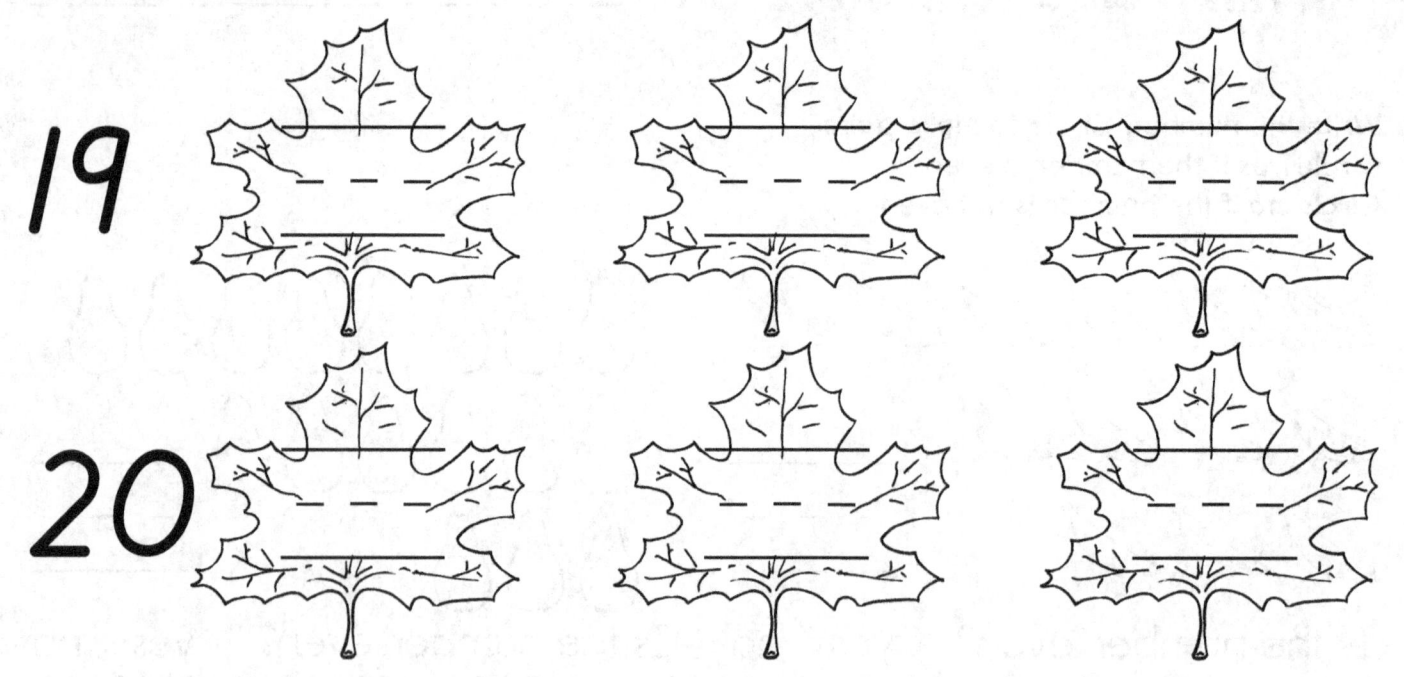

Follow the path to the pile of leaves.
Write the numbers 11–20 in order.

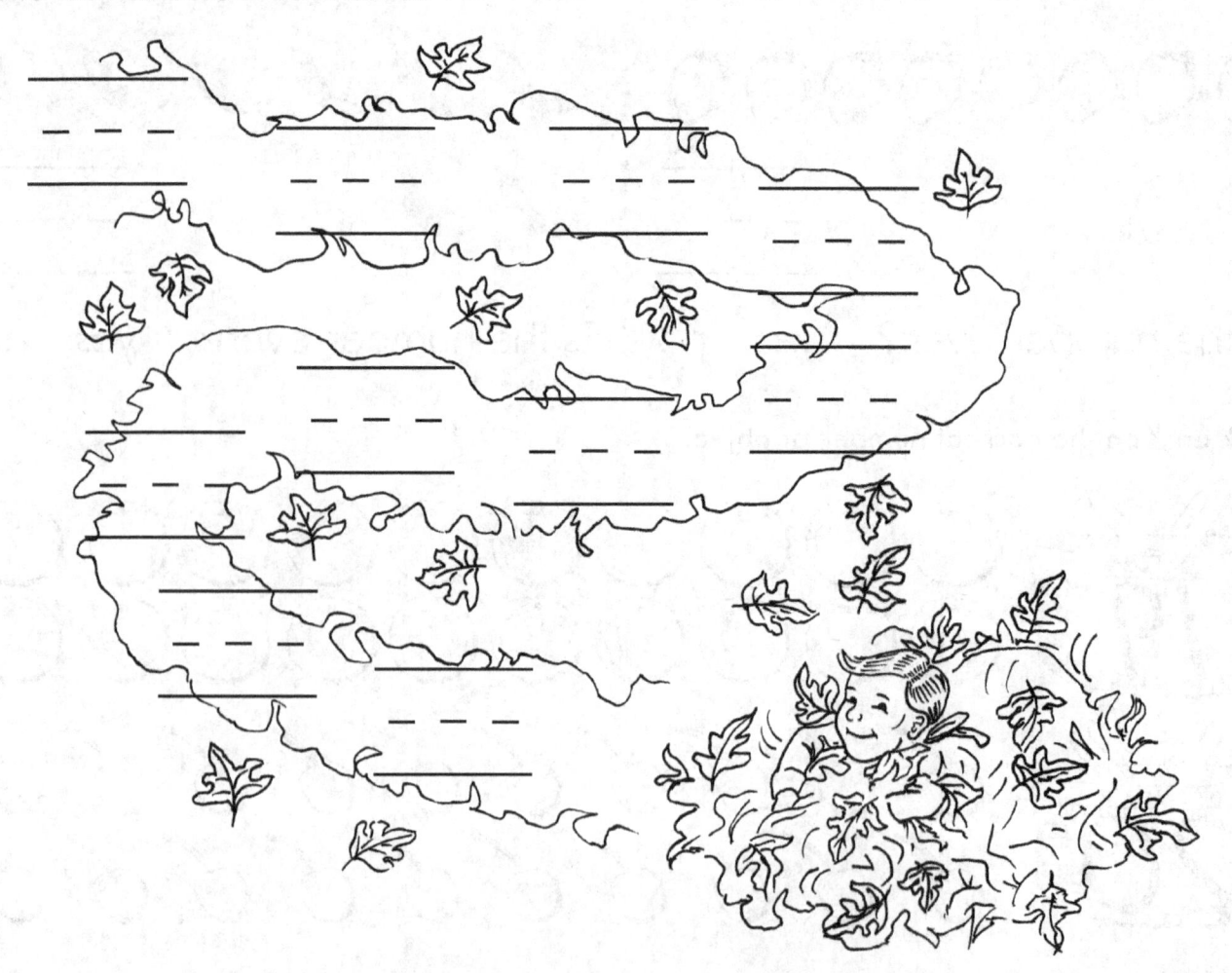

22 twenty-two

Math 1 Reviews

Chapter 2 Review

_____

Write the numbers 11–20.

_____ _____ _____ _____ _____

Write the number in each set.
Circle the correct answer.

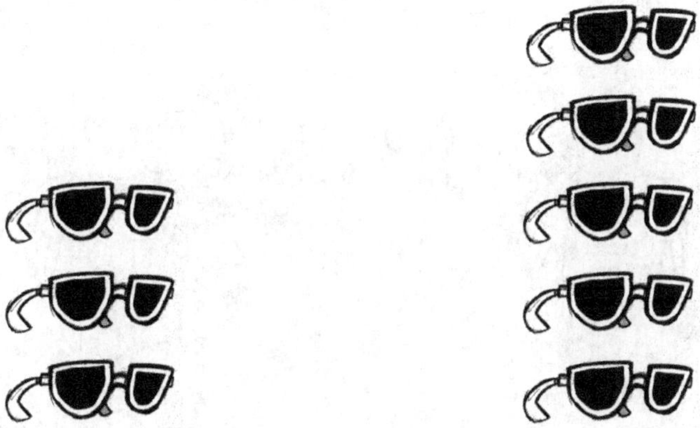

_____ is more than _____
_____ is less than _____

_____ is more than _____
_____ is less than _____

_____ is more than _____
_____ is less than _____

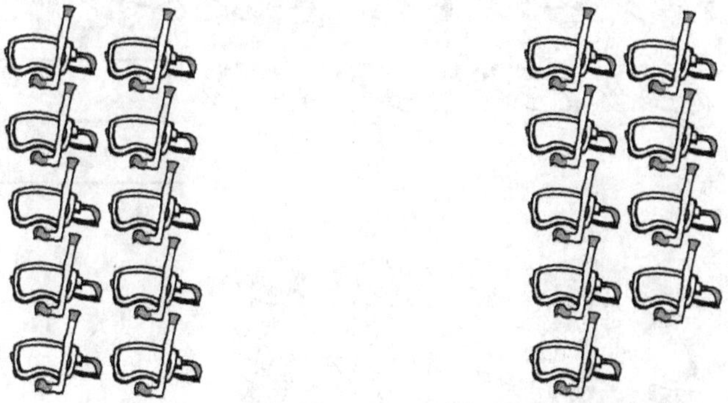

_____ is more than _____
_____ is less than _____

Chapter 2 Review   twenty-three 23

Circle the correct student.

fifth

eighth

tenth

Write the number in each set.

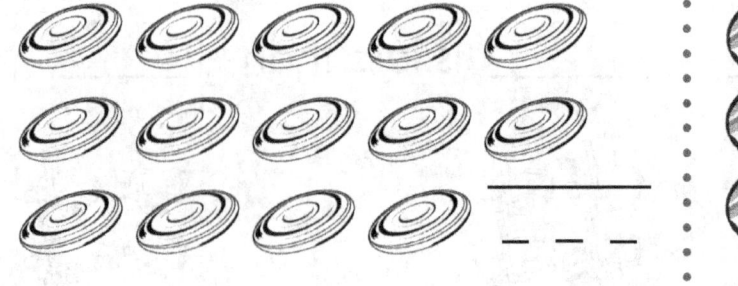

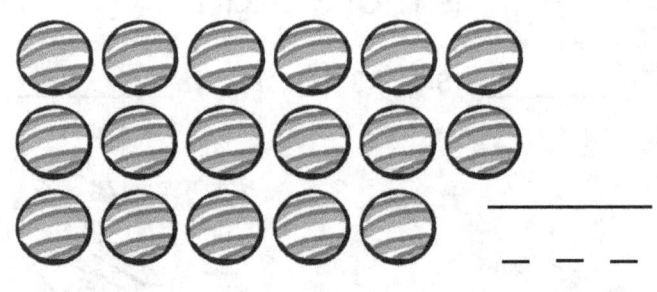

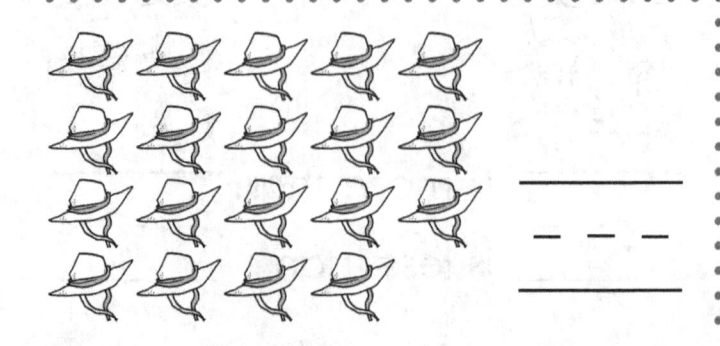

___

___

___

___

24 twenty-four

Math 1 Reviews

Cumulative Review

Color the correct book of the Bible.

sixth

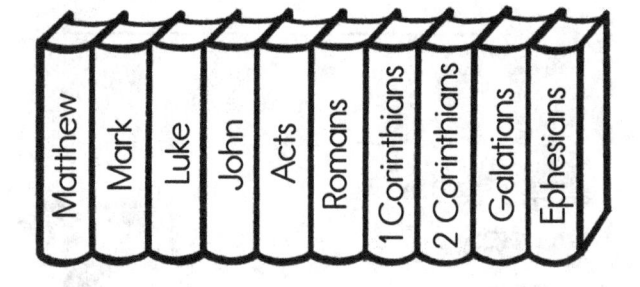

ninth

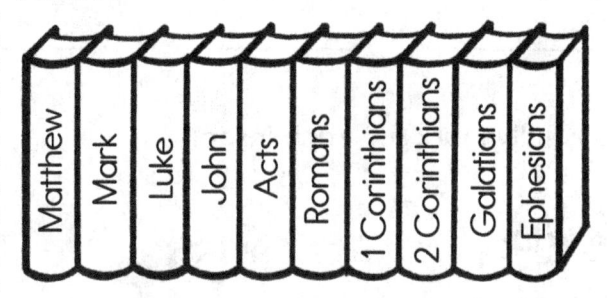

fifth

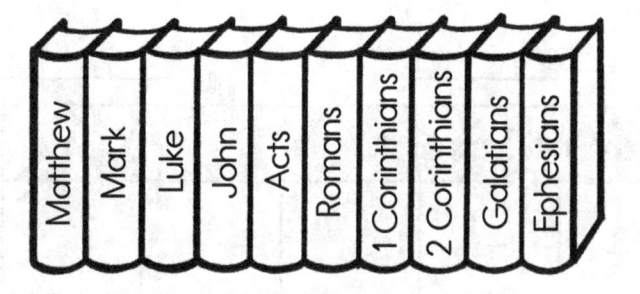

second

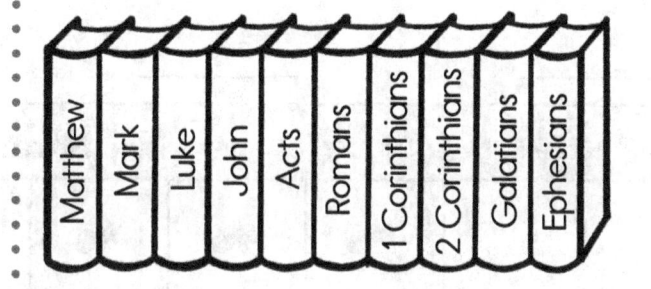

Write the number in each set.

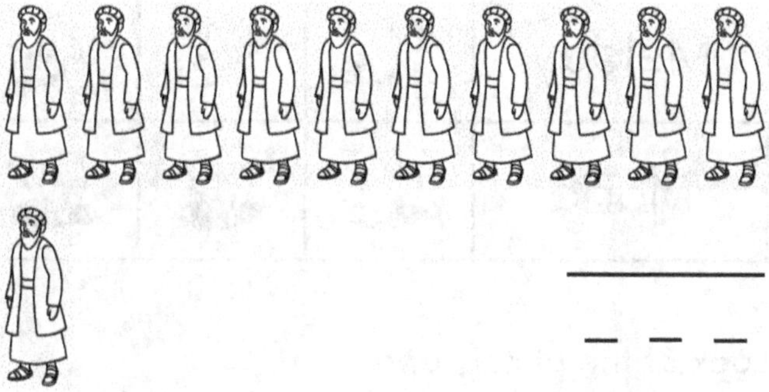

Chapter 2 Cumulative Review

twenty-five 25

Write each number.

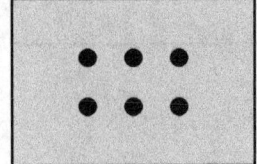

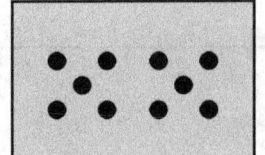

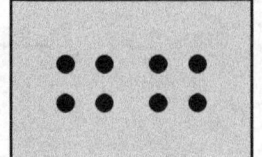

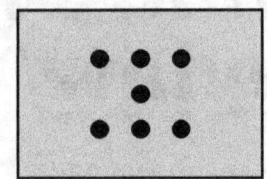

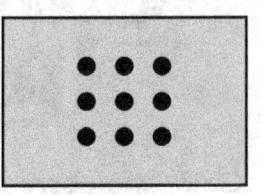

___ ___ ___ ___ ___

Make a tally mark for each insect.

___   ___   ___   ___

Name	Number of Insects
Joe	
Mark	
Melody	
Maria	

Look at the pictograph.
Write the number of insects.
Circle the name of the person who found the most.

Joe ___   Melody ___

Mark ___   Maria ___

26 twenty-six   Math 1 Reviews

Complete Addition Sentences

Complete each addition sentence.

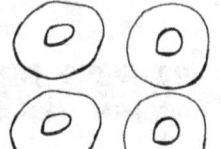

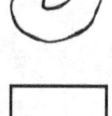

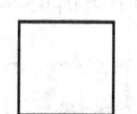

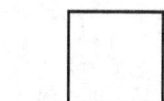

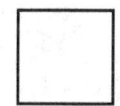

2 + 1 = ☐

2 + 4 = ☐

4 + ☐ = ☐

3 + ☐ = ☐

3 + ☐ = ☐

Chapter 3 • Lesson 14

twenty-seven **27**

Number Writing

Write each number.

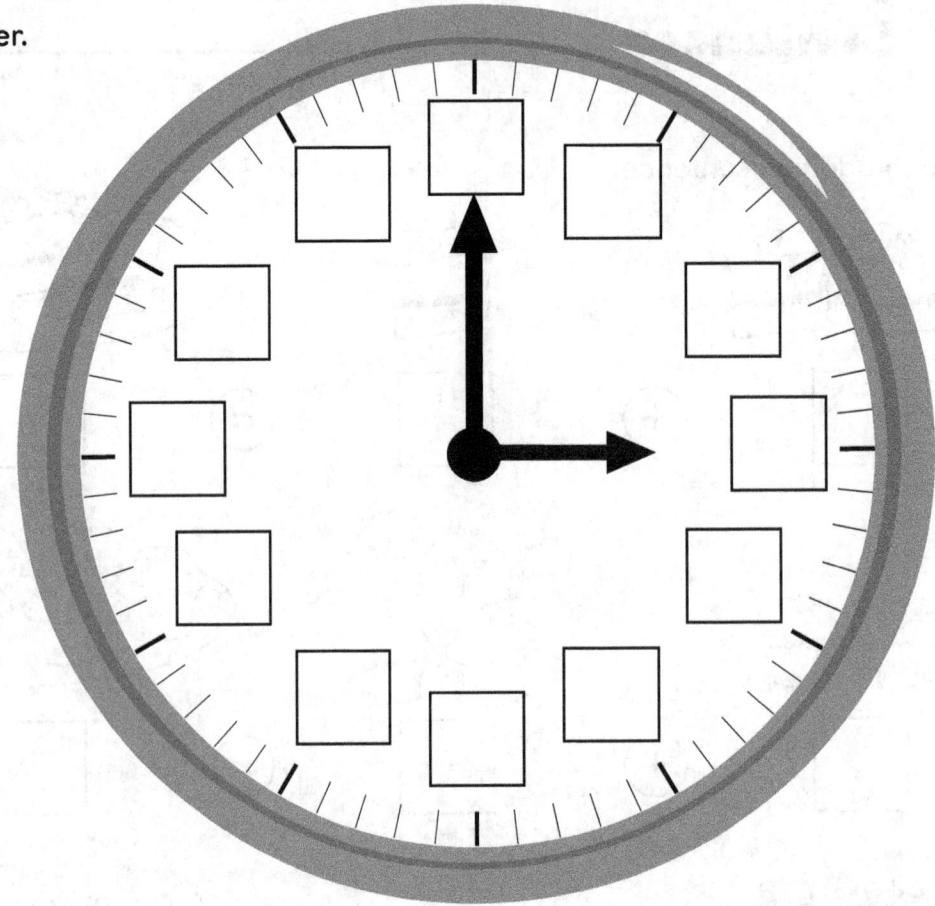

Addition Fact Review

Solve each addition problem.

$\begin{array}{r}1\\+1\\\hline\square\end{array}$ $\begin{array}{r}1\\+0\\\hline\square\end{array}$ $\begin{array}{r}2\\+1\\\hline\square\end{array}$

$\begin{array}{r}3\\+1\\\hline\square\end{array}$ $\begin{array}{r}0\\+3\\\hline\square\end{array}$ $\begin{array}{r}2\\+2\\\hline\square\end{array}$

28 twenty-eight   Math 1 Reviews

Write Addition Sentences

Complete each addition sentence.

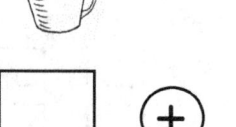

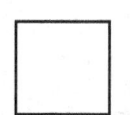

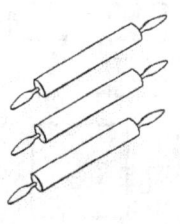

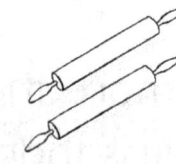

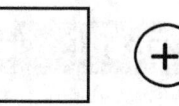

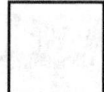

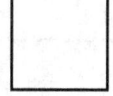

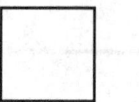

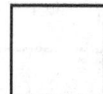

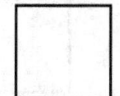

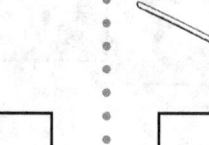

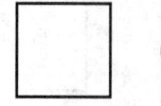

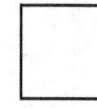

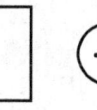

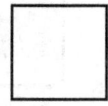

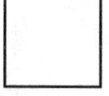

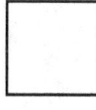

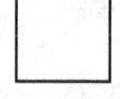

Chapter 3 • Lesson 15   twenty-nine **29**

Chapter 2 Review

Write the number in each set.
Circle the correct answer.

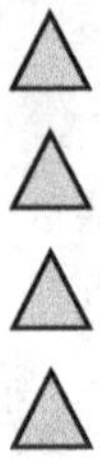

☐ is more than / is less than ☐

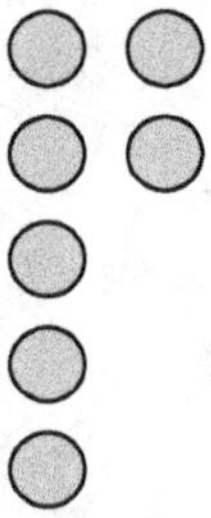

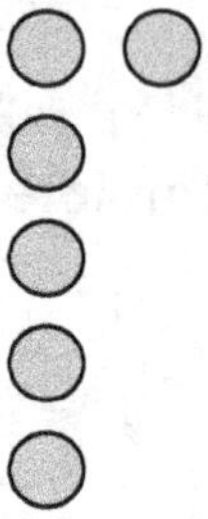

☐ is more than / is less than ☐

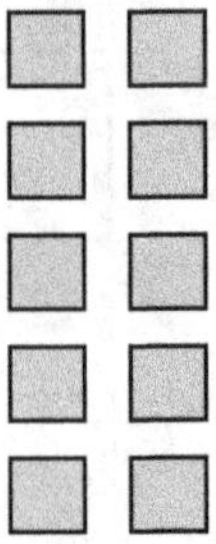

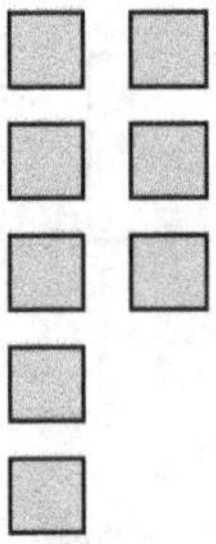

☐ is more than / is less than ☐

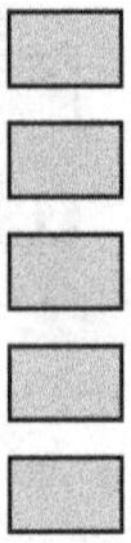

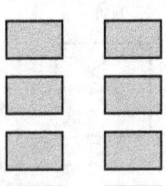

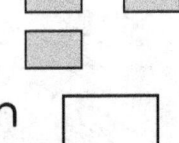

☐ is more than / is less than ☐

Addition Fact Review

Solve each addition problem.

2 + 2 = ☐

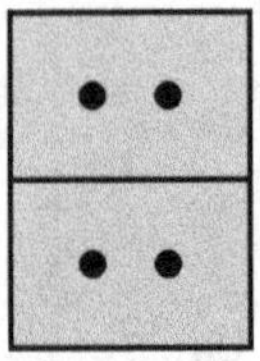

3 + 3 = ☐

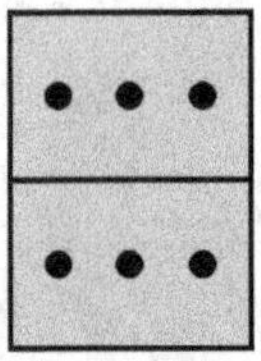

0 + 4 = ☐

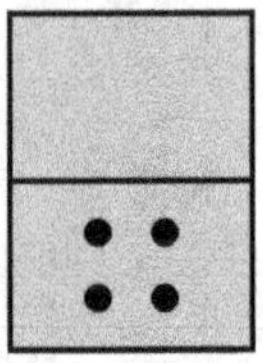

4 + 1 = ☐

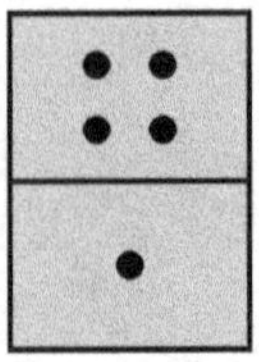

1 + 4 = ☐

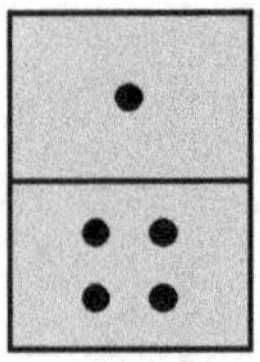

3 + 2 = ☐

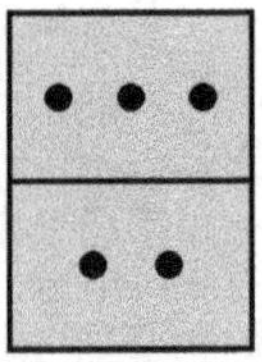

Draw Pictures to Add

Complete each addition sentence.

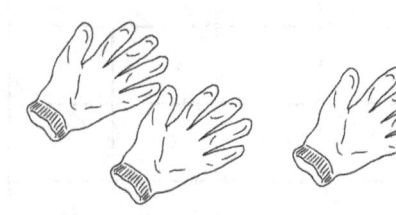

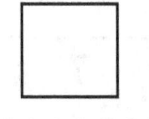

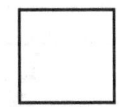

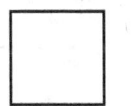

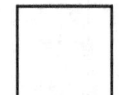

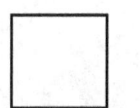

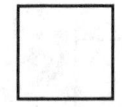

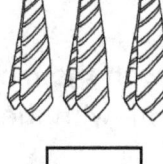

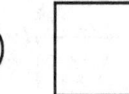

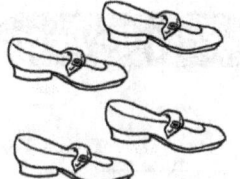

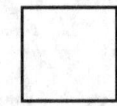

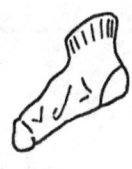

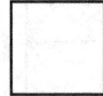

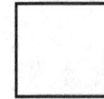

☐ + ☐ = ☐

☐ + ☐ = ☐

☐ + ☐ = ☐

☐ + ☐ = ☐

☐ + ☐ = ☐

☐ + ☐ = ☐

Draw pictures for each addition sentence. Write the answer.

1 + 5 = ☐

2 + 3 = ☐

Chapter 3 • Lesson 16

Chapter 2 Review

Use the information to make a bar graph.

△ triangles
□ squares
○ circles
▭ rectangles

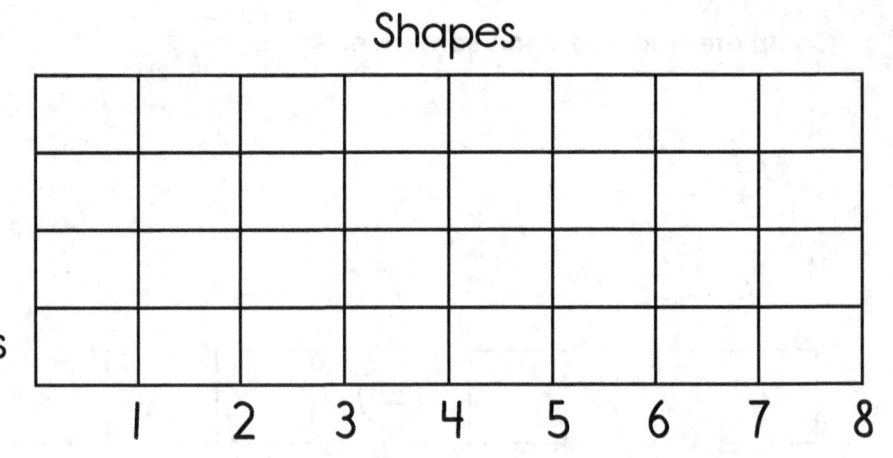

Shapes

1 2 3 4 5 6 7 8

Circle the shape with the most.
Underline the shape with the fewest.

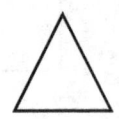

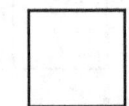

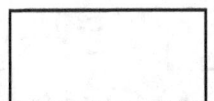

△   □   ○   ▭

Addition Fact Review

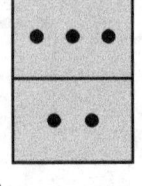

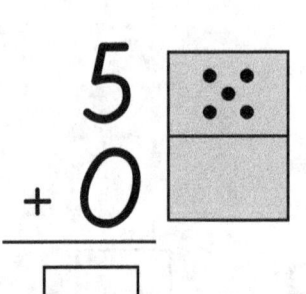

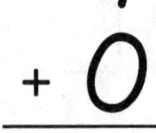

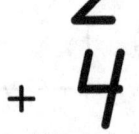

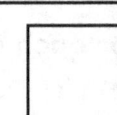

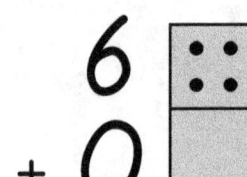

Solve each addition problem.

3
+ 2
___

4
+ 0
___

2
+ 4
___

5
+ 0
___

1
+ 5
___

6
+ 0
___

32 thirty-two

Math 1 Reviews

Addition in Vertical Form

Complete each addition problem.

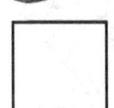

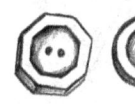

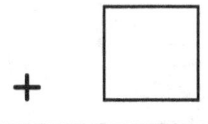

☐ + ☐ = ☐

☐
+ ☐
———
☐

Write a number sentence for each story.

Pam had 3 pins.
Pam got 2 more pins.
How many pins does she have in all?

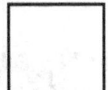

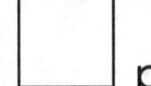

☐ + ☐ = ☐ pins

Bob had 4 tacks.
Bob got 2 more tacks.
How many tacks does he have in all?

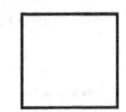

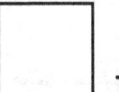

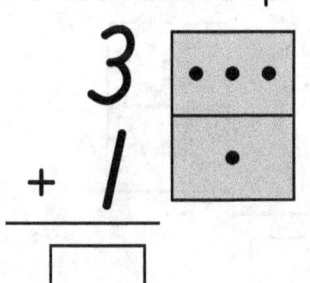

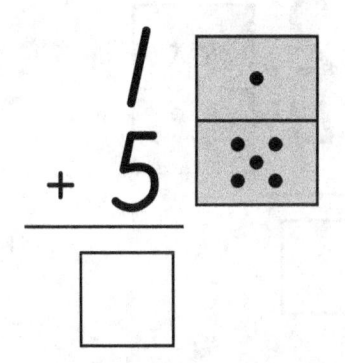

☐ + ☐ = ☐ tacks

Solve each addition problem.

3
+ 1
———
☐

4
+ 1
———
☐

0
+ 6
———
☐

1
+ 1
———
☐

3
+ 3
———
☐

1
+ 5
———
☐

Chapter 3 • Lesson 17

Number Writing

Write the numbers 1–20.

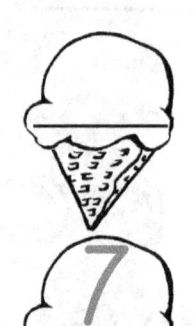

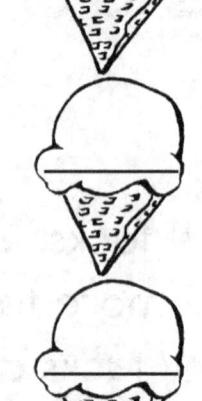

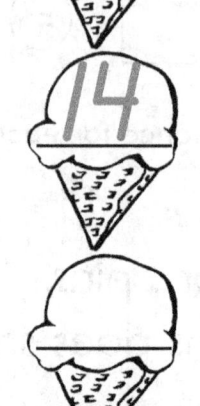

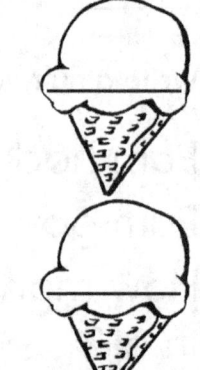

Addition Fact Review

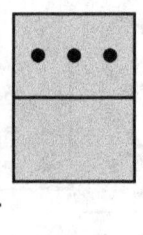

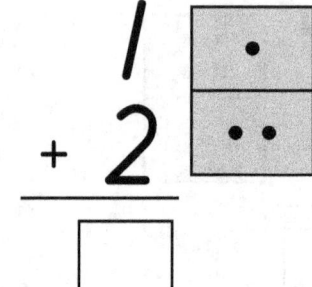

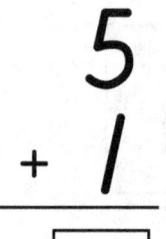

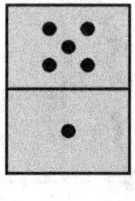

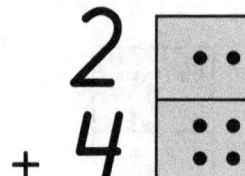

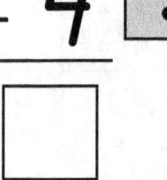

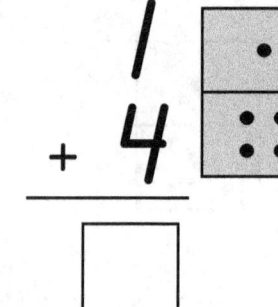

Solve each addition problem.

$\begin{array}{r}3\\+0\\\hline\end{array}$ □

$\begin{array}{r}1\\+2\\\hline\end{array}$ □

$\begin{array}{r}5\\+1\\\hline\end{array}$ □

$\begin{array}{r}2\\+4\\\hline\end{array}$ □

$\begin{array}{r}2\\+3\\\hline\end{array}$ □

$\begin{array}{r}1\\+4\\\hline\end{array}$ □

34 thirty-four

Math 1 Reviews

Addition Names for 4, 5 & 6

Color the clouds.
Complete each number sentence.

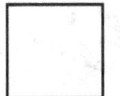

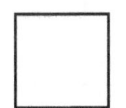

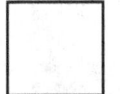

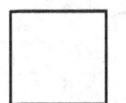

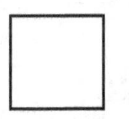

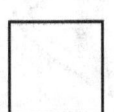

| 6 | + | 0 | = | 6 |

| 5 | + | ☐ | = | ☐ |

| ☐ | + | ☐ | = | ☐ |

| ☐ | + | ☐ | = | ☐ |

| ☐ | + | ☐ | = | ☐ |

| ☐ | + | ☐ | = | ☐ |

| ☐ | + | ☐ | = | ☐ |

Color each bow that has another name for 6.
Use the number sentences above to help.

Bows: 3+3, 1+5, 2+3, 6+0, 5+1, 2+2, 3+2, 4+2

Chapter 3 • Lesson 18   thirty-five **35**

Chapter 1 Review

Make a tally mark to show each number.

|||| 5    3    1

2    4    5

Number Writing

Write each number.

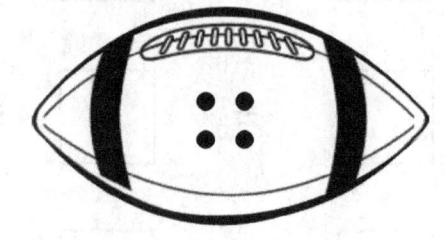

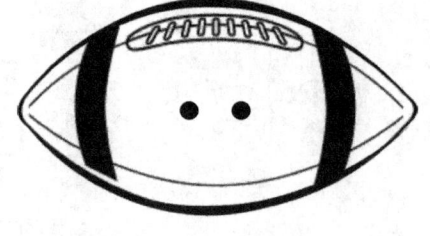

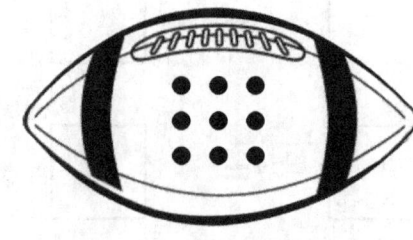

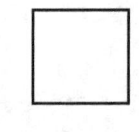

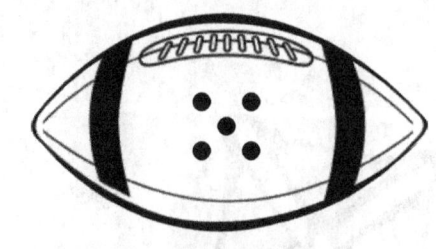

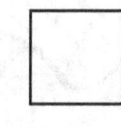

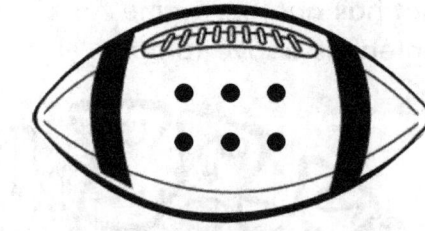

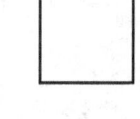

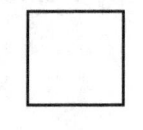

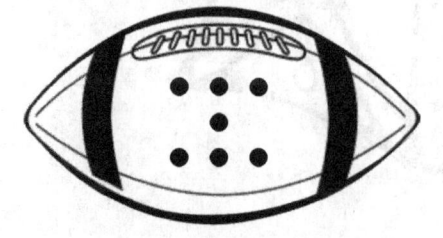

36 thirty-six

Math 1 Reviews

Order of Addends

Complete each addition problem.

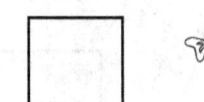

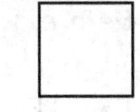

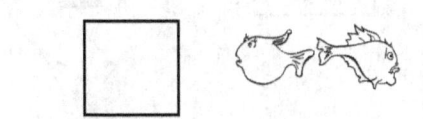

☐ + ☐ = ☐

☐ + ☐ = ☐

☐ + ☐ = ☐

☐ + ☐ = ☐

☐
+ ☐
———
☐

☐
+ ☐
———
☐

☐
+ ☐
———
☐

☐
+ ☐
———
☐

Chapter 3 • Lesson 19   thirty-seven **37**

Chapter 2 Review

Write the number that comes *between*.

18 ☐ 20   6 ☐ 8

11 ☐ 13   16 ☐ 18

Write the number that comes *after*.

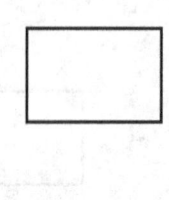

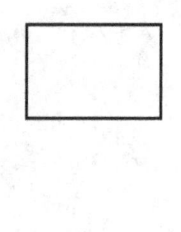

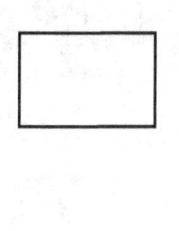

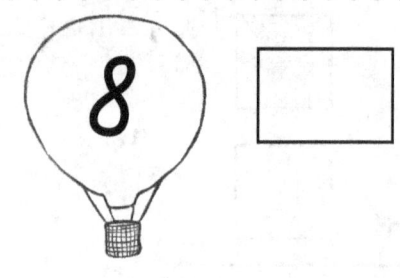

7 ☐   10 ☐   3 ☐

12 ☐   19 ☐   8 ☐

Write the number that comes *before*.

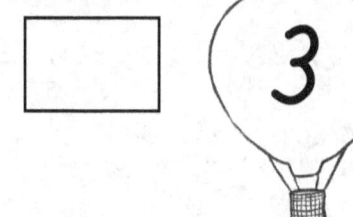

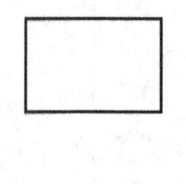

☐ 3   ☐ 7   ☐ 12

☐ 15   ☐ 18   ☐ 10

38 thirty-eight

Math 1 Reviews

Chapter 3 Review

Complete each addition problem.

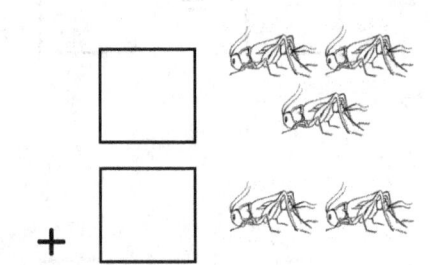

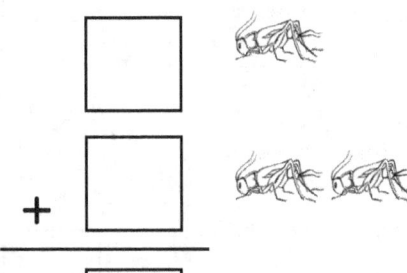

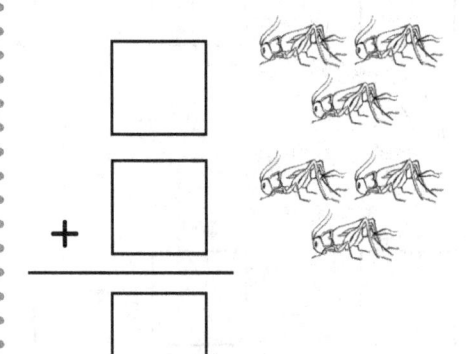

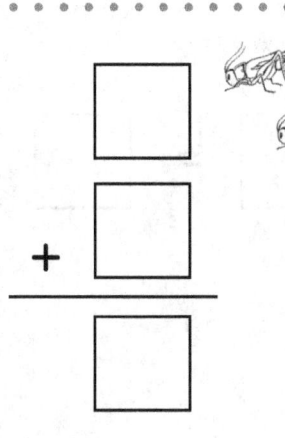

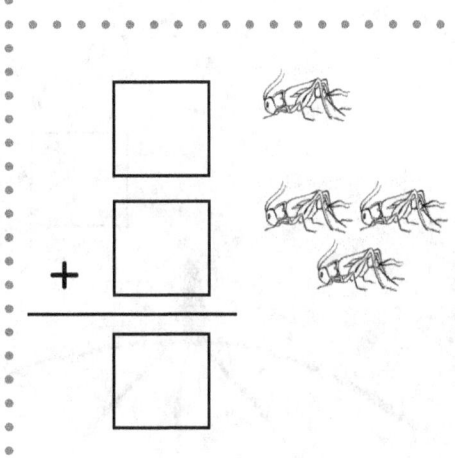

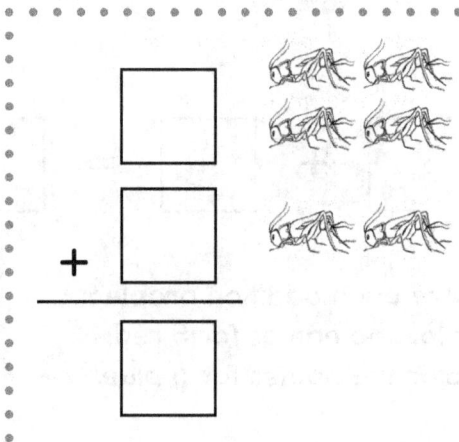

Draw pictures to show each story.
Write a number sentence for each story.

Ben had 3 bugs in the net.
He got 3 more bugs in the net.
How many bugs are in the net?

There were 4 ants on the hill.
Jon put 2 more ants on the hill.
How many ants are on the hill?

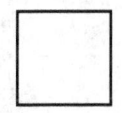

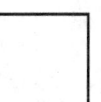

☐ + ☐ = ☐ bugs

☐ + ☐ = ☐ ants

Chapter 3 Review

thirty-nine 39

Complete each number sentence.

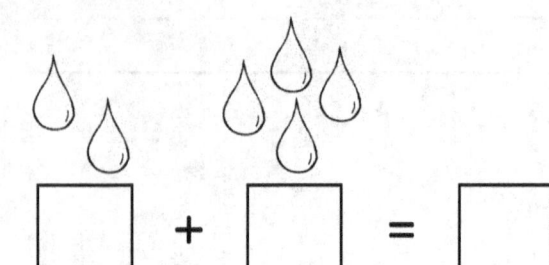

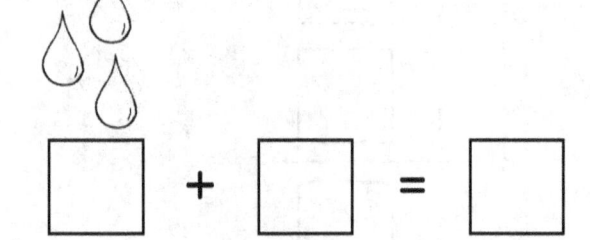

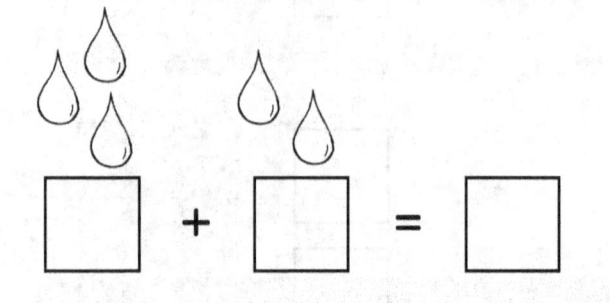

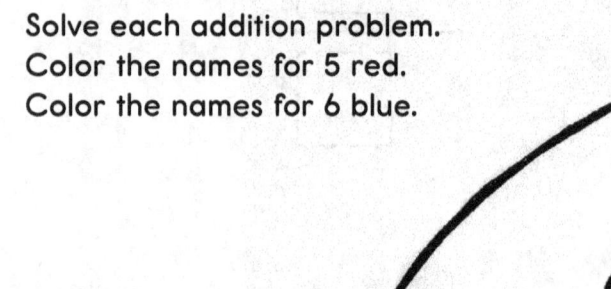

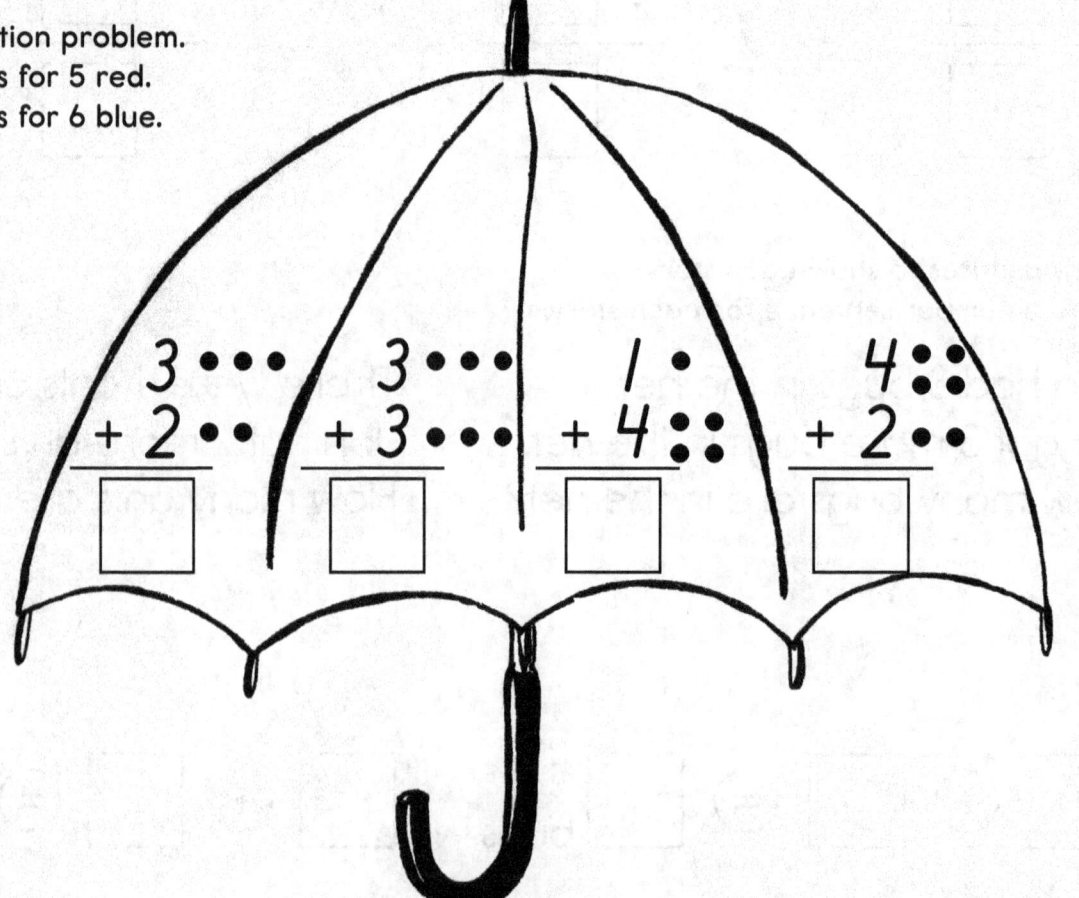

☐ + ☐ = ☐   ☐ + ☐ = ☐

☐ + ☐ = ☐   ☐ + ☐ = ☐

☐ + ☐ = ☐   ☐ + ☐ = ☐

Solve each addition problem.
Color the names for 5 red.
Color the names for 6 blue.

3
+ 2
☐

3
+ 3
☐

1
+ 4
☐

4
+ 2
☐

40 forty

Math 1 Reviews

Cumulative Review

Write the number of instruments. Circle the pairs.
Circle *yes* if the number is even.
Circle *no* if the number is not even.

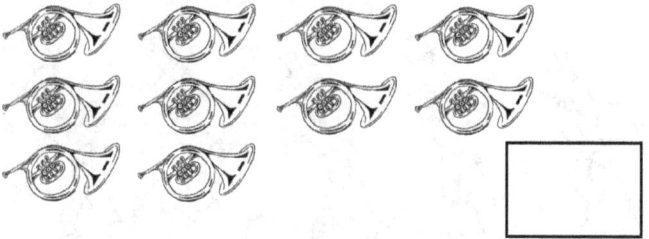

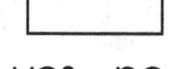

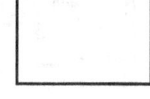

Is the number even?   yes   no

Is the number even?   yes   no

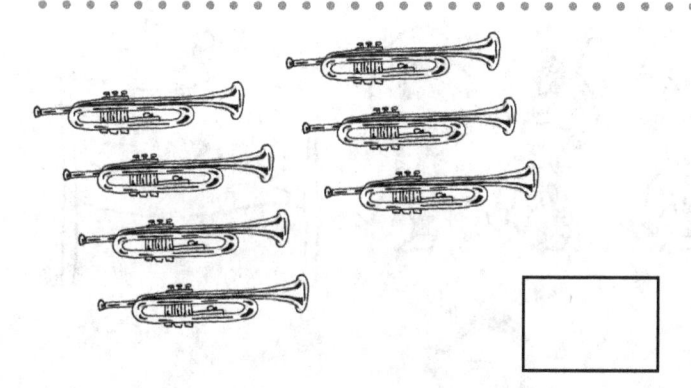

Is the number even?   yes   no

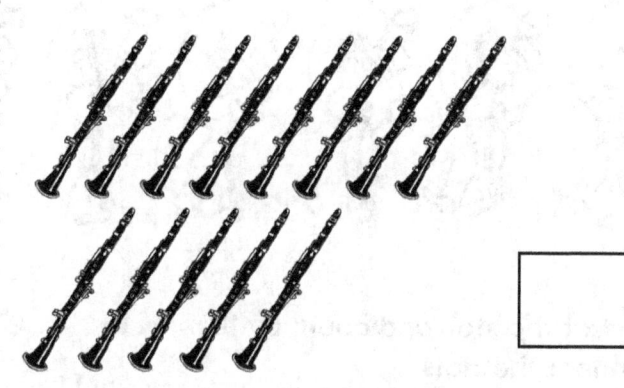

Is the number even?   yes   no

Number the instruments from smallest to largest.

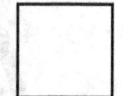

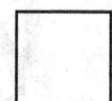

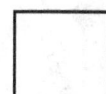

2

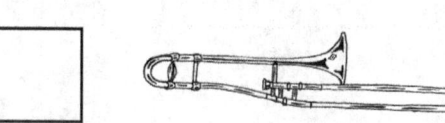

1

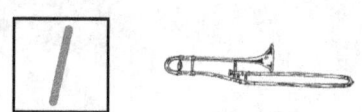

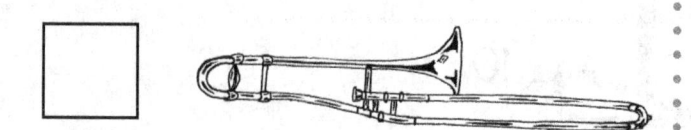

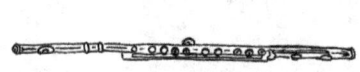

Chapter 3 Cumulative Review

forty-one **41**

Read the word. Color the player in the correct position.

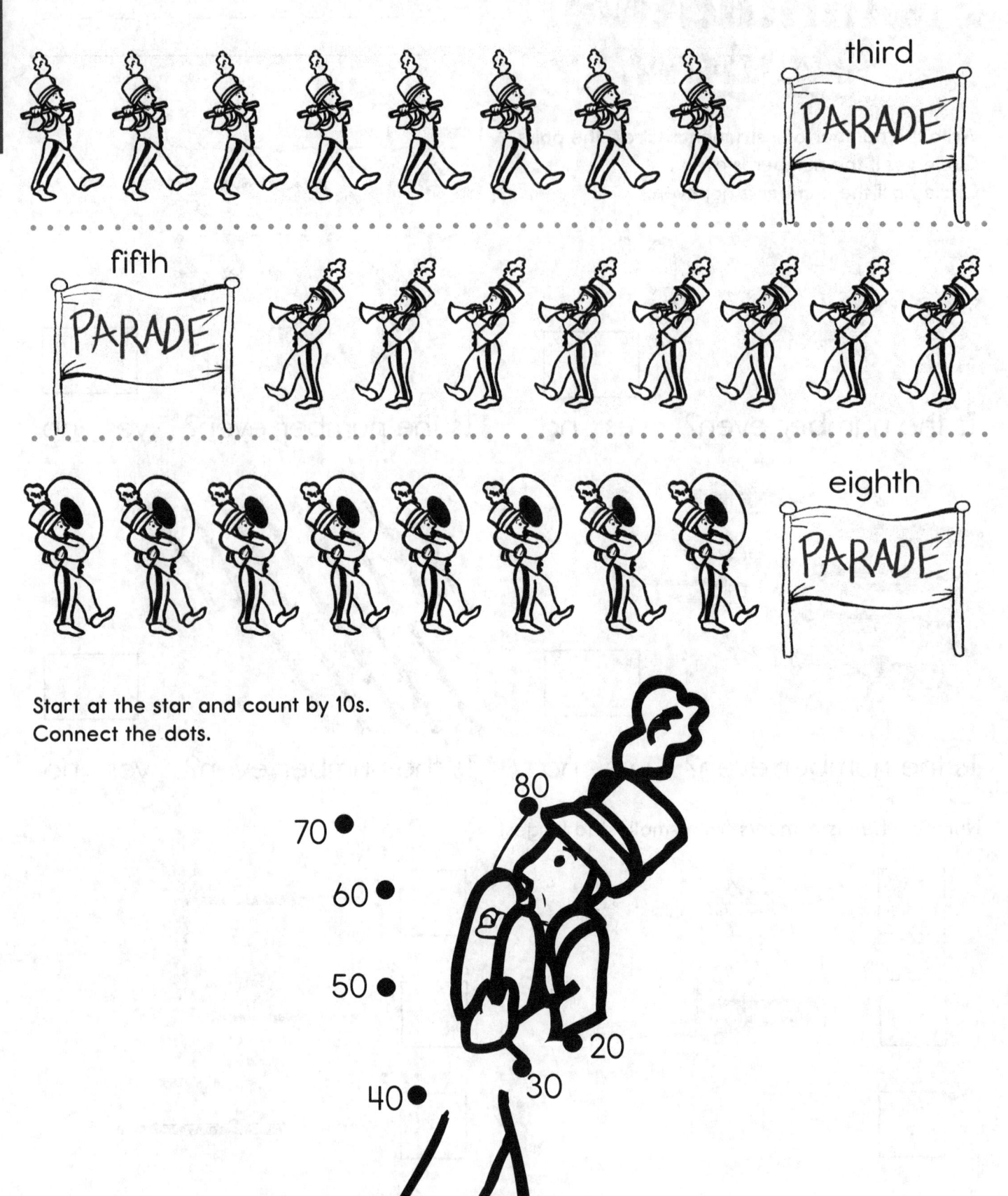

third PARADE

fifth PARADE

eighth PARADE

Start at the star and count by 10s.
Connect the dots.

42 forty-two

Math 1 Reviews

Tens & Ones in Numbers to 50

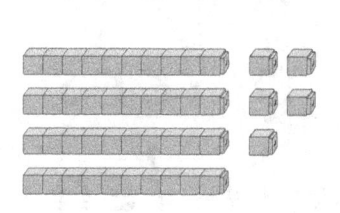

Write the number of tens and ones.
Write the number.

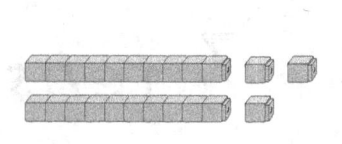

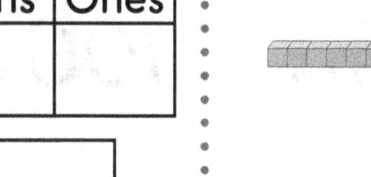

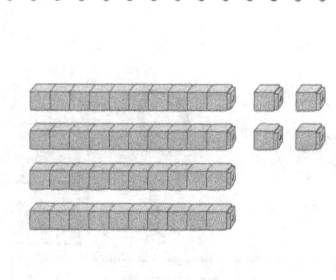

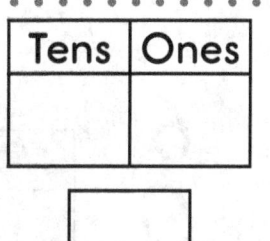

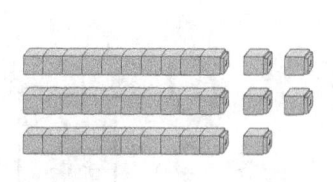

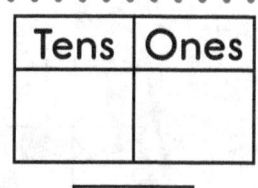

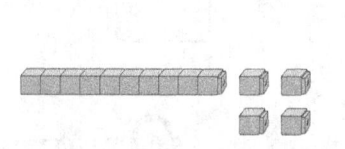

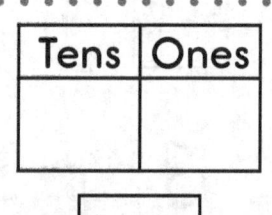

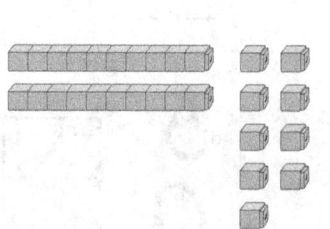

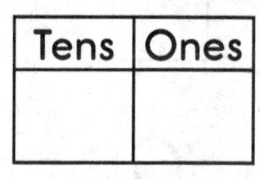

Tens	Ones
4	5

45

Write the number that comes just *after*.

38 | 24 | 43

Write the number that comes just *before*.

27 | 16 | 35

Chapter 4 • Lesson 22

Match each number to the correct number of tens.

10	3 tens	60	8 tens
20	5 tens	70	6 tens
30	1 ten	80	9 tens
40	2 tens	90	10 tens
50	4 tens	100	7 tens

(10 is matched to 1 ten)

Addition Fact Review

Add.

$1 + 0 =$ ☐

$0 + 2 =$ ☐

$6 + 0 =$ ☐

$0 + 4 =$ ☐

$0 + 0 =$ ☐

$5 + 0 =$ ☐

$3 + 0 =$ ☐

$0 + 1 =$ ☐

$0 + 6 =$ ☐

$0 + 3 =$ ☐

$2 + 0 =$ ☐

$4 + 0 =$ ☐

44 forty-four

Math 1 Reviews

Expanded Form; Even & Odd

Write the number. Circle the single cubes to make pairs.
Circle *even* if the number is even.
Circle *odd* if the number is odd.

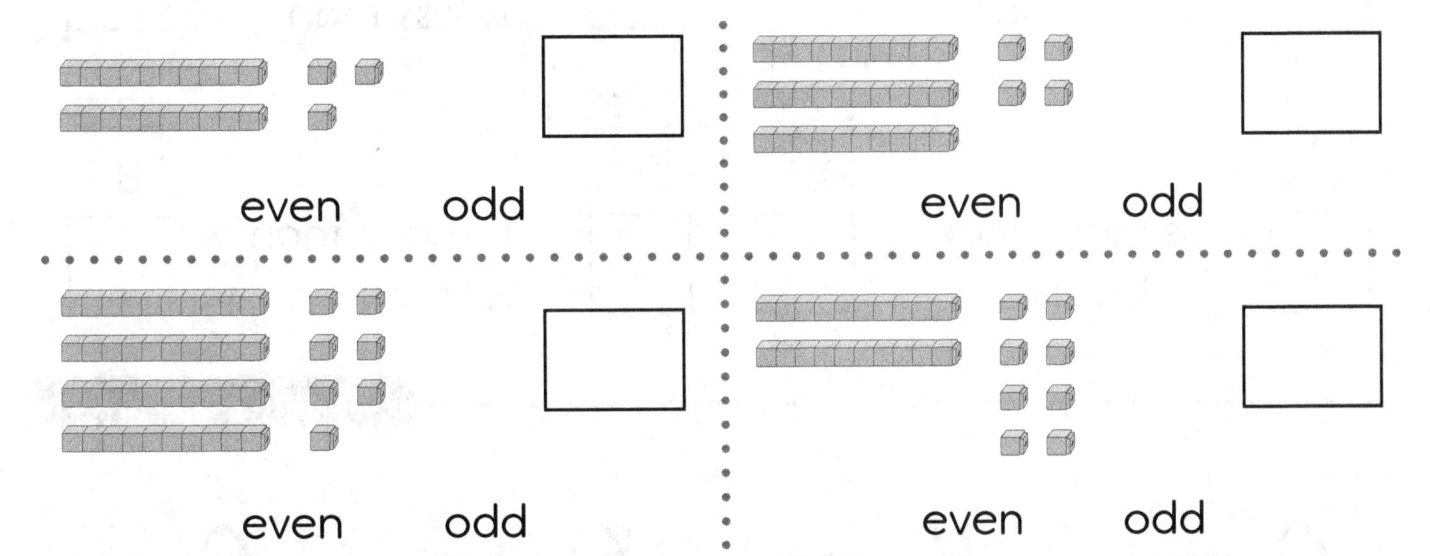

even   odd   even   odd

even   odd   even   odd

Write the number of tens and ones.
Complete the expanded form.

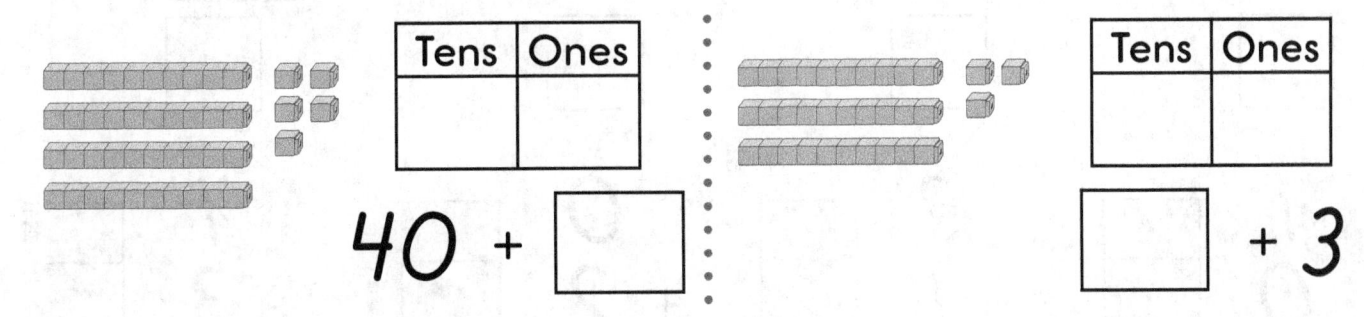

Tens	Ones

40 +

Tens	Ones

☐ + 3

Write the expanded form for each number.

38 = ☐ + ☐

42 = ☐ + ☐

21 = ☐ + ☐

Chapter 4 • Lesson 23   forty-five **45**

Chapter 2 Review

Write the number in each set.
Circle the correct answer.

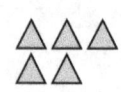

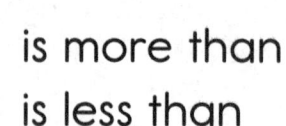

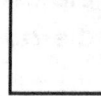

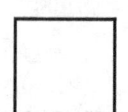

☐ 5 is more than / is less than ☐ 2

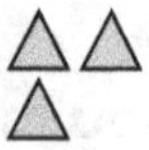

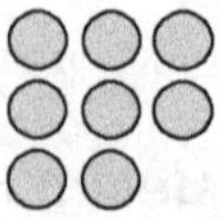

☐ 9 is more than / is less than ☐ 9

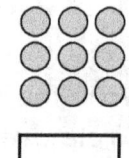

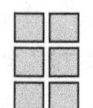

☐ is more than / is less than ☐

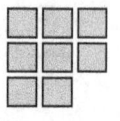

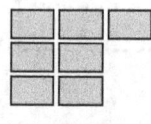

☐ is more than / is less than ☐

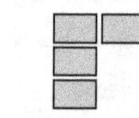

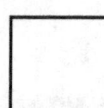

Addition Fact Review

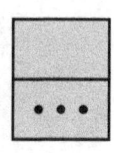

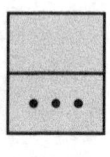

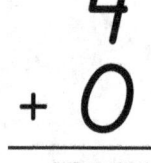

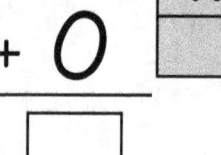

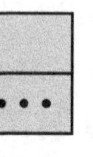

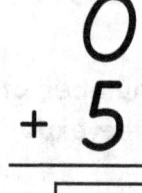

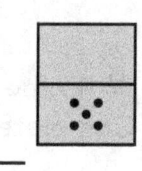

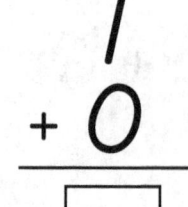

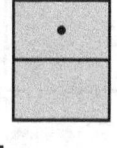

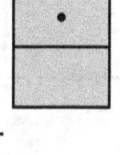

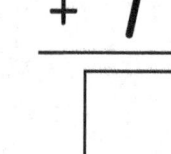

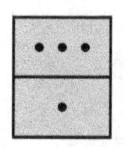

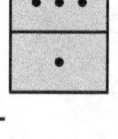

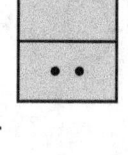

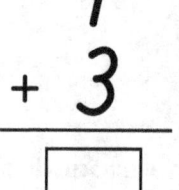

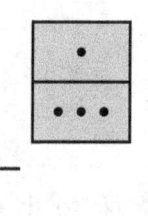

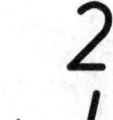

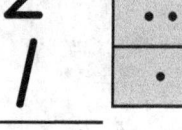

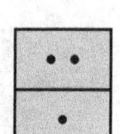

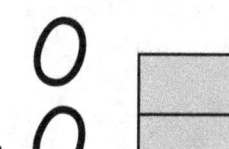

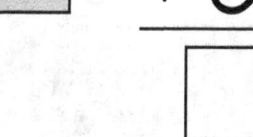

Add.

0 + 3 = ☐

4 + 0 = ☐

6 + 0 = ☐

0 + 5 = ☐

1 + 0 = ☐

3 + 1 = ☐

0 + 2 = ☐

1 + 3 = ☐

2 + 1 = ☐

0 + 0 = ☐

1 + 1 = ☐

1 + 2 = ☐

46 forty-six

Math 1 Reviews

Greater Than & Less Than

Circle the number that is *greater*.
Complete the sentence.

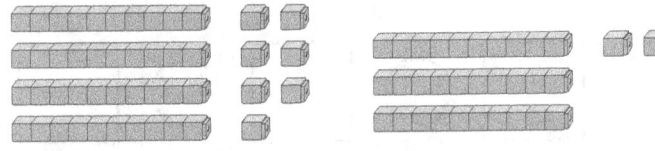

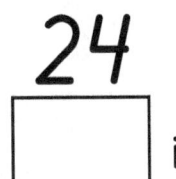

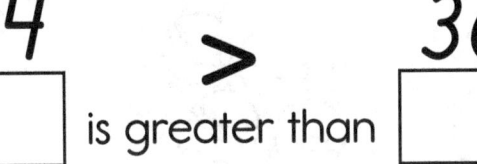

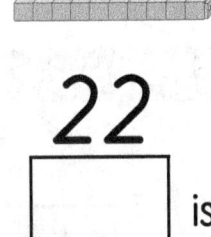

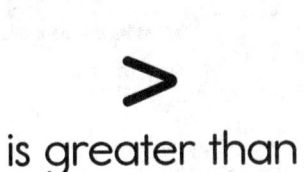

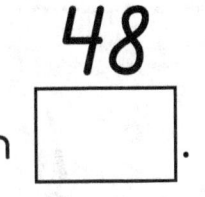

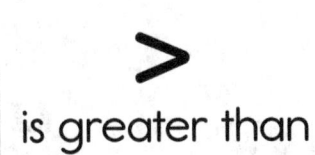

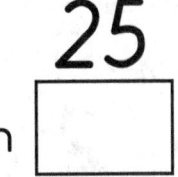

47 > **32**
[ ] is greater than [ ] .

24 > **36**
[ ] is greater than [ ] .

22 > **48**
[ ] is greater than [ ] .

32 > **25**
[ ] is greater than [ ] .

Circle the number that is *less*.
Complete the sentence.

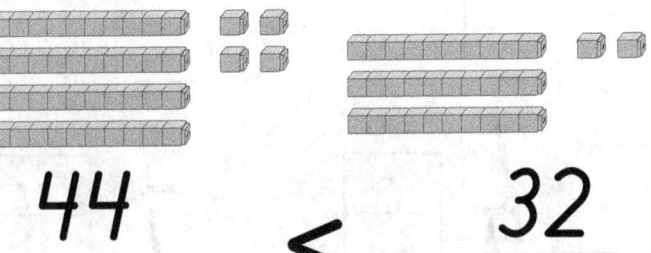

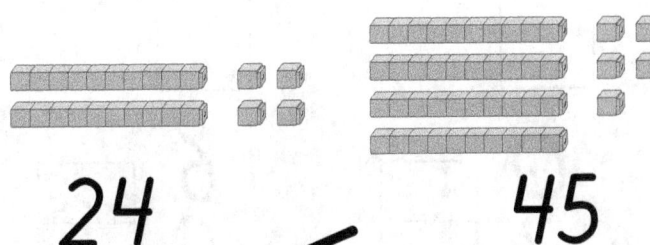

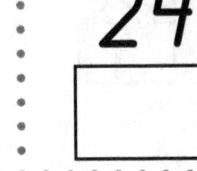

44 < **32**
[ ] is less than [ ] .

24 < **45**
[ ] is less than [ ] .

33 < **28**
[ ] is less than [ ] .

20 < **33**
[ ] is less than [ ] .

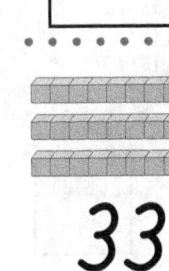

Chapter 4 • Lesson 24

Chapter 3 Review

Complete each number sentence.

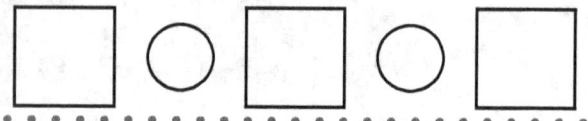

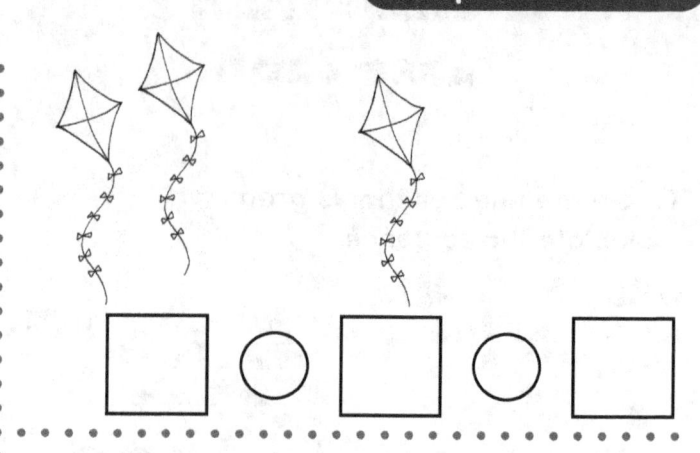

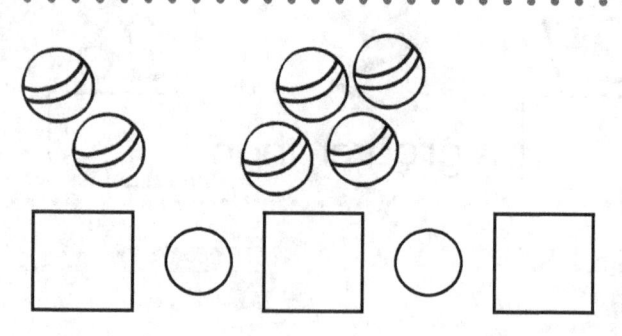

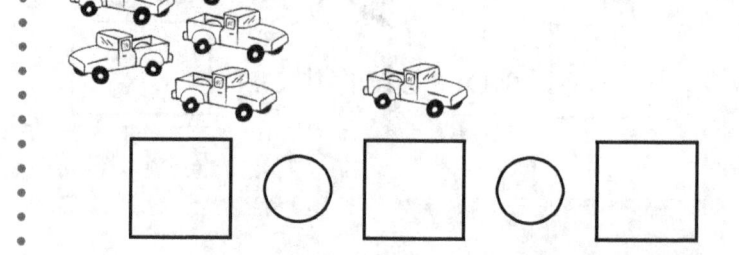

☐ ○ ☐ ○ ☐

☐ ○ ☐ ○ ☐

☐ ○ ☐ ○ ☐

☐ ○ ☐ ○ ☐

Addition Fact Review

Add.

$\begin{array}{r}2\\+1\\\hline\end{array}$ ☐

$\begin{array}{r}4\\+1\\\hline\end{array}$ ☐

$\begin{array}{r}1\\+3\\\hline\end{array}$ ☐

$\begin{array}{r}1\\+5\\\hline\end{array}$ ☐

$\begin{array}{r}0\\+3\\\hline\end{array}$ ☐

$\begin{array}{r}6\\+0\\\hline\end{array}$ ☐

$\begin{array}{r}1\\+4\\\hline\end{array}$ ☐

$\begin{array}{r}3\\+1\\\hline\end{array}$ ☐

$\begin{array}{r}5\\+1\\\hline\end{array}$ ☐

$\begin{array}{r}2\\+0\\\hline\end{array}$ ☐

$\begin{array}{r}0\\+0\\\hline\end{array}$ ☐

$\begin{array}{r}1\\+2\\\hline\end{array}$ ☐

48 forty-eight

Math 1 Reviews

Tens & Ones in Numbers to 100

Write the number of tens and ones.
Write the number.

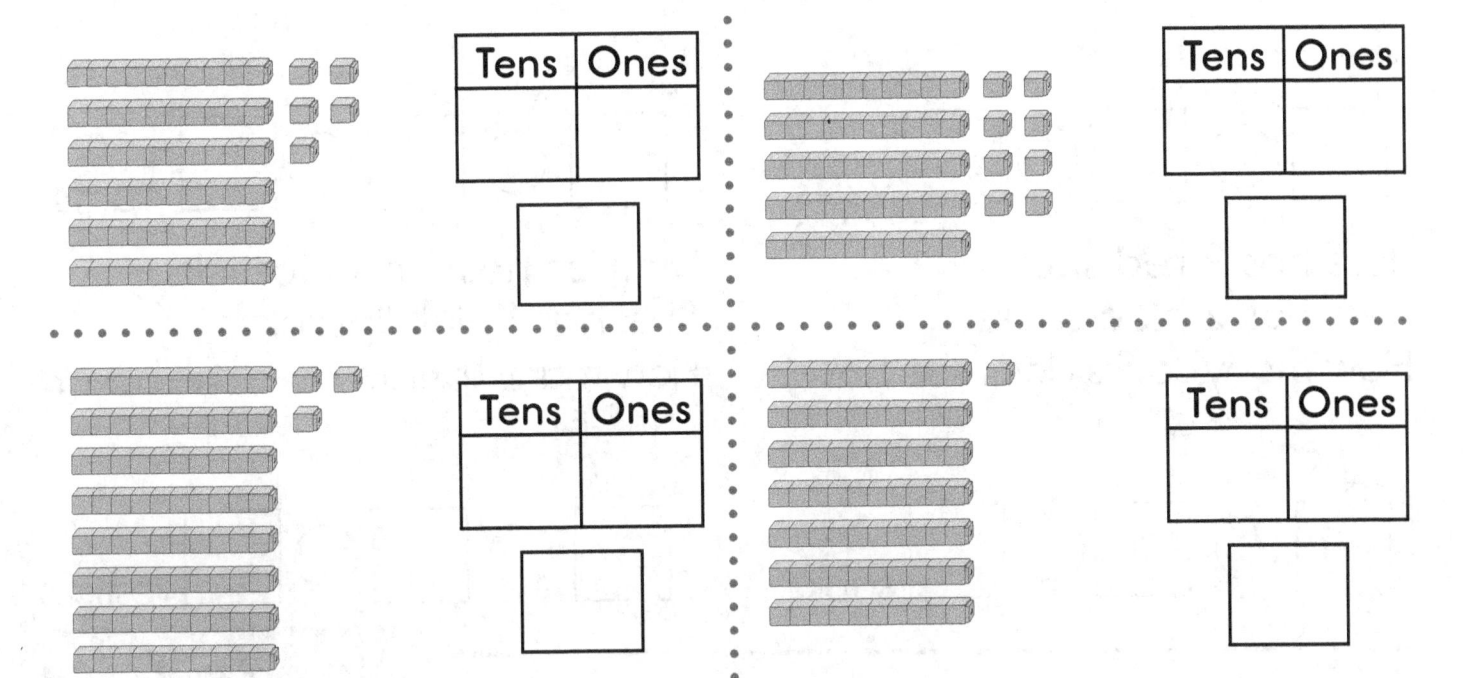

Write the number of tens and ones.
Complete the expanded form.

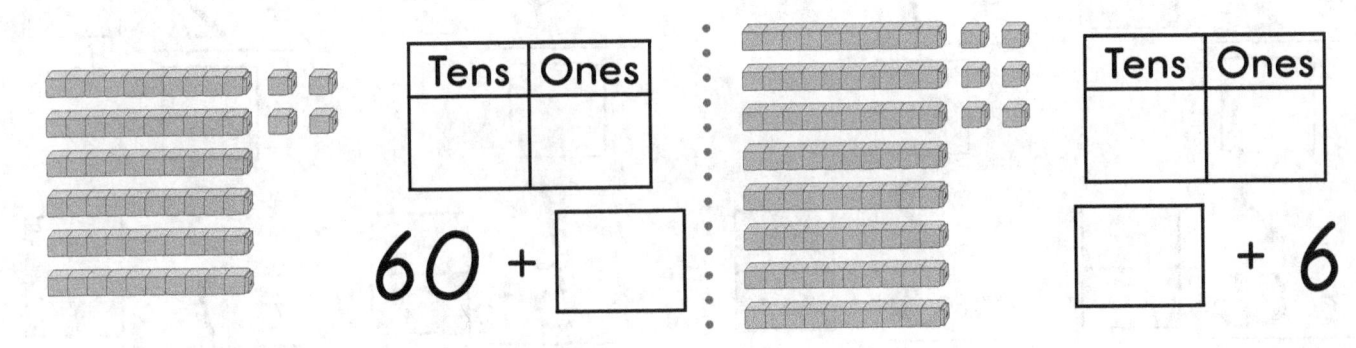

60 + ☐

☐ + 6

Write the expanded form of each number.

64 = ☐ + ☐

77 = ☐ + ☐

89 = ☐ + ☐

93 = ☐ + ☐

Chapter 4 • Lesson 25

forty-nine 49

Chapter 3 Review

Write a number sentence for each story.

Adam has 4 red vests.
Adam has 2 blue vests.
How many vests does he have?

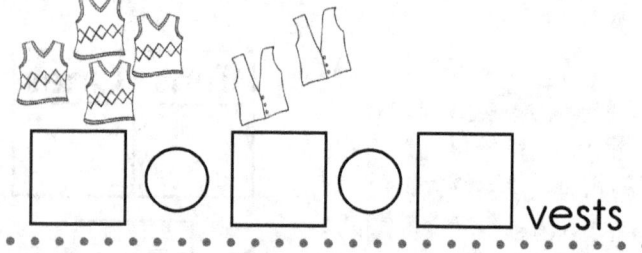

☐ ○ ☐ ○ ☐ vests

Ben has 3 tan pants.
Ben has 3 black pants.
How many pants does he have?

☐ ○ ☐ ○ ☐ pants

Jess has 2 red socks.
Jess has 2 blue socks.
How many socks does she have?

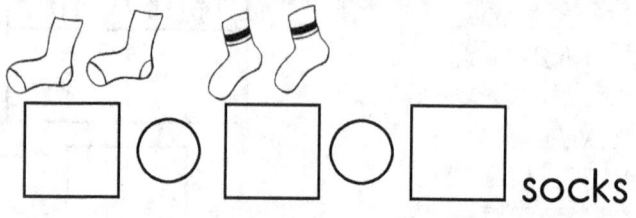

☐ ○ ☐ ○ ☐ socks

Mother has 3 red lipsticks.
She has 1 pink lipstick.
How many lipsticks does she have?

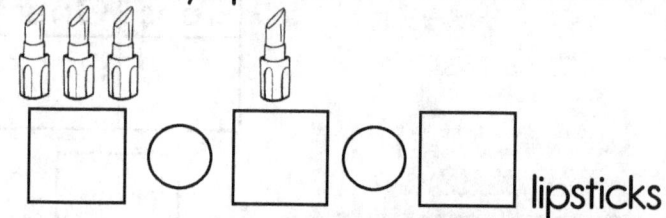

☐ ○ ☐ ○ ☐ lipsticks

Addition Fact Review

Add.

3 + 2 =

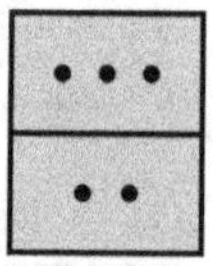

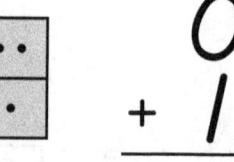

0 + 1 =

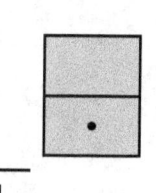

2 + 4 =

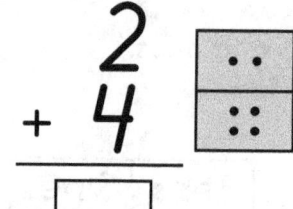

1 + 2 =

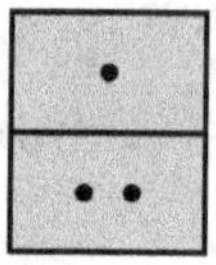

2 + 0 =

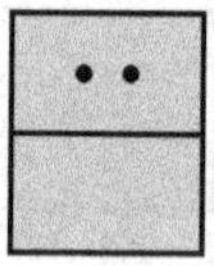

2 + 3 =

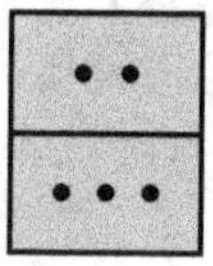

1 + 5 =

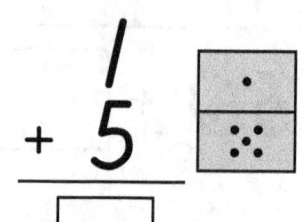

3 + 1 =

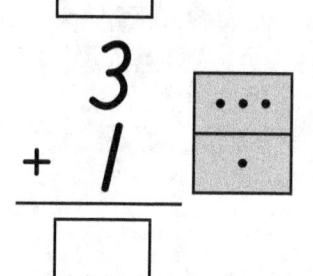

2 + 2 =

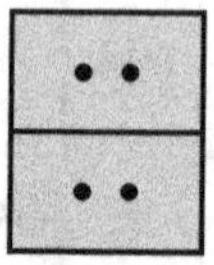

6 + 0 =

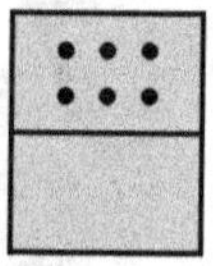

4 + 2 =

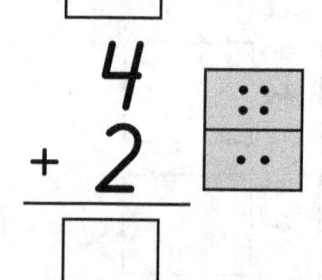

4 + 1 =

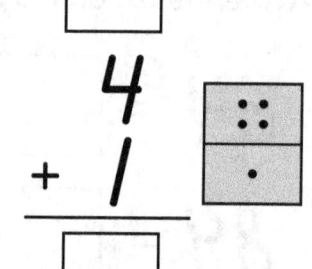

50 fifty

Math 1 Reviews

Greater Than & Less Than with Numbers to 100

Write the number. Circle the single cubes to make pairs.
Circle *even* if the number is even.
Circle *odd* if the number is odd.

even   odd

even   odd

even   odd

even   odd

Circle the correct sign.   > is greater than   < is less than

56 < 69

75 > 51

93 > 81

62 < 99

Chapter 4 • Lesson 26

Chapter 2 Review

Use the information to make a pictograph.

Don
Fred
Ed
Gary

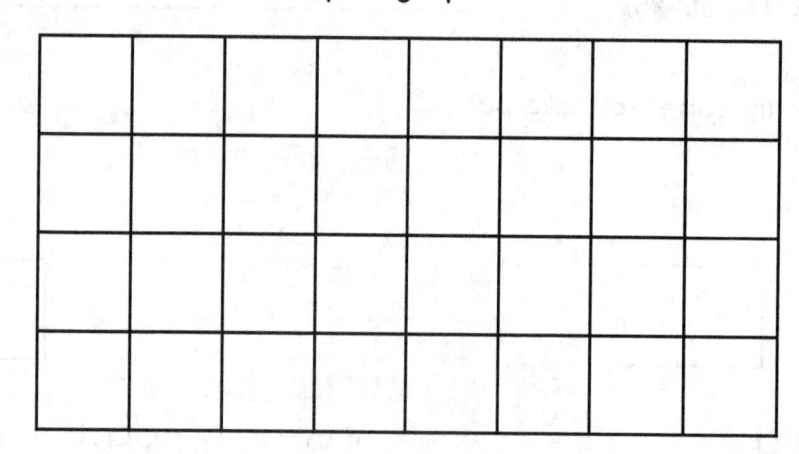

Don = 4
Fred = 6
Ed = 3
Gary = 8

Circle the person with the most balloons.
Underline the person with the fewest balloons.

Don   Fred   Ed   Gary

Addition Fact Review

Add.

$3 + 3 =$ ☐

$0 + 1 =$ ☐

$4 + 2 =$ ☐

$2 + 1 =$ ☐

$2 + 3 =$ ☐

$0 + 0 =$ ☐

$3 + 1 =$ ☐

$0 + 4 =$ ☐

$5 + 1 =$ ☐

$2 + 4 =$ ☐

$6 + 0 =$ ☐

$1 + 2 =$ ☐

52 fifty-two

Math 1 Reviews

Counting by 1s, 5s & 10s

Count by 1s on the number line.

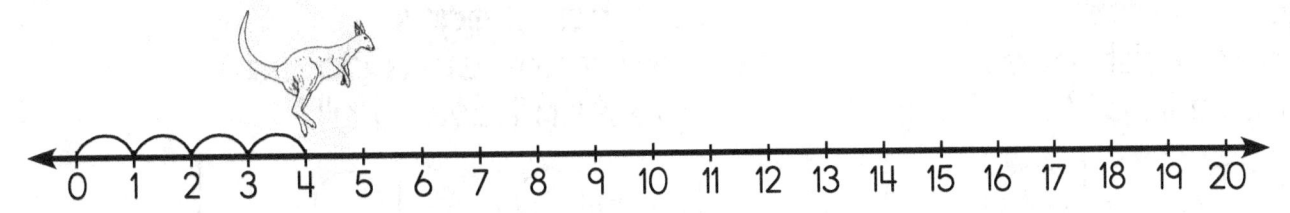

0 1 2 3 4 5 6 7 8 9 10 11 12 13 14 15 16 17 18 19 20

Count by 5s on the number line.

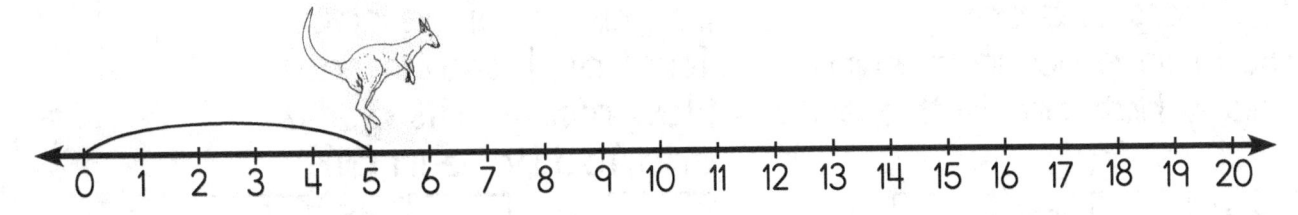

0 1 2 3 4 5 6 7 8 9 10 11 12 13 14 15 16 17 18 19 20

Complete each number pattern.

30	35	40	☐	50	65	70	75	☐	85
13	14	15	☐	17	20	30	40	☐	60
40	50	60	☐	80	41	42	43	☐	45

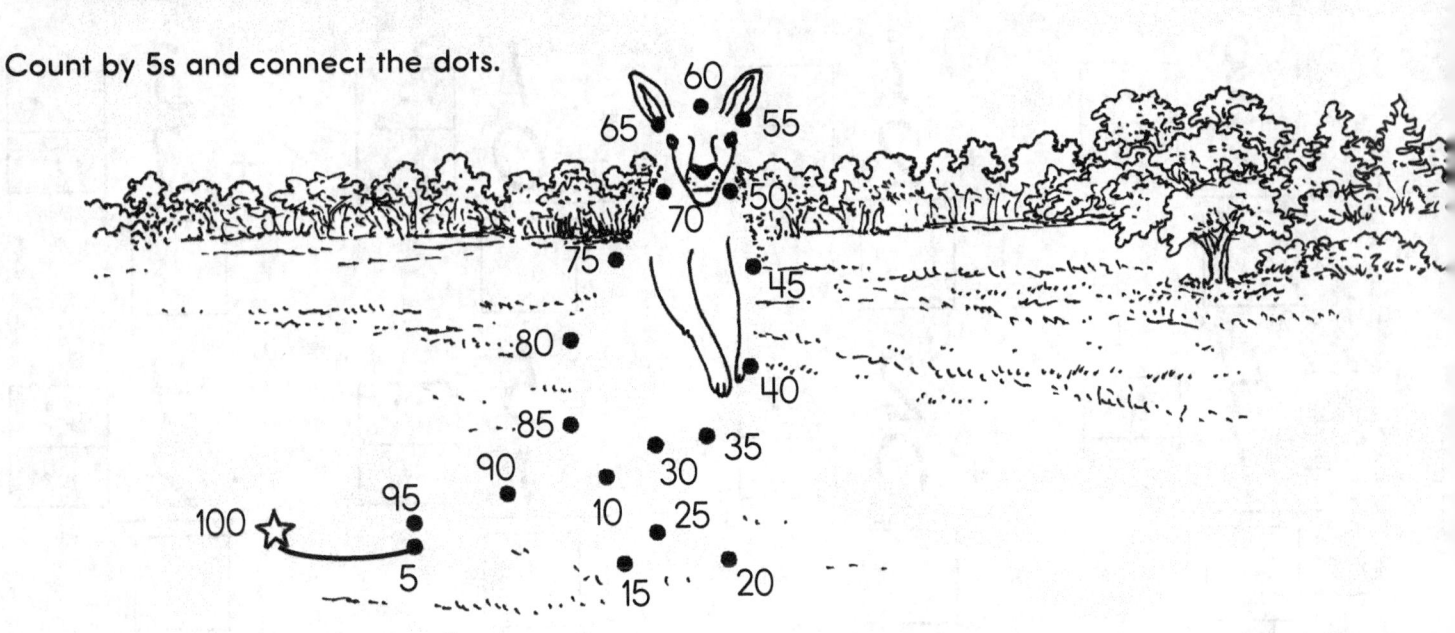

Count by 5s and connect the dots.

Chapter 4 • Lesson 27

fifty-three 53

Chapter 3 Review

Write a number sentence for each story.

Dan has 2 fish.
He gets 2 more fish.
How many fish does
Dan have in all?

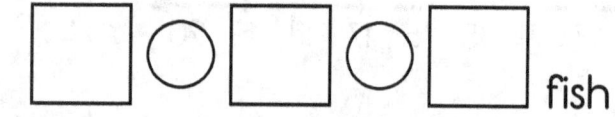

☐ ○ ☐ ○ ☐ fish

Sid has 4 eggs
Ann has 2 eggs.
How many eggs do Sid
and Ann have in all?

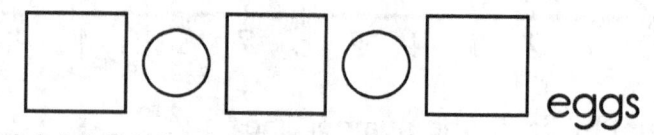

☐ ○ ☐ ○ ☐ eggs

Ed has 3 hats in a box.
He puts 1 more hat in the box.
How many hats are in the box?

☐ ○ ☐ ○ ☐ hats

Liz has 1 orange cat.
Tad has 4 brown cats.
How many cats do Liz
and Tad have in all?

☐ ○ ☐ ○ ☐ cats

Addition Fact Review

Add.

$\begin{array}{r}3\\+1\\\hline\end{array}$ $\begin{array}{r}0\\+0\\\hline\end{array}$ $\begin{array}{r}2\\+2\\\hline\end{array}$ $\begin{array}{r}0\\+3\\\hline\end{array}$

$\begin{array}{r}3\\+2\\\hline\end{array}$ $\begin{array}{r}1\\+2\\\hline\end{array}$ $\begin{array}{r}1\\+0\\\hline\end{array}$ $\begin{array}{r}3\\+3\\\hline\end{array}$

$\begin{array}{r}4\\+1\\\hline\end{array}$ $\begin{array}{r}2\\+0\\\hline\end{array}$ $\begin{array}{r}5\\+1\\\hline\end{array}$ $\begin{array}{r}4\\+2\\\hline\end{array}$

54 fifty-four

Math 1 Reviews

Introducing Hundreds

Write each missing number.

110	111	☐	☐	114	☐
142	☐	☐	145	☐	147
134	135	☐	137	☐	☐

Write the hundreds, tens, and ones.
Write the number.

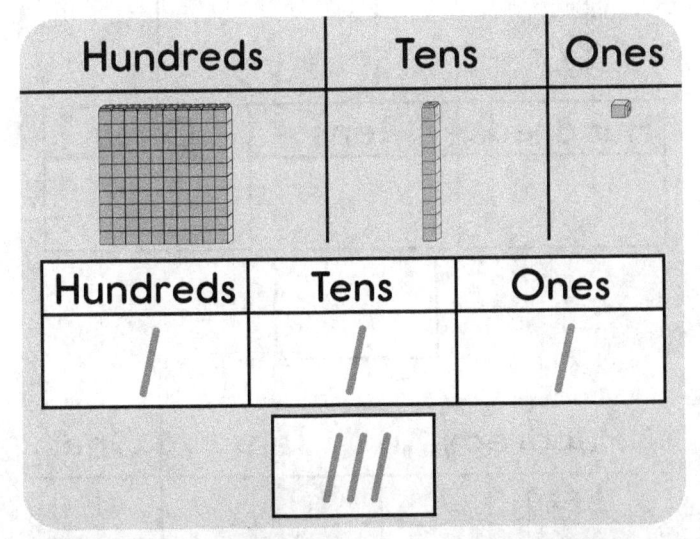

Hundreds	Tens	Ones
1	1	1

111

Hundreds	Tens	Ones

Hundreds	Tens	Ones

Hundreds	Tens	Ones

Chapter 4 • Lessons 28–29

Write the hundreds, tens, and ones.
Write the number.

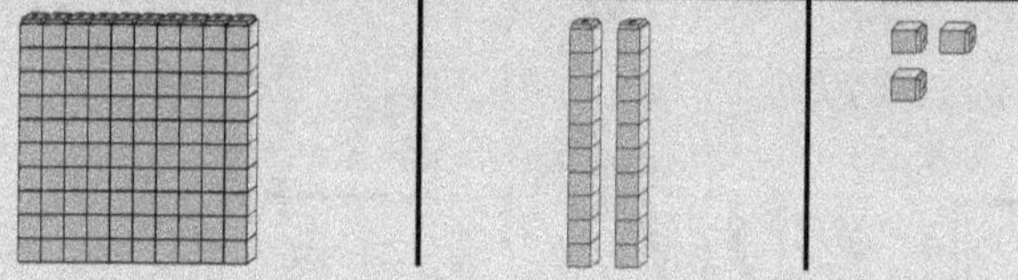

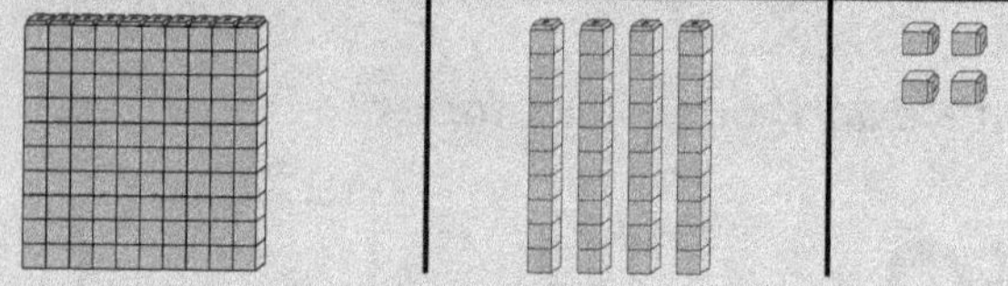

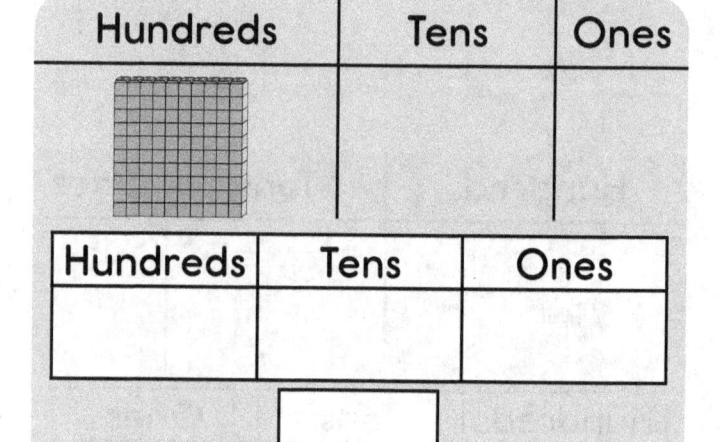

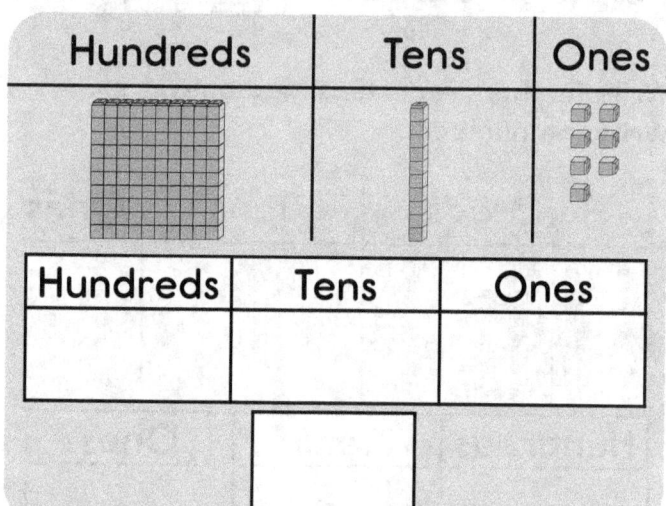

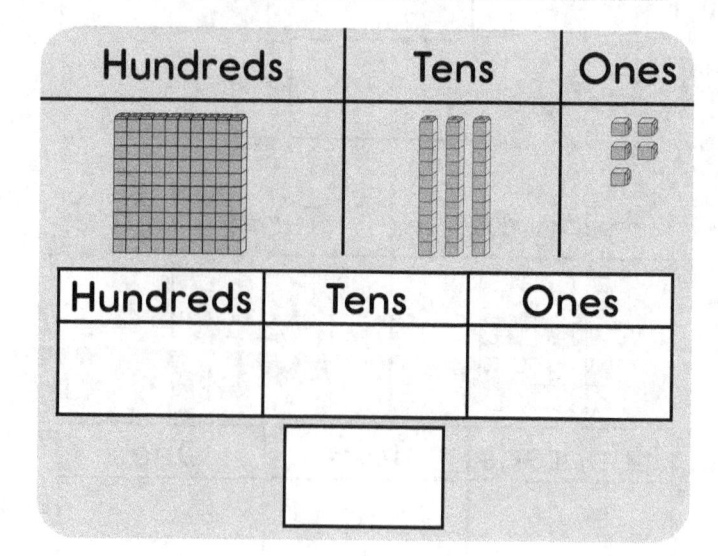

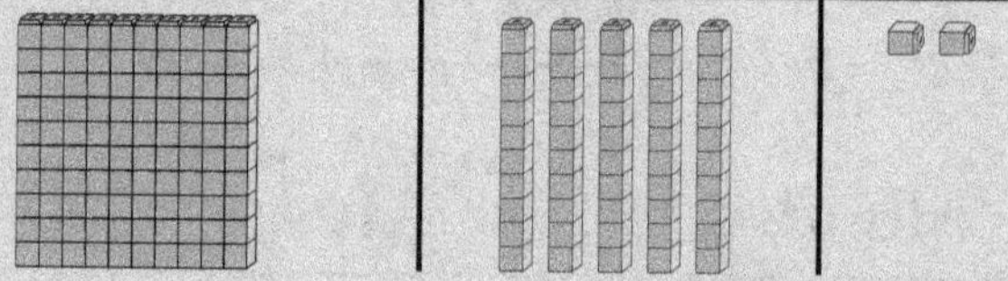

Representing Hundreds

Count by 100s to 1,000.

100 200 ☐ ☐ 500
600 ☐ 800 ☐ ☐

Write the hundreds, tens, and ones.
Write the number.

Hundreds	Tens	Ones
3	3	1

331

Hundreds	Tens	Ones

Hundreds	Tens	Ones

Hundreds	Tens	Ones

Chapter 5 • Lesson 30

Write the hundreds, tens, and ones.
Write the number.

58 fifty-eight Math 1 Reviews

Chapter 4 Review

Write the number of tens and ones.
Write the number.

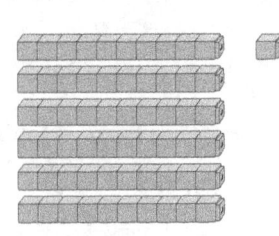

Tens	Ones

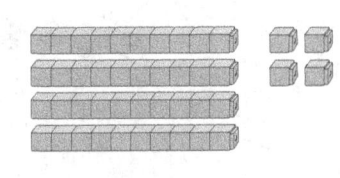

Tens	Ones

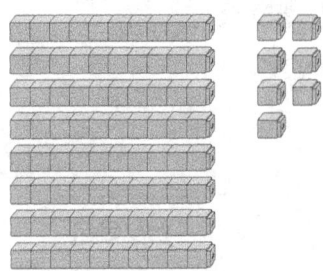

Tens	Ones

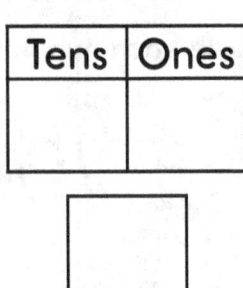

Mark the correct circle.

60 + 7

61　67　52　76
○　○　○　○

20 + 3

23　19　30　45
○　○　○　○

90 + 1

81　72　91　33
○　○　○　○

Complete the expanded form for each number.

69 = ☐ + ☐

33 = ☐ + ☐

84 = ☐ + ☐

97 = ☐ + ☐

Count by 5s.

5

55

Chapter 4 Review

fifty-nine 59

Write the number. Circle the pairs of blocks.
Circle *even* if the number is even.
Circle *odd* if the number is odd.

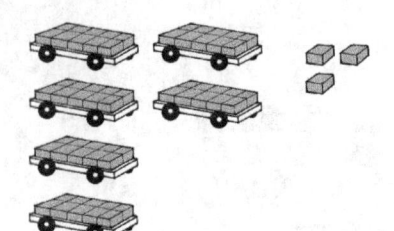

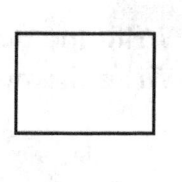

even   odd

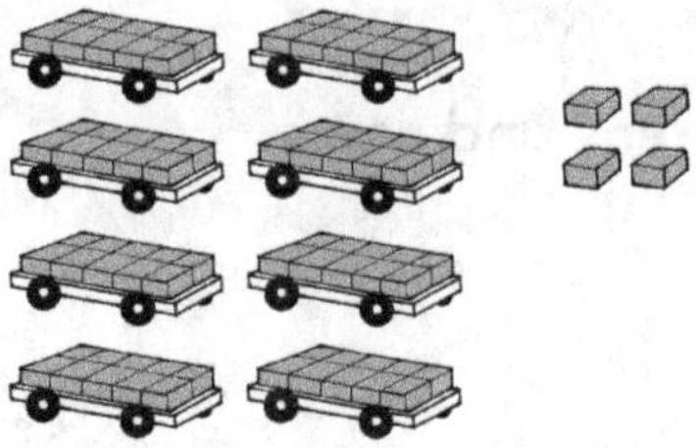

even   odd

Complete each number pattern.

8 9 10 ☐ 12   60 65 70 ☐ 80

20 30 40 ☐ 60   15 20 25 ☐ 35

Circle each correct sign.   > is greater than   < is less than

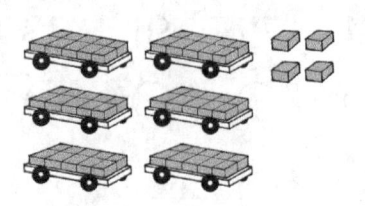

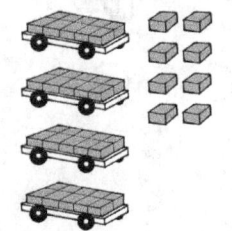

64 > < 48   73 > < 42

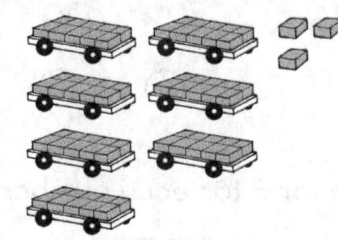

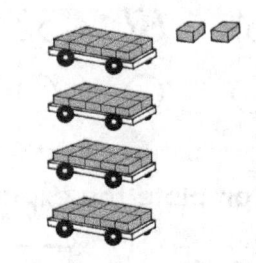

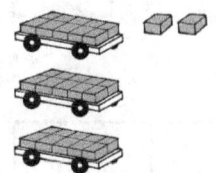

32 > < 24   56 > < 21

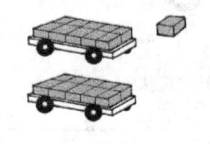

60 sixty   Math 1 Reviews

Cumulative Review

Write each missing number.

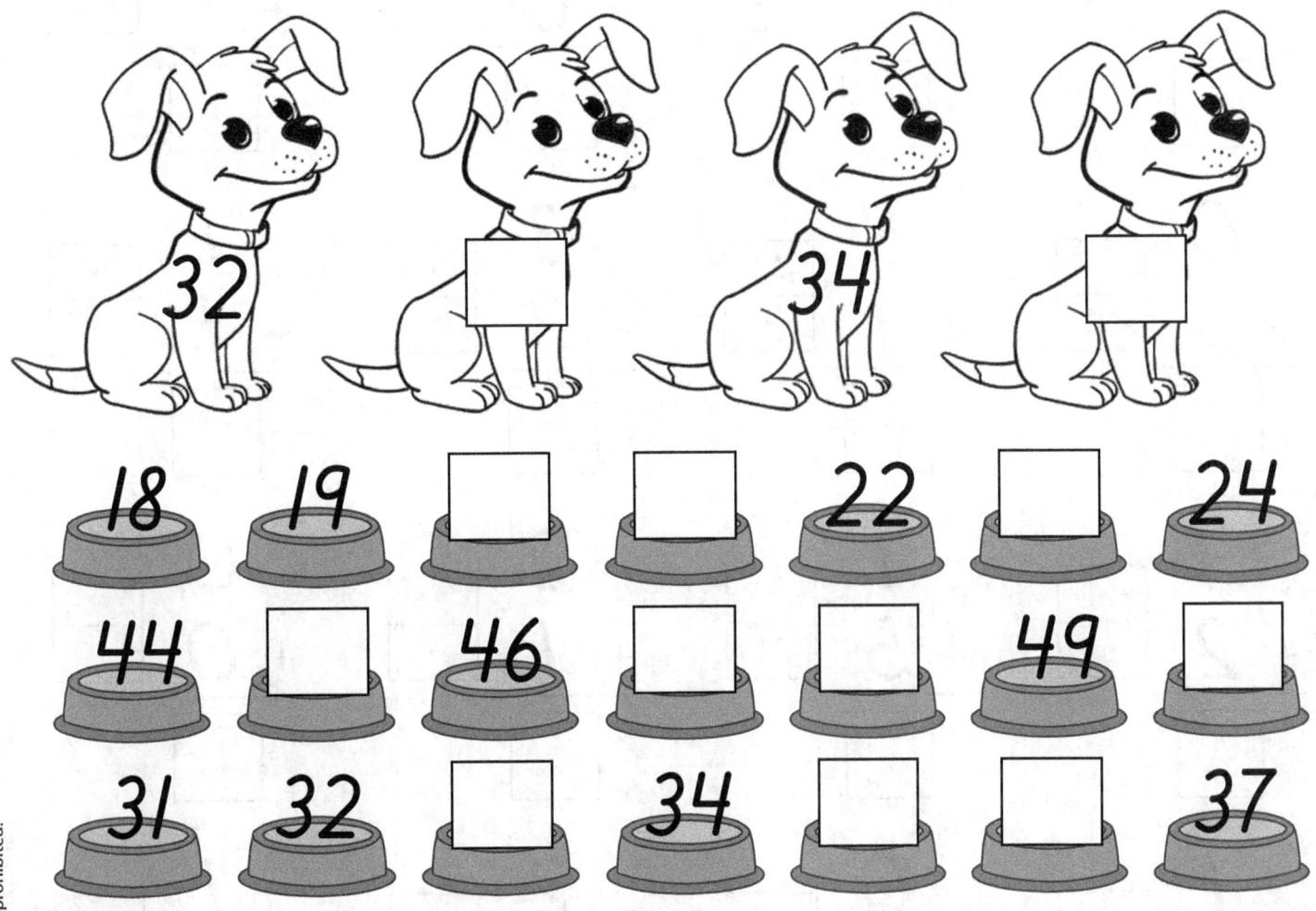

32   [ ]   34   [ ]

18 19 [ ] [ ] 22 [ ] 24

44 [ ] 46 [ ] [ ] 49 [ ]

31 32 [ ] 34 [ ] [ ] 37

Mark an *X* on the correct number of objects.

15

19

17

Chapter 4 Cumulative Review   sixty-one 61

Addition Fact Review

Add.

1 + 2 = ☐

2 + 2 = ☐

4 + 0 = ☐

3 + 1 = ☐

0 + 5 = ☐

4 + 1 = ☐

3 + 3 = ☐

1 + 1 = ☐

4 + 2 = ☐

1 + 5 = ☐

2 + 1 = ☐

0 + 0 = ☐

1 + 4 = ☐

0 + 3 = ☐

2 + 4 = ☐

1 + 3 = ☐

0 + 2 = ☐

2 + 1 = ☐

3 + 2 = ☐

6 + 0 = ☐

Beginning Subtraction

_ _ _ _ _ _ _ _ _ _ _ _ _ _ _

Complete each subtraction sentence.

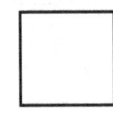

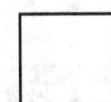

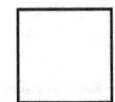

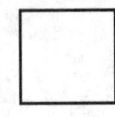

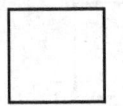

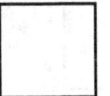

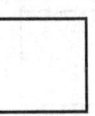

3 − 2 = ☐

☐ − ☐ = ☐

☐ − ☐ = ☐

☐ − ☐ = ☐

Chapter 5 • Lesson 34

sixty-three 63

Chapter 1 Review

Write the number of shoes. Circle the shoes to make pairs.
Circle *yes* if the number is even.
Circle *no* if the number is not even.

Is the number even?   yes   no

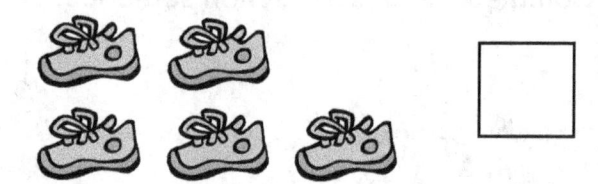

Is the number even?   yes   no

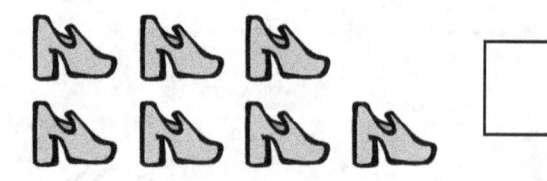

Is the number even?   yes   no

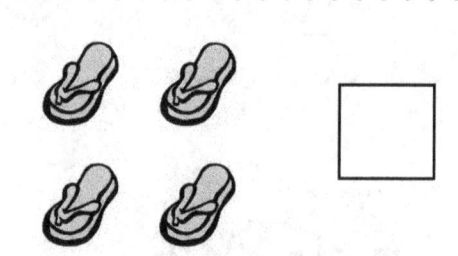

Is the number even?   yes   no

Addition Fact Review

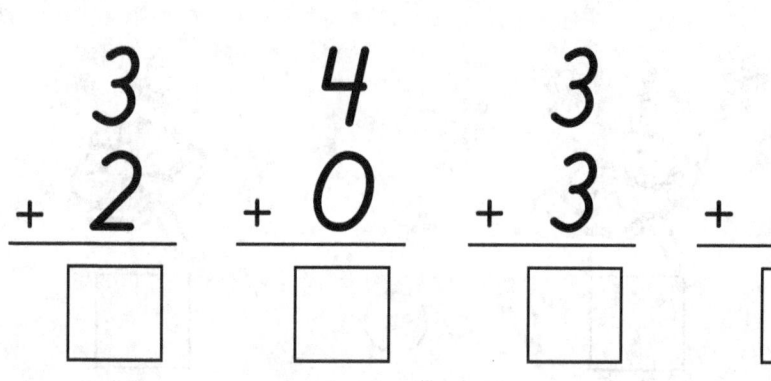

Add.

$1 + 1 =$

$1 + 4 =$

$0 + 2 =$

$3 + 1 =$

$4 + 2 =$

$5 + 0 =$

$3 + 2 =$

$4 + 0 =$

$3 + 3 =$

$2 + 1 =$

$0 + 0 =$

$2 + 2 =$

64 sixty-four

Math 1 Reviews

Subtraction Sentences

Complete each subtraction sentence.

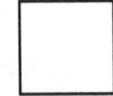

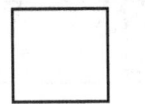

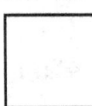

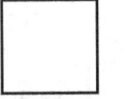

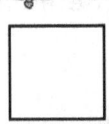

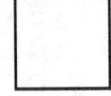

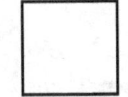

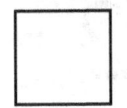

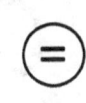

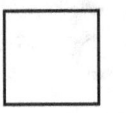

Chapter 5 • Lesson 35

sixty-five **65**

Chapter 2 Review

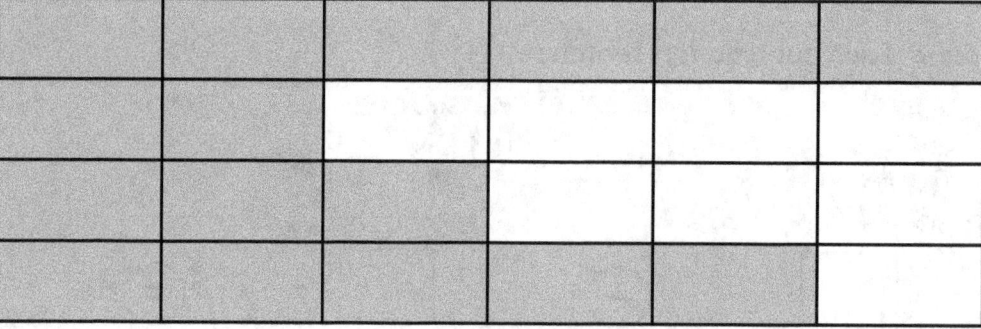

Look at the bar graph.
Write the number of each insect.

Ant
Butterfly
Ladybug
Bee

Circle the item with the most.   ant   ladybug

Circle the item with the least.   bee   butterfly

Addition Fact Review

Add.

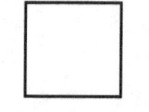

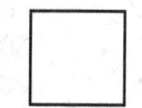

0 + 0 = ☐   1 + 5 = ☐   2 + 3 = ☐

1 + 1 = ☐   3 + 0 = ☐   4 + 1 = ☐

2 + 2 = ☐   1 + 2 = ☐   2 + 4 = ☐

3 + 3 = ☐   0 + 2 = ☐   1 + 3 = ☐

Cross Out to Subtract

Cross out the nuts to subtract.
Write the answer.

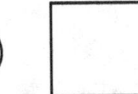

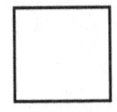

| 6 | − | 3 | = | ☐ |

| 4 | − | 2 | = | ☐ |

| 5 | − | 4 | = | ☐ |

Write a number sentence for each story.
Draw pictures to solve. Write each answer.

Nip has 6 nuts.
Nip gives 4 nuts to his mom.
How many nuts does Nip have left?

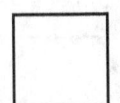

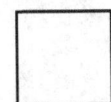

☐ − ☐ = ☐ nuts

Here are 4 nuts.
Biff taps 1 nut into his sack.
How many nuts are left?

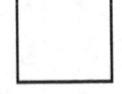

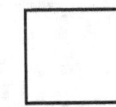

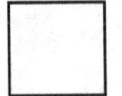

☐ − ☐ = ☐ nuts

Chapter 5 • Lesson 36

sixty-seven **67**

Chapter 4 Review

Write the number in each set.

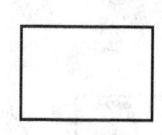

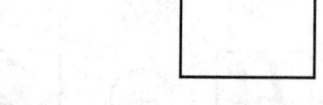

☐   ☐

☐   ☐

Addition Fact Review

Add.

3 + 0 = ☐   4 + 0 = ☐   5 + 0 = ☐

3 + 1 = ☐   4 + 1 = ☐   5 + 1 = ☐

3 + 2 = ☐   4 + 2 = ☐   1 + 2 = ☐

3 + 3 = ☐   2 + 3 = ☐   2 + 2 = ☐

Vertical Subtraction

Cross out the fish to subtract.
Write the answer.

1 - 0 = ☐

3 - 2 = ☐

5 - 4 = ☐

2 - 0 = ☐

6 - 1 = ☐

4
- 3

☐

6
- 0

☐

1
- 1

☐

2
- 2

☐

6
- 5

☐

Chapter 5 • Lesson 37

sixty-nine **69**

Chapter 2 Review

Write the number of pennies.
Write the number of dimes.

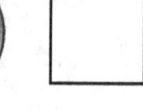

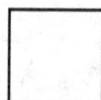

Circle the coin with the least.

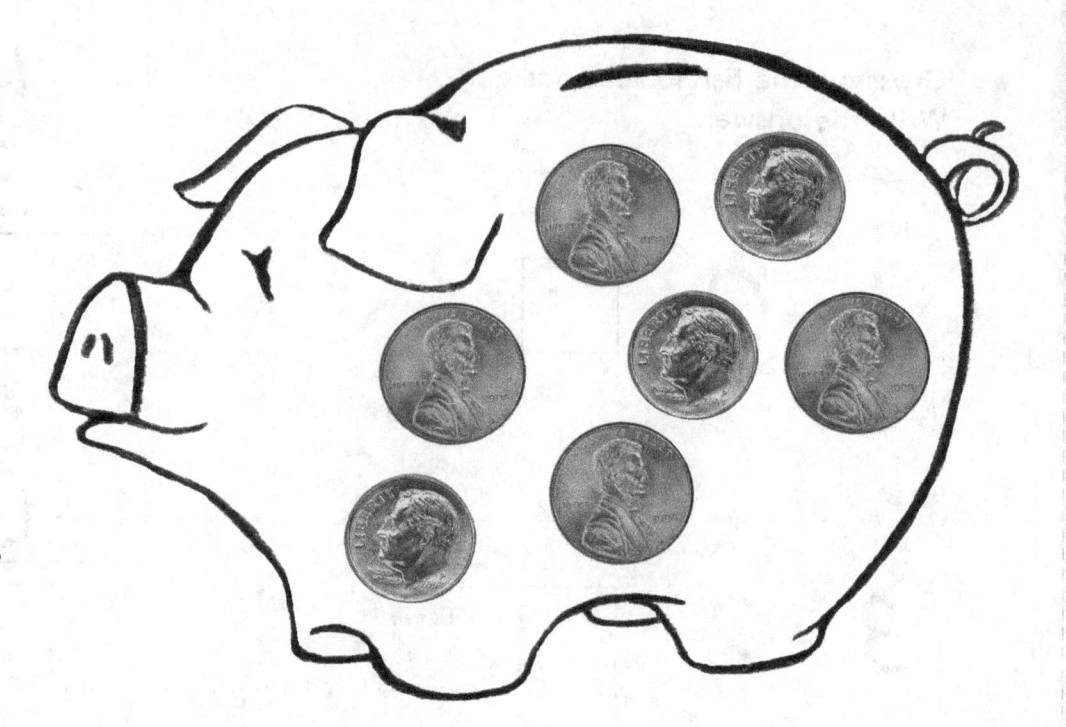

Addition Fact Review

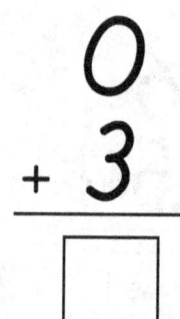

Add.

0 + 1 = ☐   1 + 2 = ☐   2 + 3 = ☐   0 + 2 = ☐   4 + 1 = ☐   1 + 4 = ☐

3 + 3 = ☐   0 + 3 = ☐   5 + 1 = ☐   1 + 3 = ☐   2 + 4 = ☐   6 + 0 = ☐

70 seventy

Math 1 Reviews

Related Subtraction Facts

Complete each related subtraction sentence.

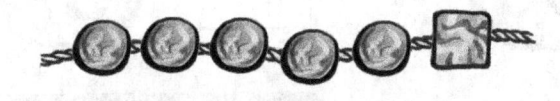

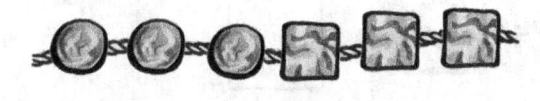

6 − 2 = ☐
6 − 4 = ☐

3 − 1 = ☐
3 − 2 = ☐

4 − 1 = ☐
4 − 3 = ☐

5 − 2 = ☐
5 − 3 = ☐

6 − 1 = ☐
6 − 5 = ☐

4 − 3 = ☐
4 − 1 = ☐

5 − 4 = ☐
5 − 1 = ☐

6 − 3 = ☐
6 − 3 = ☐

Chapter 5 • Lesson 38

Chapter 4 Review

Write the number that comes *after.*

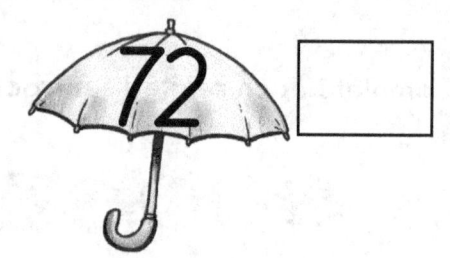

18 ☐   50 ☐   72 ☐

9 ☐   27 ☐   45 ☐

Write the number that comes *before.*

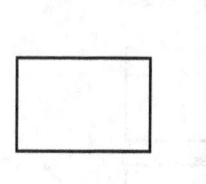

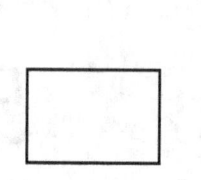

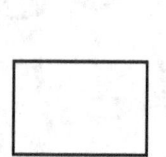

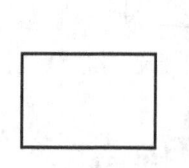

☐ 20   ☐ 43   ☐ 16

☐ 7   ☐ 68   ☐ 31

Addition Fact Review

Add.

2 + 1 = ☐   4 + 2 = ☐   2 + 0 = ☐

3 + 2 = ☐   5 + 1 = ☐   1 + 1 = ☐

2 + 2 = ☐   3 + 0 = ☐   4 + 1 = ☐

3 + 1 = ☐   1 + 0 = ☐   3 + 3 = ☐

Number Names

Cross out the dots to subtract. Write the answer.
Use the key to color the light bulb.

0 red
1 blue
2 green

5 - 4 = ☐
2 - 2 = ☐
6 - 4 = ☐
5 - 3 = ☐

4 - 4 = ☐
1 - 0 = ☐
3 - 2 = ☐
2 - 0 = ☐

Chapter 5 • Lesson 39

seventy-three 73

Chapter 2 Review

Look at the pictograph.
Write the correct number of each animal.
Circle the animal that has the most.

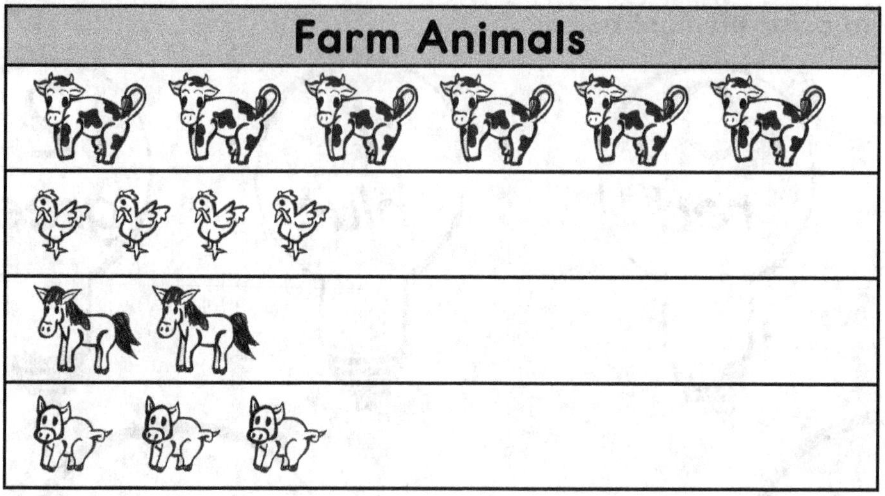

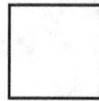

Farm Animals

Addition Fact Review

Add.

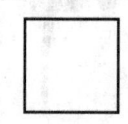

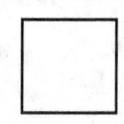

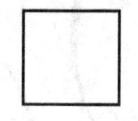

```
  1      2      3      2      0      4
+ 2    + 1    + 2    + 3    + 4    + 0
___    ___    ___    ___    ___    ___
 □      □      □      □      □      □


  2      4      1      3      3      2
+ 4    + 2    + 3    + 1    + 3    + 2
___    ___    ___    ___    ___    ___
 □      □      □      □      □      □
```

74 seventy-four

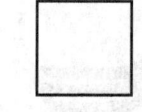

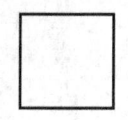

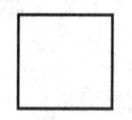

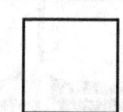

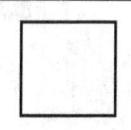

Math 1 Reviews

Chapter 5 Review

Complete each subtraction sentence.

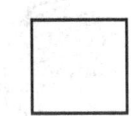

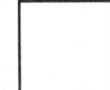

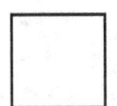

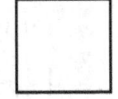

☐ — ☐ = ☐

☐ — ☐ = ☐

Cross out the balls to subtract.
Write the answer.

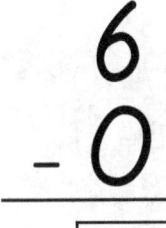

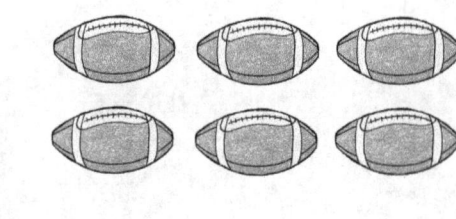

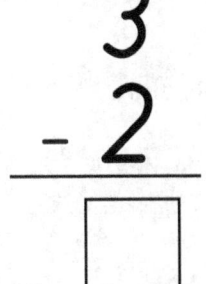

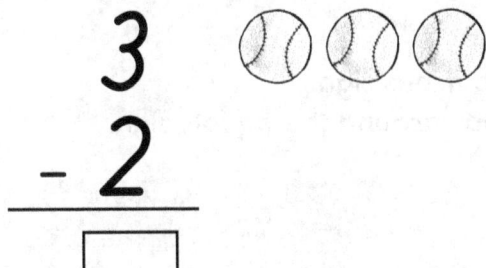

5
- 3

☐

6
- 0

☐

4
- 4

☐

3
- 2

☐

Chapter 5 Review

seventy-five **75**

Draw a line to match each pair of related facts.

4 − 1 = 3   5 − 2 = 3

6 − 2 = 4   4 − 3 = 1

5 − 3 = 2   6 − 4 = 2

Write a number sentence for each story.
Draw pictures to solve. Write each answer.

Tom has 3 flags.
He loses 1 flag.
How many flags does Tom have left?

Pam has 6 caps.
She gives 4 caps away.
How many caps does Pam have left?

☐ ⊖ ☐ ⊜ ☐ flags

☐ ⊖ ☐ ⊜ ☐ caps

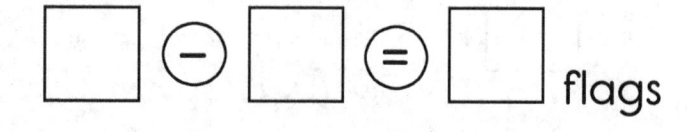

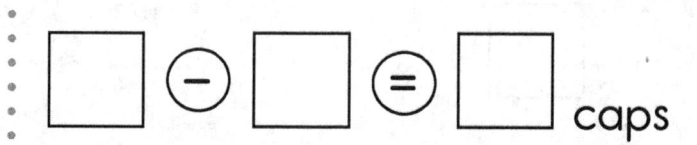

Circle the minus sign.
Draw a box around the equals sign.

=   +   ×   −

76 seventy-six

Math 1 Reviews

Cumulative Review

Count by 10s. Write each missing number.

10 ☐ ☐ 40 ☐ 60 ☐ 80 ☐ 100

Write the number of tens and ones.
Write the number.

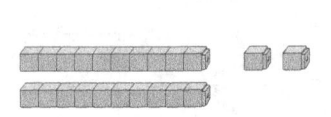

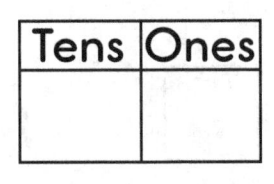

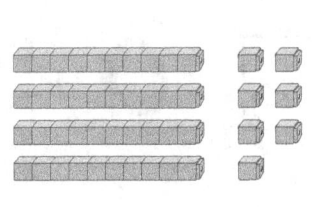

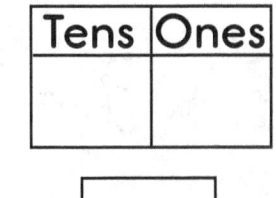

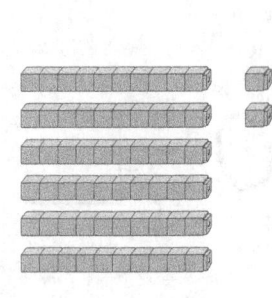

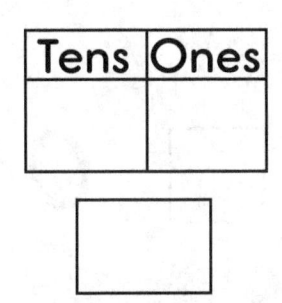

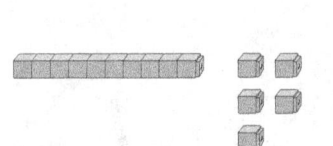

Tens	Ones

Tens	Ones

Tens	Ones

Tens	Ones

Draw a line to match the expanded form for each number.

43   20 + 5
78   40 + 3
25   60 + 2
62   70 + 8

Chapter 5 Cumulative Review   seventy-seven 77

Addition Fact Review

Add.

0 + 3 = ☐   4 + 1 = ☐   3 + 3 = ☐

1 + 1 = ☐   4 + 2 = ☐   2 + 1 = ☐

2 + 3 = ☐   0 + 4 = ☐   1 + 5 = ☐

5 + 1 = ☐   1 + 2 = ☐   3 + 2 = ☐

2 + 2 = ☐   1 + 3 = ☐   0 + 0 = ☐

3 + 1 = ☐   0 + 5 = ☐   2 + 4 = ☐

2 + 0 = ☐   1 + 4 = ☐   0 + 6 = ☐

Penny, Nickel & Dime

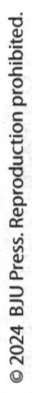

Write the value of each coin.

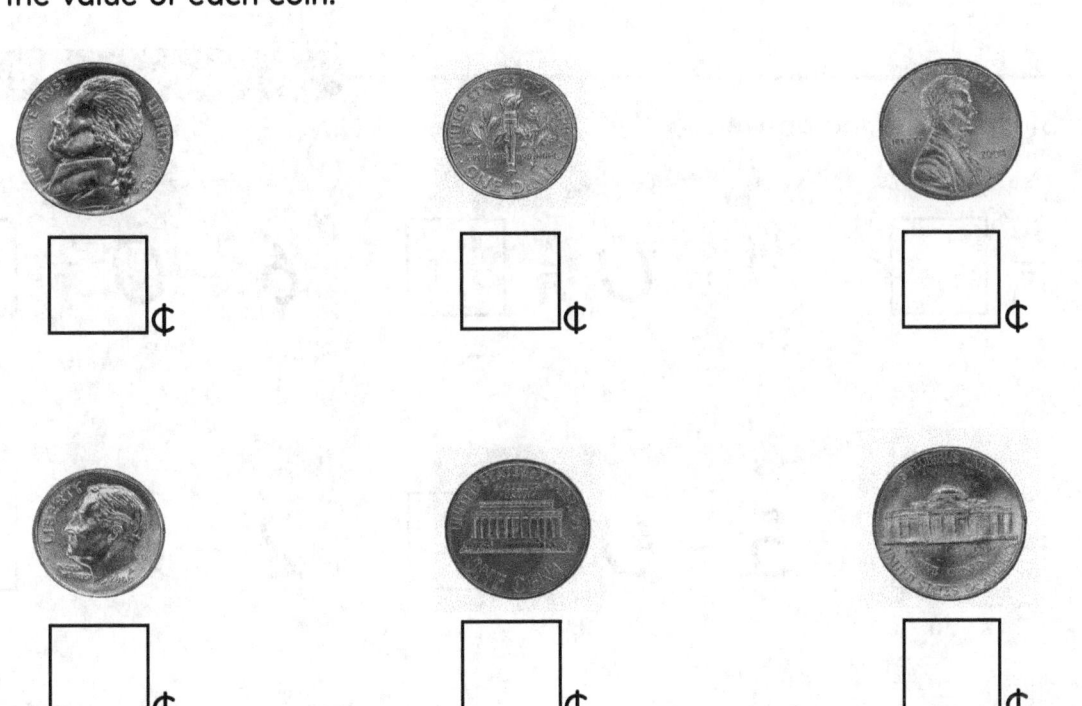

☐ ¢   ☐ ¢   ☐ ¢

☐ ¢   ☐ ¢   ☐ ¢

Chapter 6 • Lesson 42

seventy-nine **79**

Chapter 3 Review

Complete each addition sentence.
Cross out the sentence that does not belong.

4

2 + 2 = ☐

3 + 1 = ☐

3 + 2 = ☐

0 + 4 = ☐

5

3 + 2 = ☐

2 + 2 = ☐

5 + 0 = ☐

1 + 4 = ☐

6

6 + 0 = ☐

5 + 1 = ☐

3 + 3 = ☐

2 + 3 = ☐

Subtraction Fact Review

Cross out to subtract. Write the answer.

1 − 0 = ☐   4 − 0 = ☐   6 − 0 = ☐

3 − 0 = ☐   5 − 0 = ☐   2 − 0 = ☐

Counting with Coins

Write the total value on the pocket.

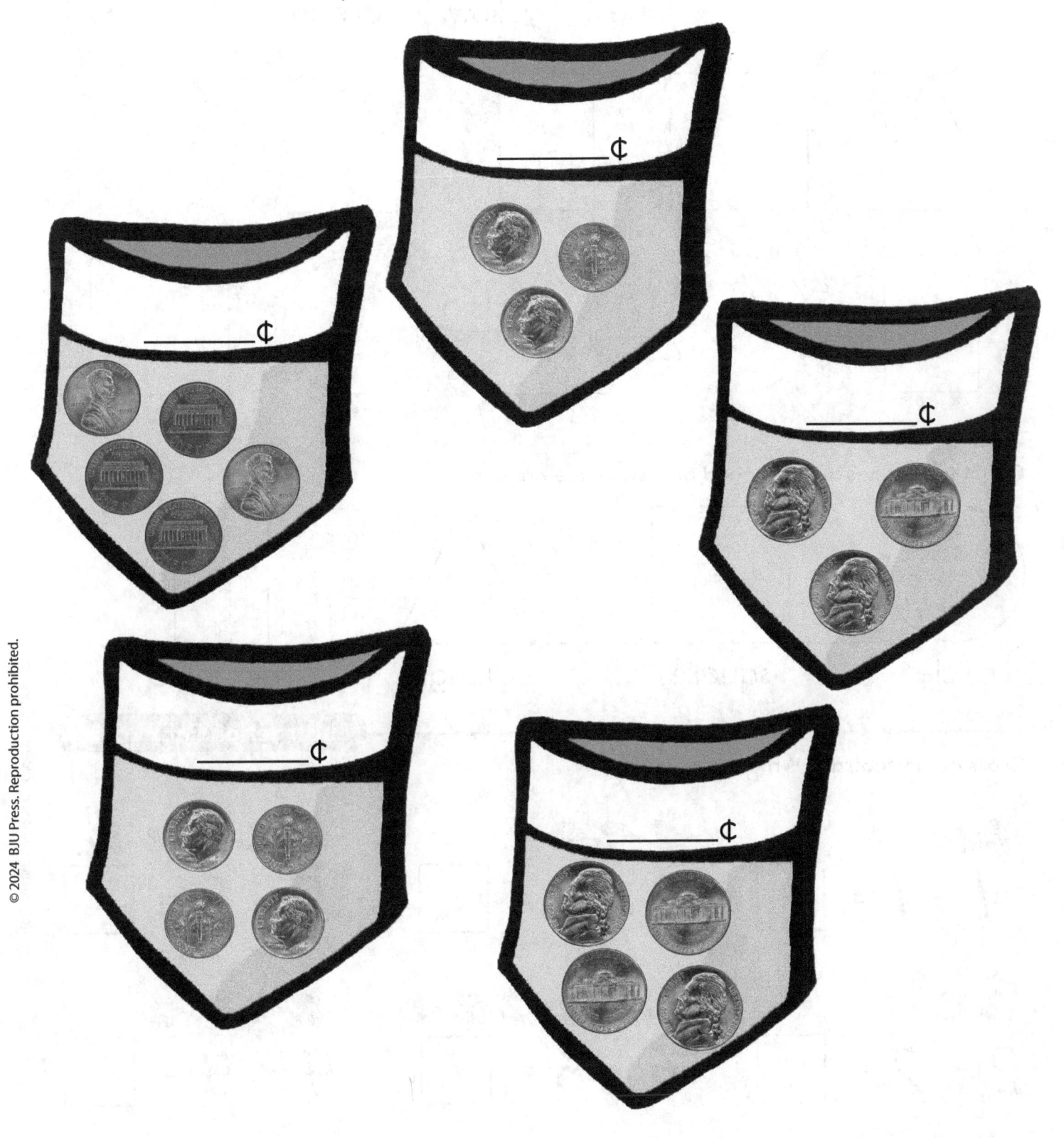

_____ ¢

_____ ¢

_____ ¢

_____ ¢

_____ ¢

Chapter 6 • Lesson 43

eighty-one **81**

Chapter 1 Review

Use the key to color each shape.

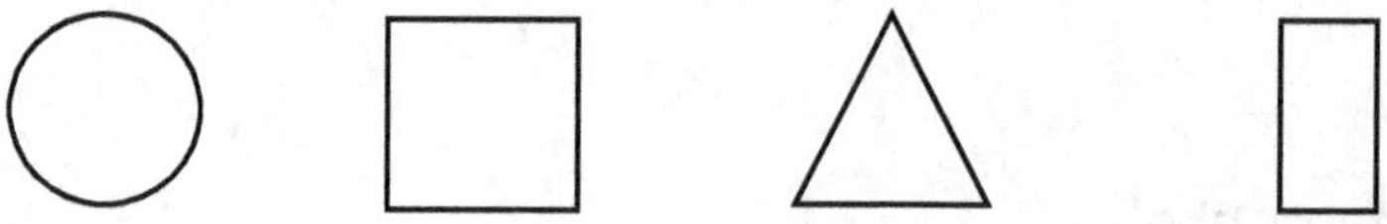

○ red   □ blue   △ yellow   ▭ green

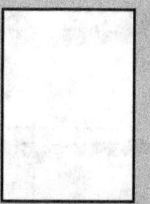

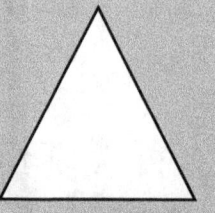

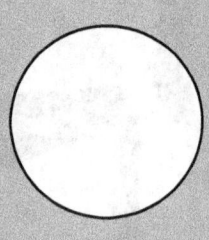

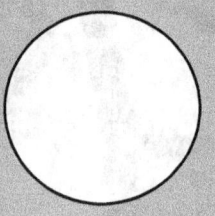

Count the shapes in the shaded box. Write the answer.

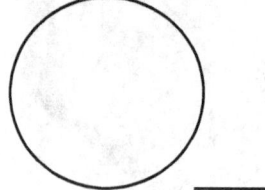

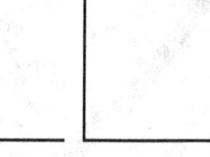

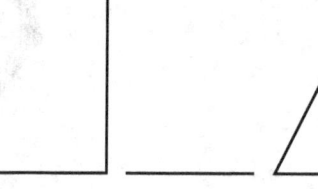

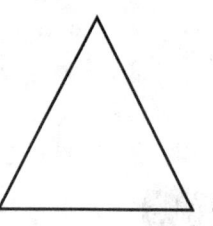

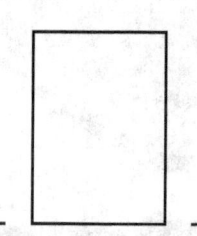

○ _____   □ _____   △ _____   ▭ _____
circle   square   triangle   rectangle

Subtraction Fact Review

Cross out to subtract. Write the answer.

1 − 1 = ☐   3 − 3 = ☐   5 − 5 = ☐

2 − 2 = ☐   6 − 6 = ☐   4 − 4 = ☐

82 eighty-two   Math 1 Reviews

Counting Dimes & Pennies

Write the total value as you *count on*.
Draw a line to match each set of coins with the correct item.

____¢ ____¢ ____¢ ____¢ ____¢ ____¢

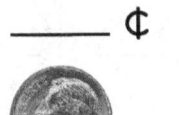

41¢

____¢ ____¢ ____¢ ____¢ ____¢

33¢

____¢ ____¢ ____¢ ____¢ ____¢

23¢

Write the number of dimes and pennies.
Write the total value.

Tens 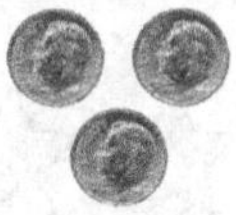	Ones 

____ dimes ____ pennies

____¢

Tens 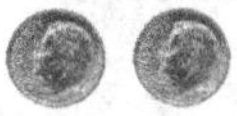	Ones 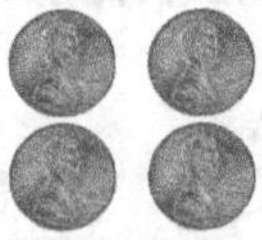

____ dimes ____ pennies

____¢

Tens 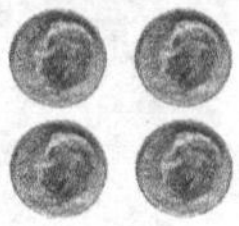	Ones 

____ dimes ____ pennies

____¢

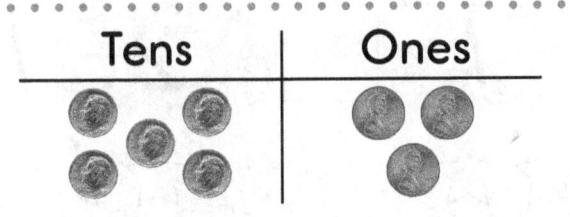

Tens	Ones

____ dimes ____ pennies

____¢

Chapter 6 • Lesson 44

Chapter 2 Review

Draw a line under the third duck.
Draw a circle around the ninth duck.
Draw a box around the seventh duck.

Chapter 1 Review

Write tally marks for each number.

1	2	3	4	5
6	7	8	9	10

Subtraction Fact Review

Cross out to subtract. Write the answer.

3 − 0 = ☐   2 − 0 = ☐   4 − 0 = ☐

3 − 1 = ☐   2 − 1 = ☐   4 − 4 = ☐

3 − 2 = ☐   2 − 2 = ☐   0 − 0 = ☐

84 eighty-four   Math 1 Reviews

Probability; Counting Nickels & Pennies

Write the total value as you *count on*.

___¢ ___¢ ___¢ ___¢

___¢ ___¢ ___¢ ___¢ ___¢ ___¢

___¢ ___¢ ___¢ ___¢ ___¢ ___¢

___¢ ___¢ ___¢ ___¢ ___¢

___¢ ___¢ ___¢ ___¢ ___¢

Chapter 6 • Lesson 45

Chapter 4 Review

Write the number that comes *before*.

☐ 23
☐ 46
☐ 70

Write the number that comes *after*.

35 ☐
59 ☐
98 ☐

Write the number that comes *between*.

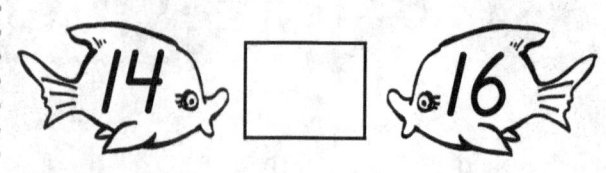

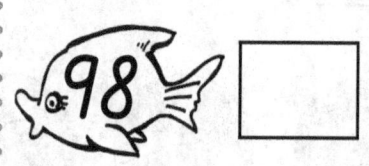

29 ☐ 31
14 ☐ 16
83 ☐ 85

Subtraction Fact Review

Cross out to subtract. Write the answer.

4 − 1 = ☐

4 − 3 = ☐

3 − 2 = ☐

3 − 1 = ☐

5 − 0 = ☐

5 − 5 = ☐

6 − 6 = ☐

6 − 0 = ☐

Chapter 6 Review

Write the value of each coin.

_____ ¢   _____ ¢   _____ ¢

Write the value for each set of coins.

_____ ¢   _____ ¢

_____ ¢   _____ ¢

Chapter 6 Review   eighty-seven 87

Write the total value as you *count on*.

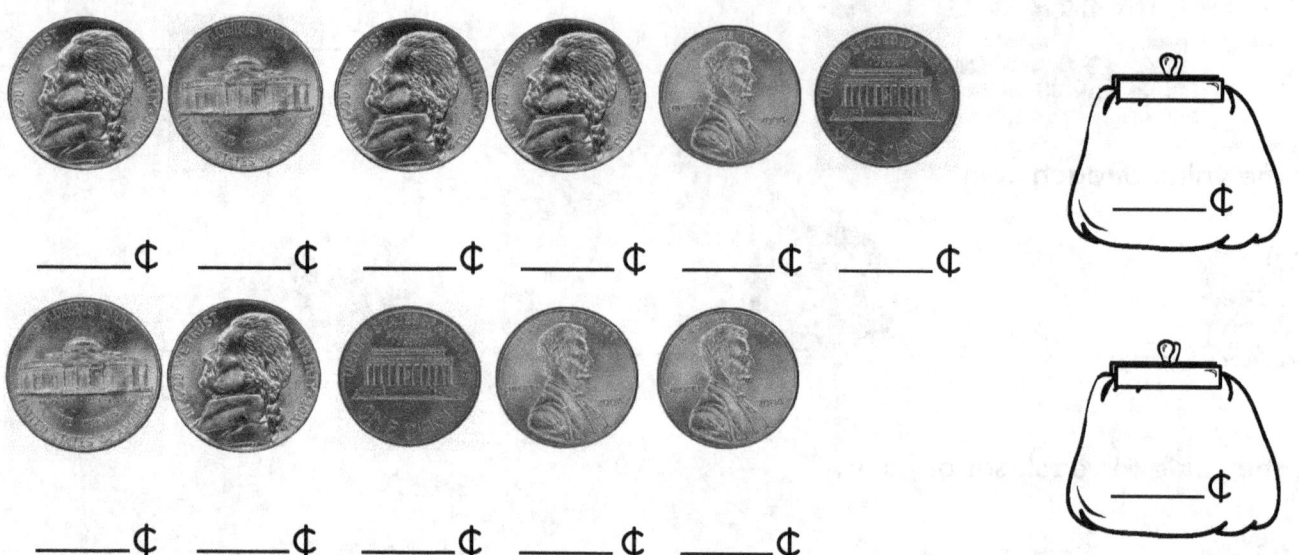

____¢ ____¢ ____¢ ____¢ ____¢ ____¢

____¢

____¢ ____¢ ____¢ ____¢ ____¢

____¢

Write the total value as you *count on*.
Do you have enough money to buy each item? Circle the answer.

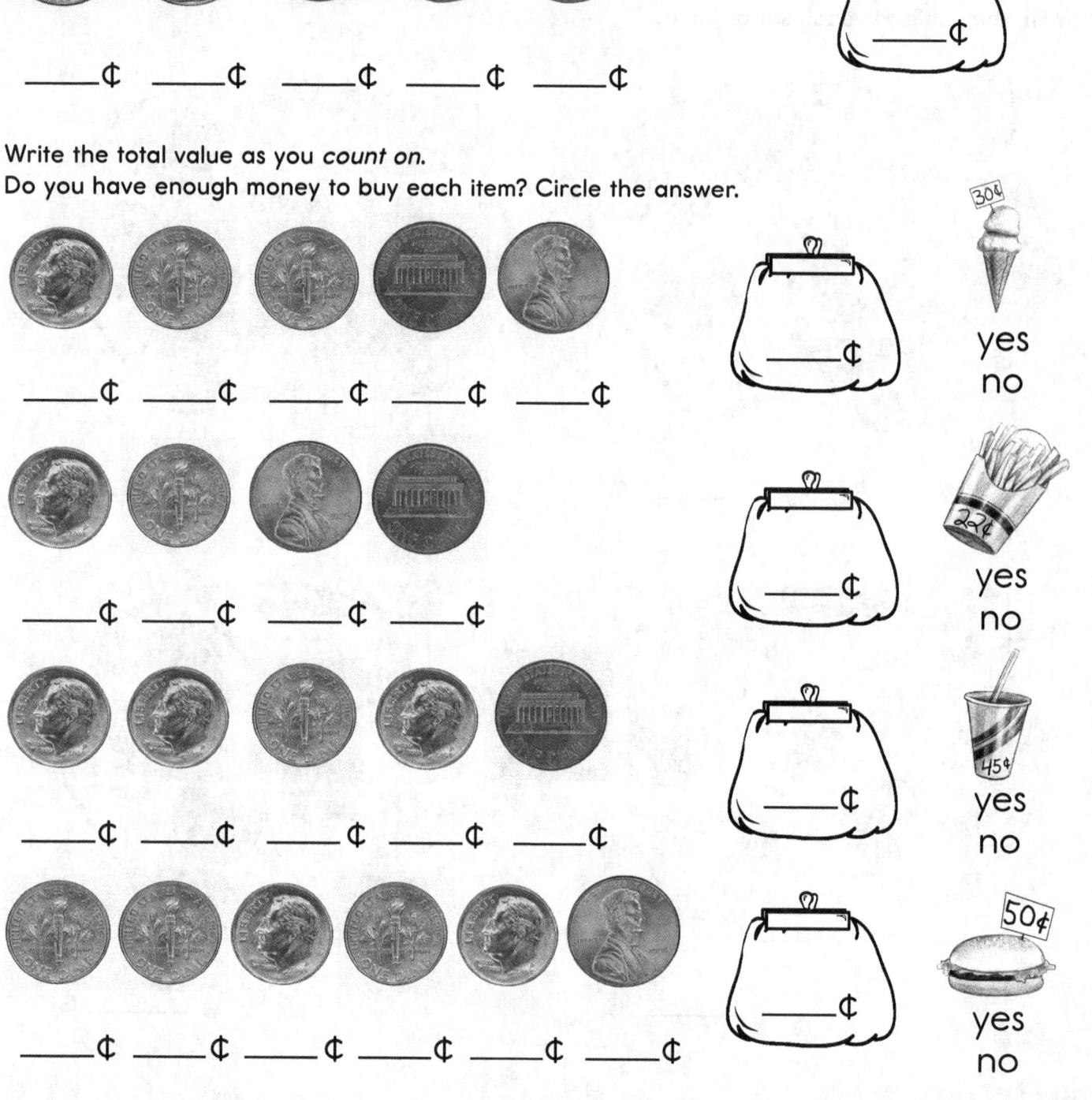

____¢ ____¢ ____¢ ____¢ ____¢

____¢ 30¢ ice cream yes / no

____¢ ____¢ ____¢ ____¢

____¢ 22¢ fries yes / no

____¢ ____¢ ____¢ ____¢ ____¢

____¢ 45¢ drink yes / no

____¢ ____¢ ____¢ ____¢ ____¢ ____¢

____¢ 50¢ burger yes / no

88 eighty-eight

Math 1 Reviews

Cumulative Review

Write the number of tens and ones.
Complete the expanded form.

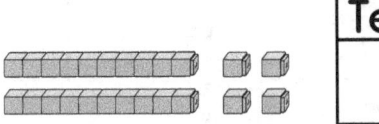

Tens	Ones

20 + ☐

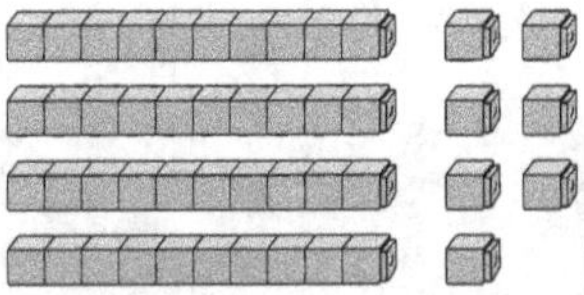

Tens	Ones

☐ + 7

Draw a line to match each expanded form to the correct number.

30 + 7   23

50 + 9   14

20 + 3   37

10 + 4   72

70 + 2   59

Write the number. Circle to make pairs.
Circle *even* if the number is even.
Circle *odd* if the number is odd.

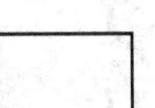

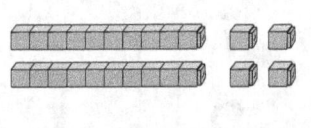

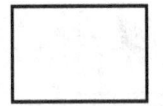

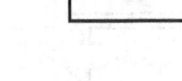

even   odd

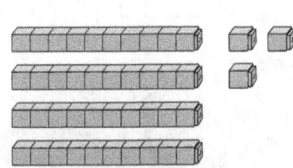

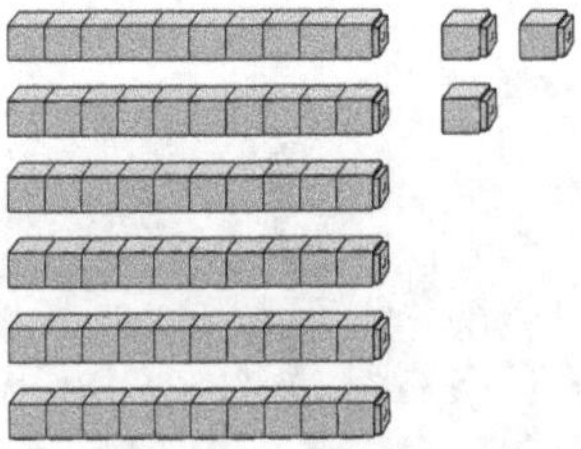

even   odd

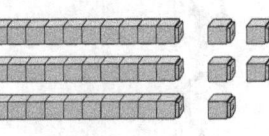

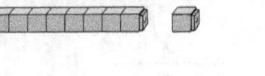

even   odd

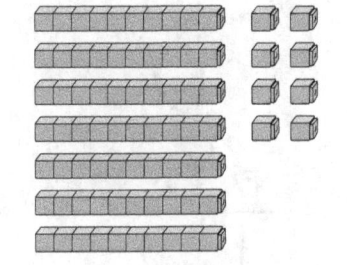

even   odd

Chapter 6 Cumulative Review   eighty-nine **89**

Addition Fact Review

Add.

2 + 0 = ☐	1 + 3 = ☐	4 + 2 = ☐	2 + 1 = ☐	0 + 0 = ☐
5 + 1 = ☐	3 + 3 = ☐	5 + 0 = ☐	1 + 1 = ☐	3 + 2 = ☐
6 + 0 = ☐	2 + 2 = ☐	0 + 1 = ☐	4 + 1 = ☐	0 + 4 = ☐
1 + 5 = ☐	3 + 1 = ☐	2 + 4 = ☐	1 + 2 = ☐	0 + 2 = ☐
2 + 3 = ☐	0 + 6 = ☐	1 + 4 = ☐	4 + 0 = ☐	0 + 5 = ☐

Count on 1 or 2

Count on 1

$8 + 1 = \boxed{}$

Count on 2

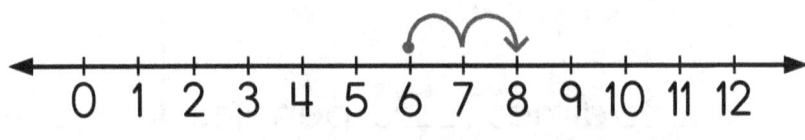

$6 + 2 = \boxed{}$

Use the number line to add.
Draw ⤵ to find the sum.

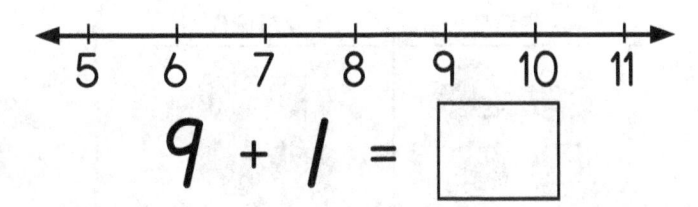

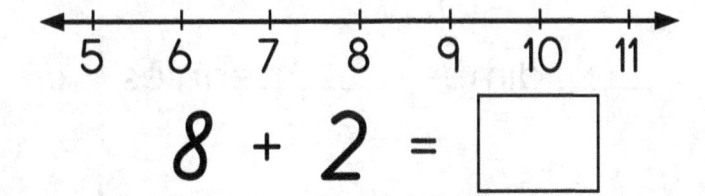

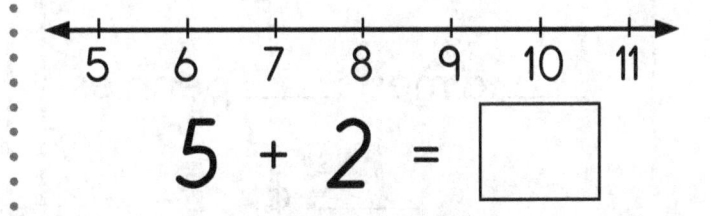

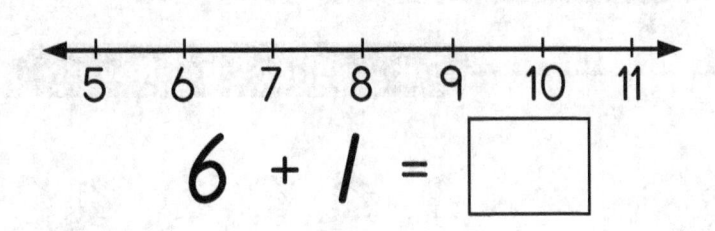

$9 + 1 = \boxed{}$

$7 + 1 = \boxed{}$

$8 + 2 = \boxed{}$

$5 + 2 = \boxed{}$

$6 + 1 = \boxed{}$

$5 + 1 = \boxed{}$

Circle each addend.
Draw a square around each sum.

$8 + 3 = 11$ $3 + 9 = 12$ $5 + 5 = 10$

Chapter 7 • Lesson 48

Chapter 6 Review

Write the number of dimes and pennies.
Write the total value.

Tens	Ones
(3 dimes)	(4 pennies)

___ dimes ___ pennies

___ ¢

Tens	Ones
(6 dimes)	

___ dimes ___ pennies

___ ¢

Tens	Ones
(1 dime)	(5 pennies)

___ dimes ___ pennies

___ ¢

Tens	Ones
(4 dimes)	(5 pennies)

___ dimes ___ pennies

___ ¢

Subtraction Fact Review

Cross out to subtract. Write the answer.

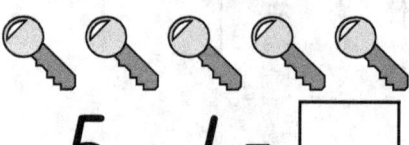

5 − 1 = ☐

3 − 2 = ☐

5 − 4 = ☐

3 − 1 = ☐

4 − 3 = ☐

1 − 1 = ☐

Zero & Order Principles

Read the principle. Complete the facts.

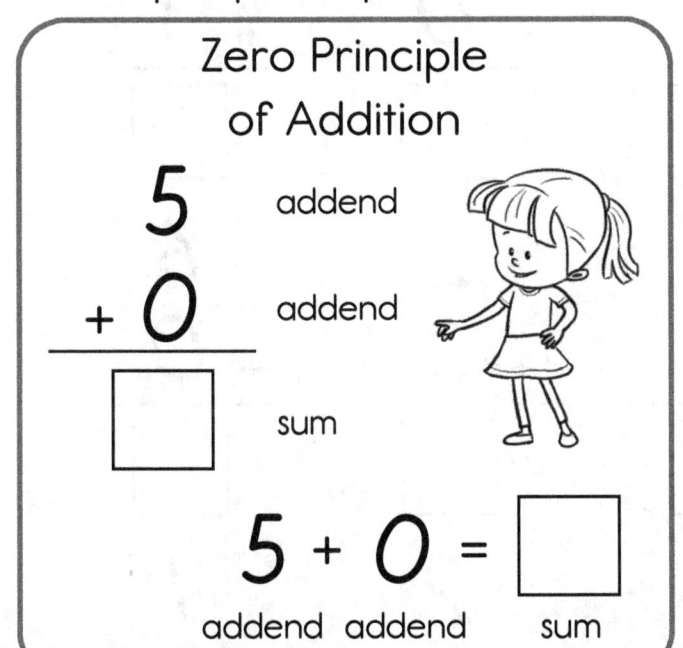

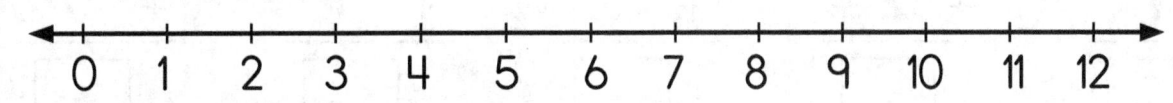

Zero Principle of Addition

5 addend
+ 0 addend
☐ sum

5 + 0 = ☐
addend addend   sum

Order Principle of Addition

2 + 3 = ☐

3 + 2 = ☐

0 1 2 3 4 5 6 7 8 9 10 11 12

Add. Use the number line if needed.
Use the Order Principle to write the new fact.

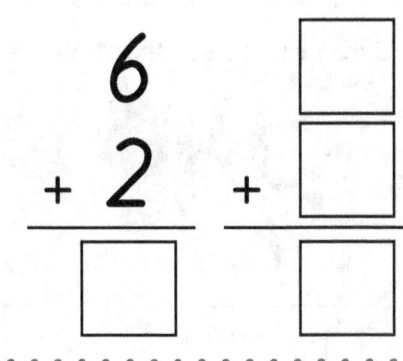

5   ☐   9   ☐   6   ☐
+ 3   + ☐   + 0   + ☐   + 2   + ☐
──   ──   ──   ──   ──   ──
☐   ☐   ☐   ☐   ☐   ☐

7   ☐   8   ☐   0   ☐
+ 3   + ☐   + 1   + ☐   + 7   + ☐
──   ──   ──   ──   ──   ──
☐   ☐   ☐   ☐   ☐   ☐

Chapter 7 • Lesson 49

Chapter 3 Review

Add.

```
  2        5        1        4        3
+ 2      + 1      + 3      + 0      + 3
─────    ─────    ─────    ─────    ─────
 □        □        □        □        □

  1        4        1        2        6
+ 1      + 2      + 4      + 3      + 0
─────    ─────    ─────    ─────    ─────
 □        □        □        □        □

  2        3        0        2        1
+ 1      + 2      + 5      + 4      + 5
─────    ─────    ─────    ─────    ─────
 □        □        □        □        □
```

Subtraction Fact Review

Cross out to subtract. Write the answer.

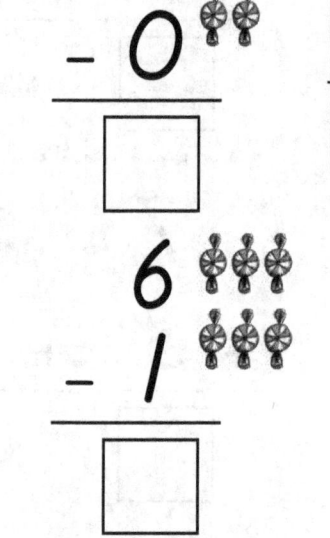

```
  3        6        5        4
- 1      - 0      - 4      - 0
─────    ─────    ─────    ─────
 □        □        □        □

  4        3        6        6
- 4      - 2      - 5      - 1
─────    ─────    ─────    ─────
 □        □        □        □
```

94 ninety-four

Math 1 Reviews

Add on a Number Line; Double Facts

Use the number line to add.
Draw ⌒↘. Write the sum.

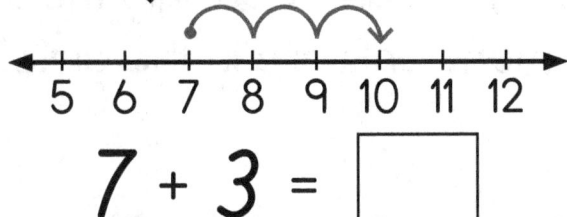

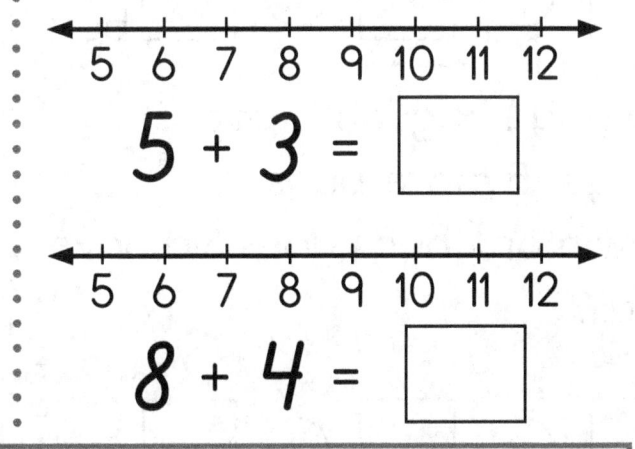

7 + 3 = ☐

9 + 3 = ☐

5 + 4 = ☐

5 + 3 = ☐

8 + 4 = ☐

Double Facts
Find the sum by joining two sets with the same number of objects in each set.

Circle each double fact.
Use the number line to add.
Write each sum.

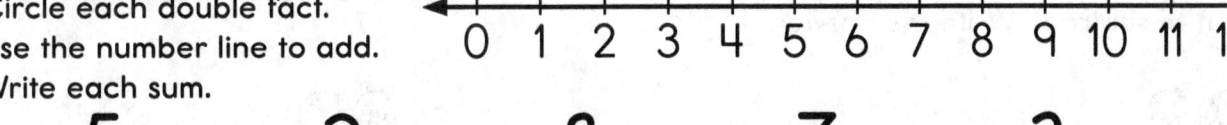

5	2	8	7	3	6
+ 3	+ 2	+ 3	+ 3	+ 3	+ 3
☐	☐	☐	☐	☐	☐

4	7	6	9	1	5
+ 4	+ 1	+ 6	+ 2	+ 1	+ 5
☐	☐	☐	☐	☐	☐

Chapter 7 • Lesson 50

Chapter 3 Review

Draw pictures to show each word problem.
Write an equation. Solve.

Annie had 4 bugs in the can.
She put 1 more bug in the can.
How many bugs are in the can?

☐ ○ ☐ ○ ☐ bugs

There were 3 ants in the glass.
Then 2 more ants went into the glass.
How many ants are in the glass?

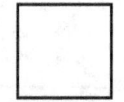

☐ ○ ☐ ○ ☐ ants

Sid had 1 bug in his cap.
He got 5 more bugs.
How many bugs does Sid have in all?

☐ ○ ☐ ○ ☐ bugs

Dad hit 6 bugs with his hand.
He did not hit any more bugs.
How many bugs did Dad hit in all?

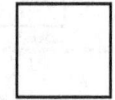

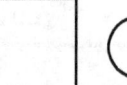

☐ ○ ☐ ○ ☐ bugs

Subtraction Fact Review

Cross out to subtract. Write the answer.

6 − 5 = ☐   3 − 2 = ☐   5 − 1 = ☐

5 − 2 = ☐   6 − 1 = ☐   4 − 3 = ☐

4 − 1 = ☐   5 − 3 = ☐   2 − 1 = ☐

Add with 3 Addends

Write each sum.

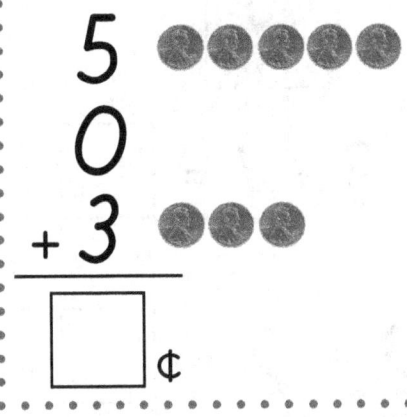

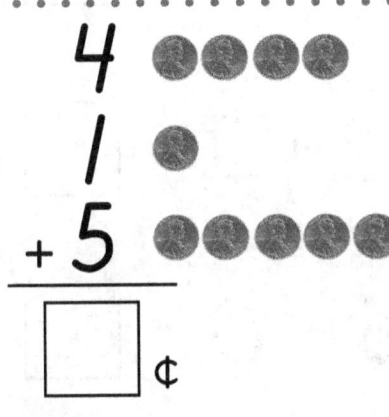

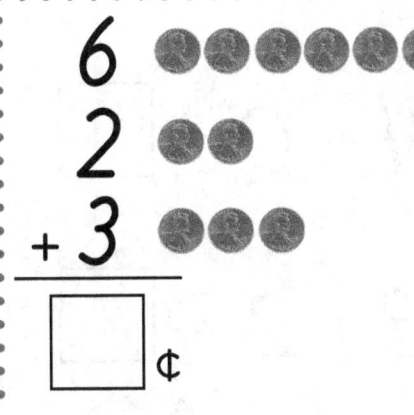

```
  3
  3
+ 1
-----
 □ ¢
```

```
  2
  4
+ 3
-----
 □ ¢
```

```
  5
  0
+ 3
-----
 □ ¢
```

```
  4
  1
+ 5
-----
 □ ¢
```

```
  5
  0
+ 6
-----
 □ ¢
```

```
  6
  2
+ 3
-----
 □ ¢
```

Add.
Use pennies or counters if needed.

```
  2
  8
+ 2
-----
  □
```

```
  6
  4
+ 1
-----
  □
```

```
  1
  3
+ 5
-----
  □
```

```
  9
  0
+ 3
-----
  □
```

Write an equation for the picture. Solve.

How many cupcakes are there?

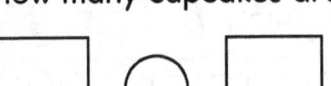

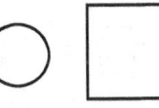

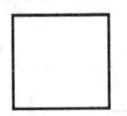

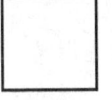

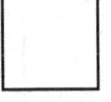

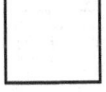

□ ○ □ ○ □ ○ □ cupcakes

Chapter 7 • Lesson 51   ninety-seven **97**

Chapter 3 Review

Complete each addition sentence.
Cross out the sentence that does not belong.

4

3 + 1 = ☐

2 + 2 = ☐

4 + 0 = ☐

3 + 2 = ☐

5

0 + 5 = ☐

1 + 3 = ☐

3 + 2 = ☐

1 + 4 = ☐

6

3 + 3 = ☐

5 + 1 = ☐

2 + 3 = ☐

6 + 0 = ☐

Subtraction Fact Review

Cross out to subtract. Write each answer.

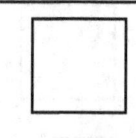

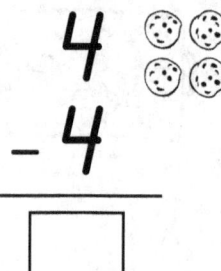

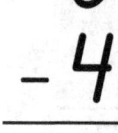

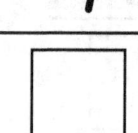

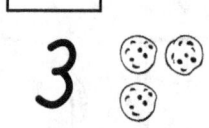

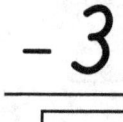

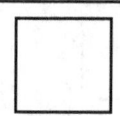

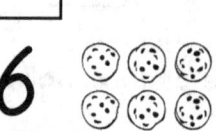

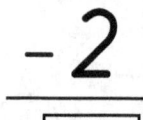

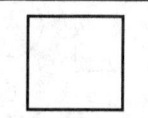

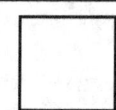

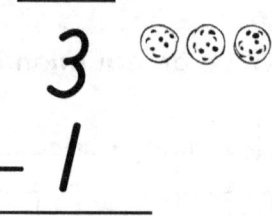

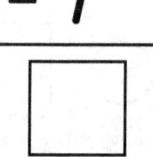

5 − 1 = ☐

6 − 4 = ☐

5 − 3 = ☐

6 − 5 = ☐

4 − 4 = ☐

3 − 2 = ☐

6 − 2 = ☐

3 − 1 = ☐

98 ninety-eight

Add with 5

Color. *Count on* from 5. Write the sum.

				5
6	7	8	9	

5 + 4 = ☐

5 + 3 = ☐

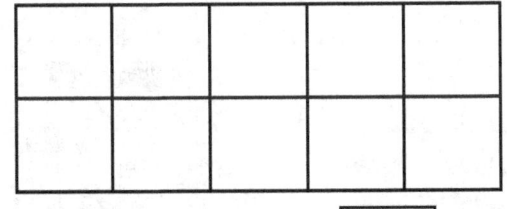

Color. *Count on* from 10. Write the sum.

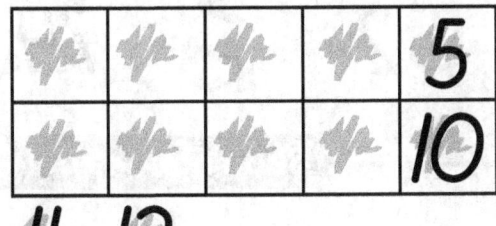

				5
				10

11 12

5 + 7 = ☐

5 + 1 = ☐

5 + 5 = ☐

5 + 6 = ☐

Mark each correct equation.

William ran 4 laps.
He rested.
Then he ran 2 more laps.
How many laps did William run in all?

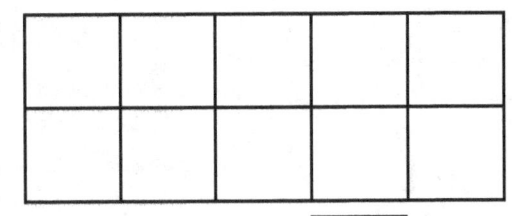

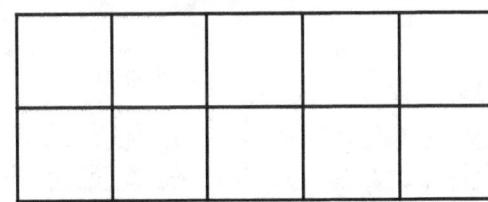

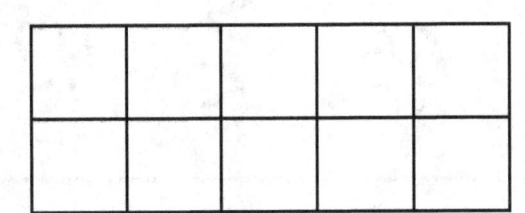

○ 4 + 2 = 6 laps

○ 4 − 2 = 2 laps

Liz had 6 gumdrops.
She gave 4 gumdrops to Meg.
How many gumdrops does Liz have left?

○ 6 + 4 = 10 gumdrops

○ 6 − 4 = 2 gumdrops

Chapter 7 • Lesson 52

Chapter 4 Review

Circle the correct sign.

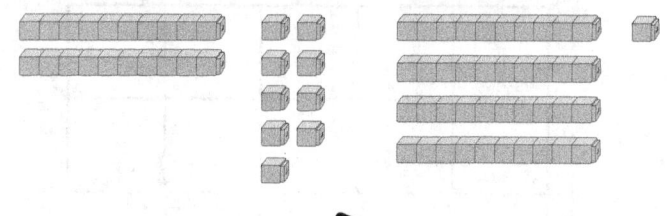

> is greater than   < is less than

29 < 41   57 > 23

65 > 43   52 < 74

Subtraction Fact Review

Cross out to subtract. Write the answer.

6 − 3 = ☐   5 − 2 = ☐   6 − 5 = ☐

4 − 2 = ☐   6 − 4 = ☐   5 − 3 = ☐

2 − 1 = ☐   6 − 2 = ☐   5 − 4 = ☐

Names for 10

_____

0 1 2 3 4 5 6 7 8 9 10 11 12

Add.
Use the number line if needed.
Color according to the key.

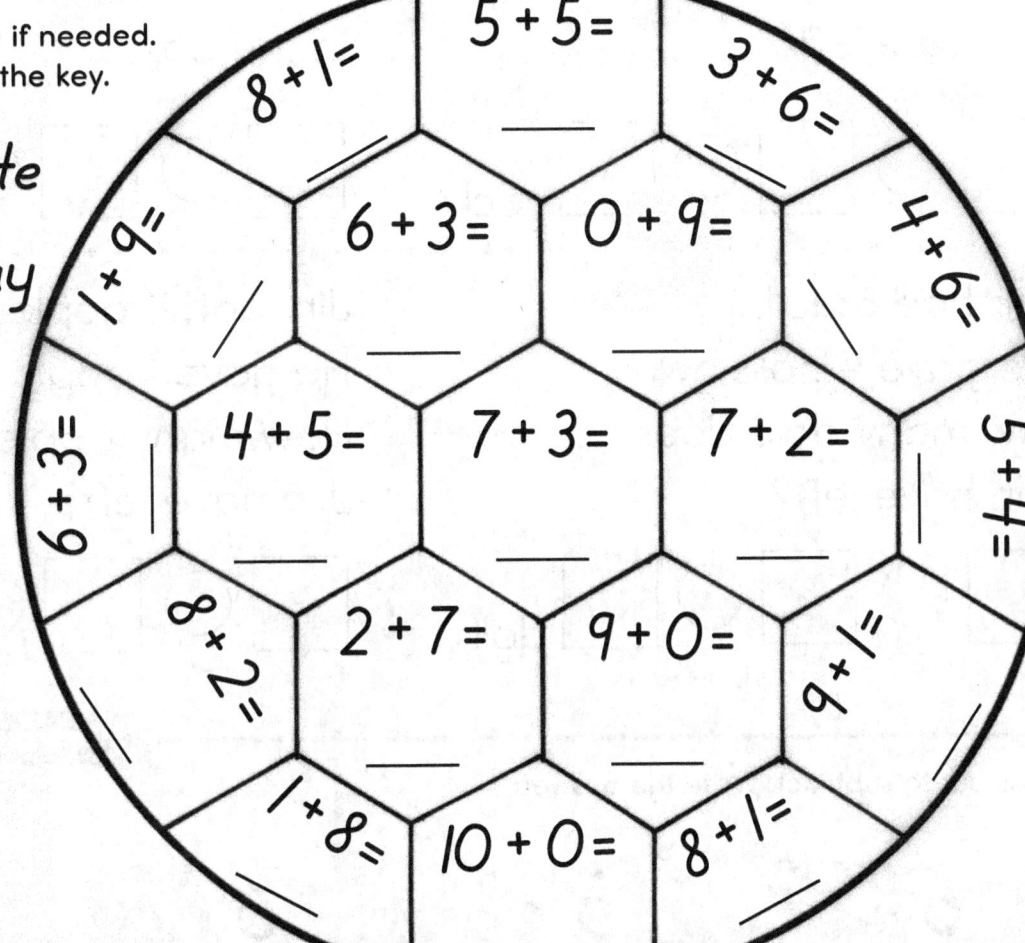

9 = ☐ white
10 = ■ gray

Hexagons:
- 8 + 1 =
- 5 + 5 =
- 3 + 6 =
- 1 + 9 =
- 6 + 3 =
- 0 + 9 =
- 4 + 6 =
- 6 + 3 =
- 4 + 5 =
- 7 + 3 =
- 7 + 2 =
- 5 + 4 =
- 8 + 2 =
- 2 + 7 =
- 9 + 0 =
- 9 + 1 =
- 1 + 8 =
- 10 + 0 =
- 8 + 1 =

Write an equation for the word problem.
Solve.

Kim has 4 hats.
Jill has 2 hats.
How many hats do they have in all?

☐ ○ ☐ ○ ☐ hats

Add.
Use the Order Principle to write the new fact.

8 + 2 = ☐
☐ + ☐ = ☐

7 + 3 = ☐
☐ + ☐ = ☐

Chapter 7 • Lesson 53   one hundred one **101**

Chapter 5 Review

Write an equation for each word problem. Solve.

Jack gave Ben 4 rocks.
Ben lost 4 rocks.
How many rocks does
Ben have left?

☐ ○ ☐ ○ ☐ rocks

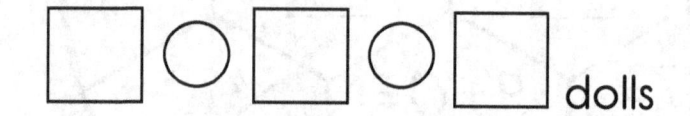

Don had 6 caps in the box.
He got 3 caps from the box.
How many caps are left
in the box?

☐ ○ ☐ ○ ☐ caps

Ann had 5 dolls.
She gave 4 dolls away.
How many dolls does
Ann have left?

☐ ○ ☐ ○ ☐ dolls

Jim had 4 apples.
He gave 1 apple to Ken.
How many apples does
Jim have left?

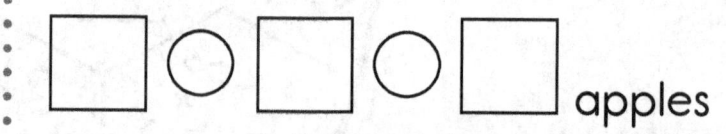

☐ ○ ☐ ○ ☐ apples

Subtraction Fact Review

Cross out to subtract. Write the answer.

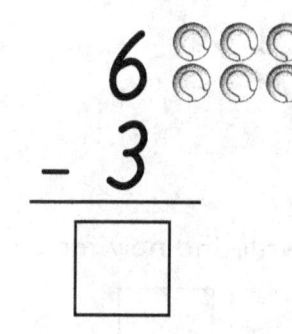

6 − 3 = ☐

5 − 3 = ☐

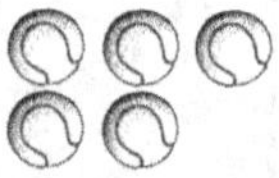

6 − 5 = ☐

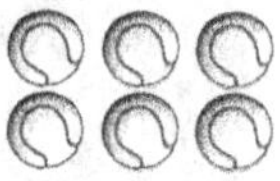

5 − 1 = ☐

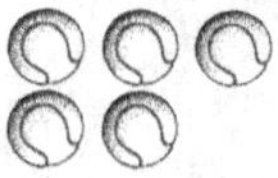

4 − 2 = ☐

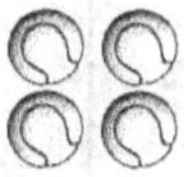

5 − 2 = ☐

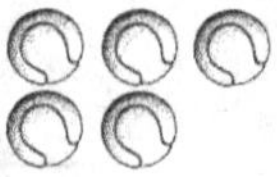

6 − 1 = ☐

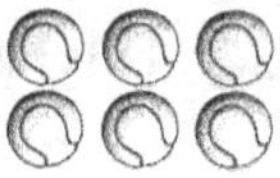

5 − 4 = ☐

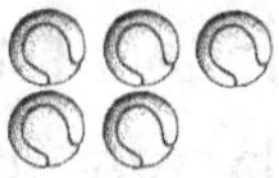

Chapter 7 Review

Use the number line if needed.

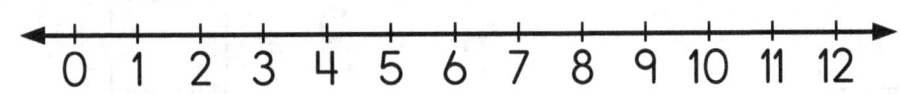

0 1 2 3 4 5 6 7 8 9 10 11 12

Add. Circle each double fact.

3	6	6	5	4	5
+3	+3	+4	+5	+4	+4
☐	☐	☐	☐	☐	☐

Use the Zero Principle or *count on* 1 or 2 to add.

7	6	5	9	8	7
+2	+1	+0	+1	+2	+0
☐	☐	☐	☐	☐	☐

Add.

0	4	3	6	2	9
8	3	7	1	5	0
+2	+2	+1	+3	+2	+3
☐	☐	☐	☐	☐	☐

Chapter 7 Review one hundred three 103

Color. Use the Ten Bar to add.

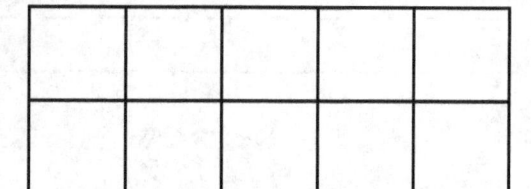

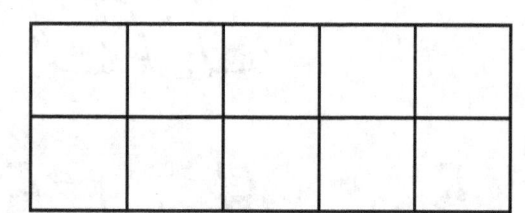

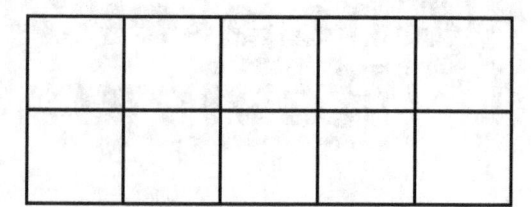

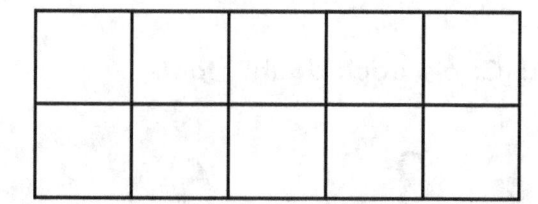

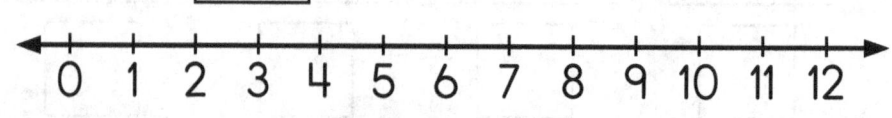

$5 + 2 = \boxed{}$

$5 + 3 = \boxed{}$

$5 + 6 = \boxed{}$

$5 + 5 = \boxed{}$

0 1 2 3 4 5 6 7 8 9 10 11 12

Add. Use the number line if needed.
Use the Order Principle to write each new fact.

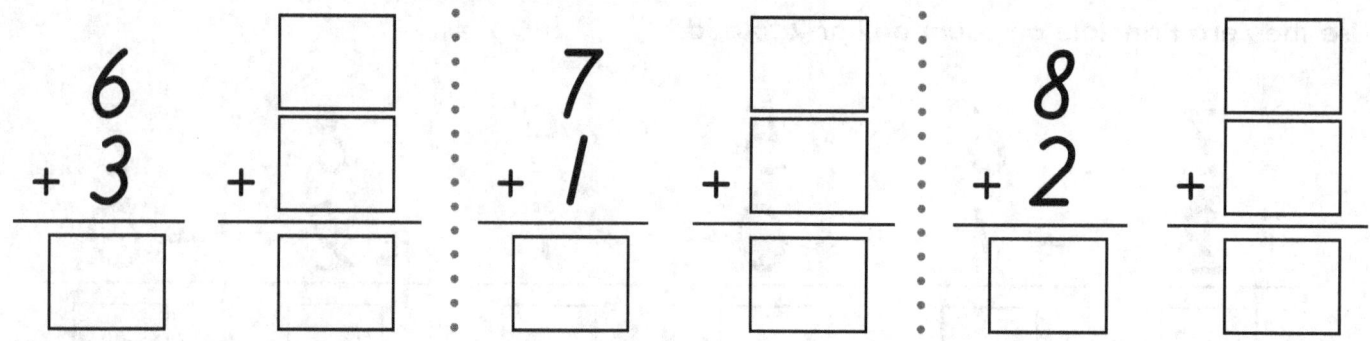

$$\begin{array}{r}6\\+\ 3\\\hline\end{array}$$

$$\begin{array}{r}\ \\+\ \\\hline\end{array}$$

$$\begin{array}{r}7\\+\ 1\\\hline\end{array}$$

$$\begin{array}{r}\ \\+\ \\\hline\end{array}$$

$$\begin{array}{r}8\\+\ 2\\\hline\end{array}$$

$$\begin{array}{r}\ \\+\ \\\hline\end{array}$$

Mark the correct equation.

There were 4 bears napping in the sun. Then 2 bears got up. How many bears are still napping?

○ $4 + 2 = 6$ bears
○ $4 - 2 = 2$ bears

The 3 little bears sat on a hill. The 3 big bears sat on a hill. How many bears sat on a hill?

○ $3 + 3 = 6$ bears
○ $3 - 3 = 0$ bears

104 one hundred four

Math 1 Reviews

Cumulative Review

Complete each pattern.

10 11 12 ☐ 14   15 20 25 ☐ 35

20 30 40 ☐ 60   45 50 55 ☐ 65

Write each missing number.

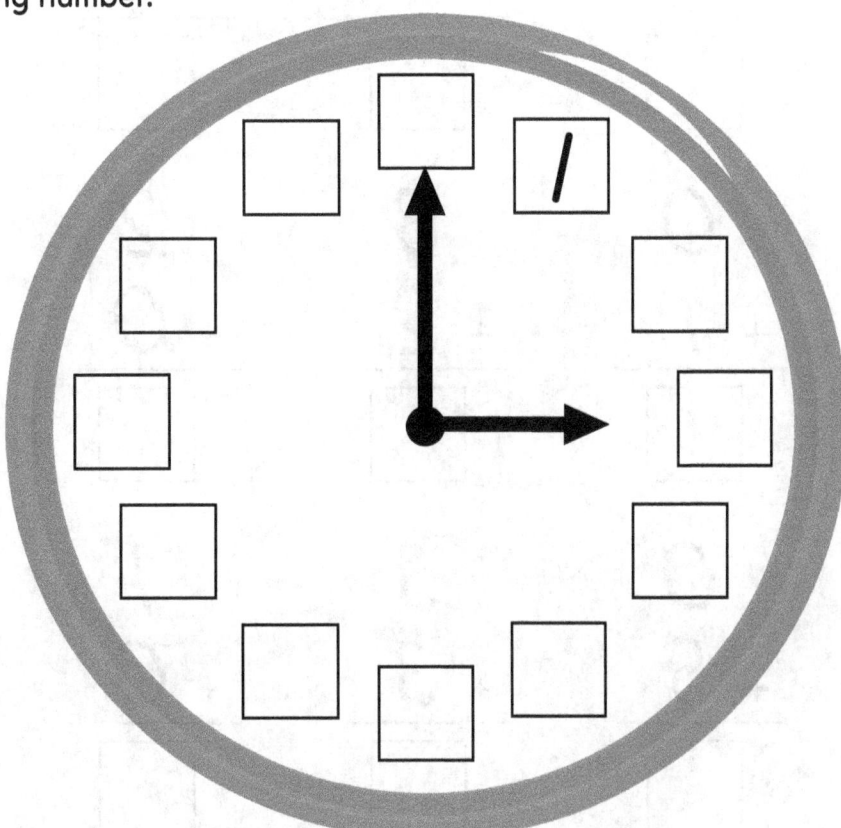

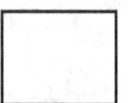

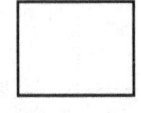

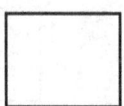

Write the number that comes *between*.

36 ☐ 38   72 ☐ 74

49 ☐ 51   88 ☐ 90

Chapter 7 Cumulative Review   one hundred five 105

Addition Fact Review

Add.

$1 + 1 = \square$ $2 + 0 = \square$ $1 + 4 = \square$ $6 + 0 = \square$ $1 + 5 = \square$

$4 + 0 = \square$ $1 + 3 = \square$ $5 + 1 = \square$ $2 + 1 = \square$ $0 + 3 = \square$

$2 + 2 = \square$ $0 + 1 = \square$ $3 + 2 = \square$ $2 + 0 = \square$ $4 + 1 = \square$

$3 + 1 = \square$ $0 + 5 = \square$ $3 + 3 = \square$ $1 + 0 = \square$ $0 + 2 = \square$

$1 + 2 = \square$ $2 + 3 = \square$ $0 + 0 = \square$ $4 + 2 = \square$ $3 + 0 = \square$

Telling Time

- - - - - - - - - - - - - -

Number the pictures in order from 1 to 3 for each job.

Circle the correct time.

12 o'clock
5 o'clock

8 o'clock
9 o'clock

2 o'clock
1 o'clock

11 o'clock
10 o'clock

Write the time.

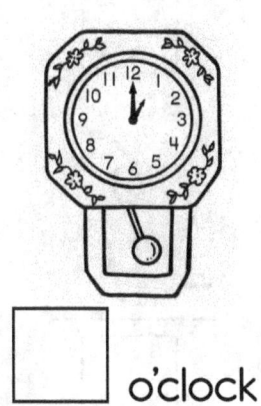

☐ o'clock

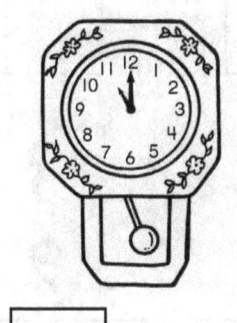

☐ o'clock

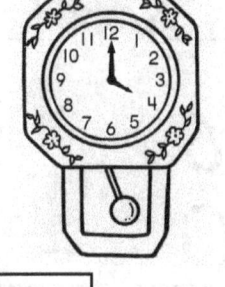

☐ o'clock

Chapter 8 • Lesson 56 — one hundred seven — **107**

Chapter 1 Review

Color the birds over the tree blue.
Color the birds under the tree red.
Color the picnic table on the right brown.
Color the picnic table on the left black.
Color the flowers on the right yellow.
Color the flowers on the left orange.

Subtraction Fact Review

Subtract.

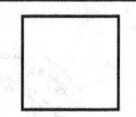

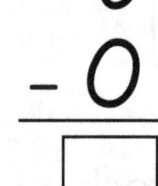

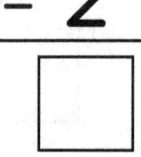

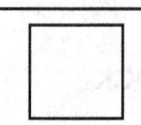

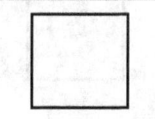

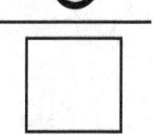

5	5	6	3	4	2
-4	-3	-6	-1	-3	-0
☐	☐	☐	☐	☐	☐

6	4	5	3	4	6
-0	-2	-1	-2	-0	-4
☐	☐	☐	☐	☐	☐

one hundred eight   Math 1 Reviews

Using Clocks to Write Time

Draw a line to match each analog clock to the correct digital clock.

Write each time.

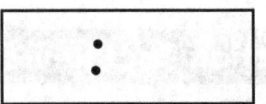

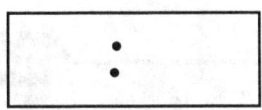

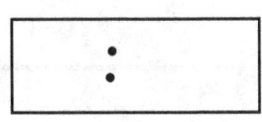

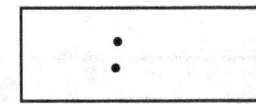

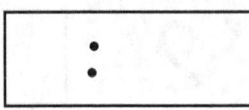

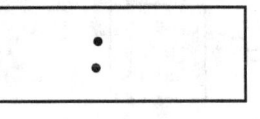

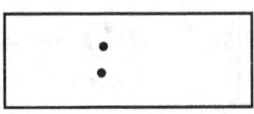

Chapter 8 • Lesson 57 — one hundred nine — **109**

Chapter 6 Review

Write the total value as you *count on*.

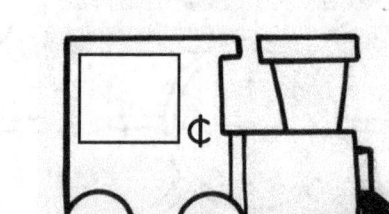

___¢ ___¢ ___¢ ___¢ ___¢ ___¢

___¢ ___¢ ___¢ ___¢ ___¢ ___¢

___¢ ___¢ ___¢ ___¢ ___¢ ___¢

___¢ ___¢ ___¢ ___¢ ___¢

Subtraction Fact Review

Subtract.

5 − 5 = ☐   6 − 2 = ☐   6 − 1 = ☐

4 − 1 = ☐   5 − 2 = ☐   6 − 3 = ☐

3 − 3 = ☐   2 − 2 = ☐   4 − 4 = ☐

2 − 1 = ☐   3 − 0 = ☐   5 − 2 = ☐

Using a Schedule

_ _ _ _ _ _ _ _ _ _ _ _ _

Write the time under each clock.
Fill in the hours passed for each activity.

Weekday Schedule

Activity	Start	Finish	Time Passed
 (boy reading)	 (clock showing 2:00) ☐:☐	 (clock showing 4:00) ☐:☐	☐ hour(s) passed
 (girl on swing)	 (clock showing 11:00) ☐:☐	 (clock showing 1:00) ☐:☐	☐ hour(s) passed
 (boy writing)	 6:00 ☐:☐	 7:00 ☐:☐	☐ hour(s) passed

Look at the table and write each time.

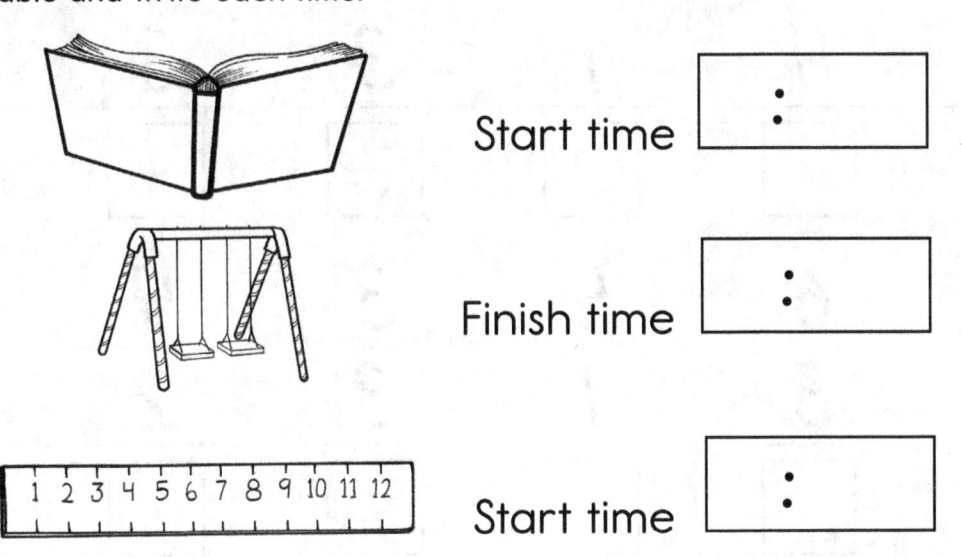

Start time ☐:☐

Finish time ☐:☐

Start time ☐:☐

Chapter 8 • Lesson 58   one hundred eleven   **111**

Chapter 6 Review

Write each total value as you *count on*.

____¢ ____¢ ____¢ ____¢ ____¢ ____¢   ____¢

____¢ ____¢ ____¢ ____¢ ____¢ ____¢   ____¢

____¢ ____¢ ____¢ ____¢ ____¢ ____¢   ____¢

____¢ ____¢ ____¢ ____¢ ____¢ ____¢   ____¢

Subtraction Fact Review

Subtract.

$\begin{array}{r}6\\-5\\\hline\end{array}$ $\begin{array}{r}5\\-4\\\hline\end{array}$ $\begin{array}{r}3\\-1\\\hline\end{array}$ $\begin{array}{r}4\\-3\\\hline\end{array}$ $\begin{array}{r}5\\-3\\\hline\end{array}$ $\begin{array}{r}2\\-2\\\hline\end{array}$

$\begin{array}{r}3\\-2\\\hline\end{array}$ $\begin{array}{r}6\\-3\\\hline\end{array}$ $\begin{array}{r}4\\-1\\\hline\end{array}$ $\begin{array}{r}3\\-3\\\hline\end{array}$ $\begin{array}{r}4\\-2\\\hline\end{array}$ $\begin{array}{r}5\\-1\\\hline\end{array}$

one hundred twelve   Math 1 Reviews

Time to the Half-Hour

Count by 5s and write the minutes.

minute hand

Write the hour. Count by 5s and write the minutes.

hour hand

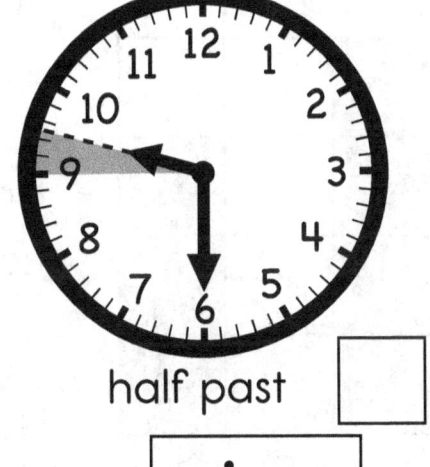

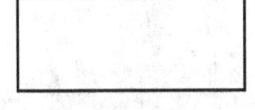

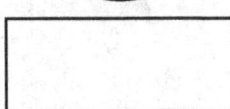

half past ☐

☐ : ☐

Write each time.

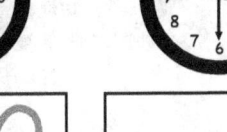

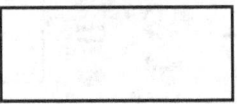

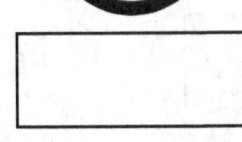

3:00

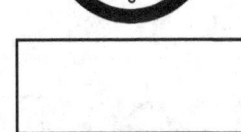

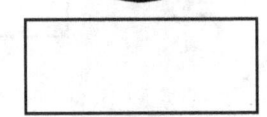

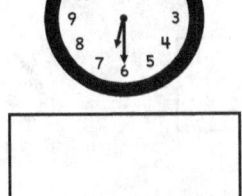

Chapter 8 • Lessons 59–60

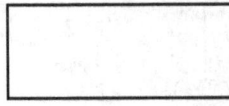

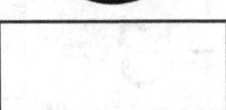

one hundred thirteen **113**

Chapter 2 Review

Pets

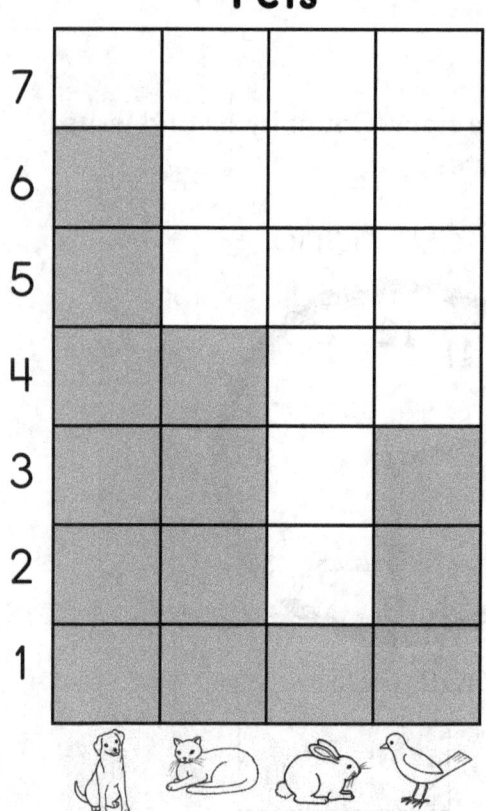

Use the graph to solve each problem.

How many kids have dogs and birds?

☐ + ☐ = ☐ kids

How many kids have rabbits and cats?

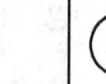

☐ + ☐ = ☐ kids

How many more dogs are there than cats?

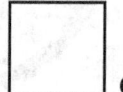

☐ − ☐ = ☐ dogs

How many more birds are there than rabbits?

☐ − ☐ = ☐ birds

Subtraction Fact Review

Subtract.

6 − 0 = ☐   4 − 4 = ☐   5 − 0 = ☐

2 − 1 = ☐   3 − 2 = ☐   6 − 5 = ☐

4 − 2 = ☐   6 − 3 = ☐   5 − 4 = ☐

3 − 3 = ☐   5 − 2 = ☐   4 − 3 = ☐

Time to the Hour & Half-Hour

Arrival   Departure

4:00   4:30

Write each time.

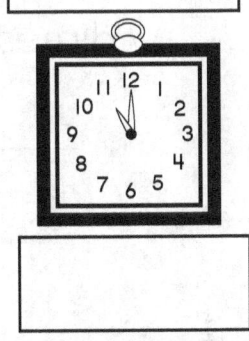

Circle the time passed.

Plane	Arrival	Departure	Time Passed
243	11:00	11:30	30 minutes or 1 hour
691	4:00	4:30	30 minutes or 1 hour
514	6:00	7:00	30 minutes or 1 hour

Chapter 8 • Lessons 61–62   one hundred fifteen   115

Chapter 6 Review

Write the number of dimes and pennies.
Write the total value.

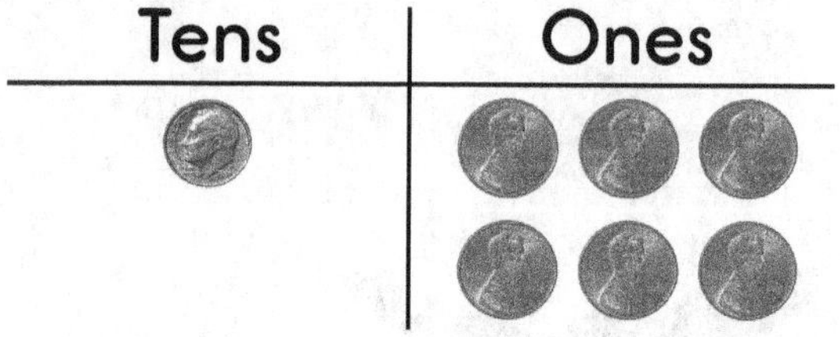

Tens	Ones
(1 dime)	(6 pennies)

___ dime ___ pennies

_____ ¢

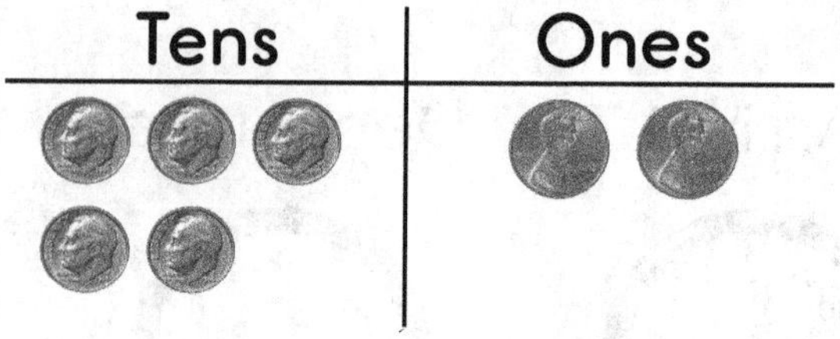 Tens	Ones
(4 dimes)	(2 pennies)

___ dimes ___ pennies

_____ ¢

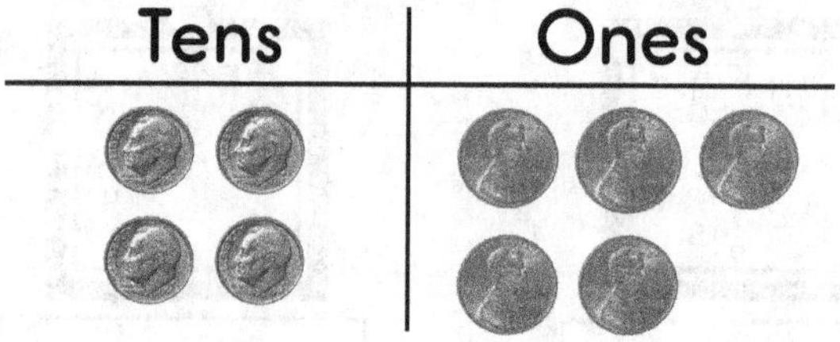 Tens	Ones
(4 dimes)	(5 pennies)

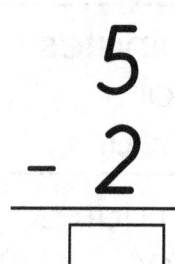

___ dimes ___ pennies

_____ ¢

Tens	Ones
(3 dimes)	(6 pennies)

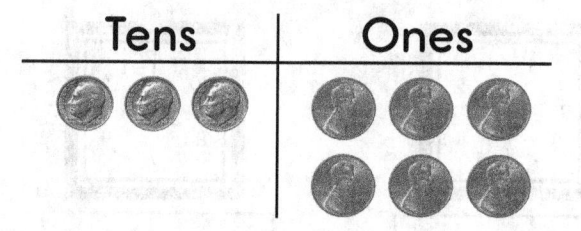

___ dimes ___ pennies

_____ ¢

Subtraction Fact Review

Subtract.

$$\begin{array}{c} 6 \\ -4 \\ \hline \Box \end{array} \qquad \begin{array}{c} 4 \\ -0 \\ \hline \Box \end{array} \qquad \begin{array}{c} 3 \\ -2 \\ \hline \Box \end{array} \qquad \begin{array}{c} 5 \\ -1 \\ \hline \Box \end{array} \qquad \begin{array}{c} 4 \\ -2 \\ \hline \Box \end{array} \qquad \begin{array}{c} 6 \\ -0 \\ \hline \Box \end{array}$$

$$\begin{array}{c} 5 \\ -2 \\ \hline \Box \end{array} \qquad \begin{array}{c} 1 \\ -1 \\ \hline \Box \end{array} \qquad \begin{array}{c} 5 \\ -0 \\ \hline \Box \end{array} \qquad \begin{array}{c} 2 \\ -1 \\ \hline \Box \end{array} \qquad \begin{array}{c} 2 \\ -0 \\ \hline \Box \end{array} \qquad \begin{array}{c} 3 \\ -3 \\ \hline \Box \end{array}$$

Using Calendars

Color all the Wednesdays red.
Find July 27 and mark an *X* on it.

Find the day after July 27 and mark a *T* on it.
Find the day before July 27 and mark a *Y* on it.

July						
Sunday	Monday	Tuesday	Wednesday	Thursday	Friday	Saturday
			1	2	3	4
5	6	7	8	9	10	11
12	13	14	15	16	17	18
19	20	21	22	23	24	25
26	27	28	29	30	31	

Number the days of the week in order.

___ Tuesday
___ Thursday
___ Saturday
___ Sunday
___ Wednesday
___ Friday
___ Monday

Number the months of the year in order.

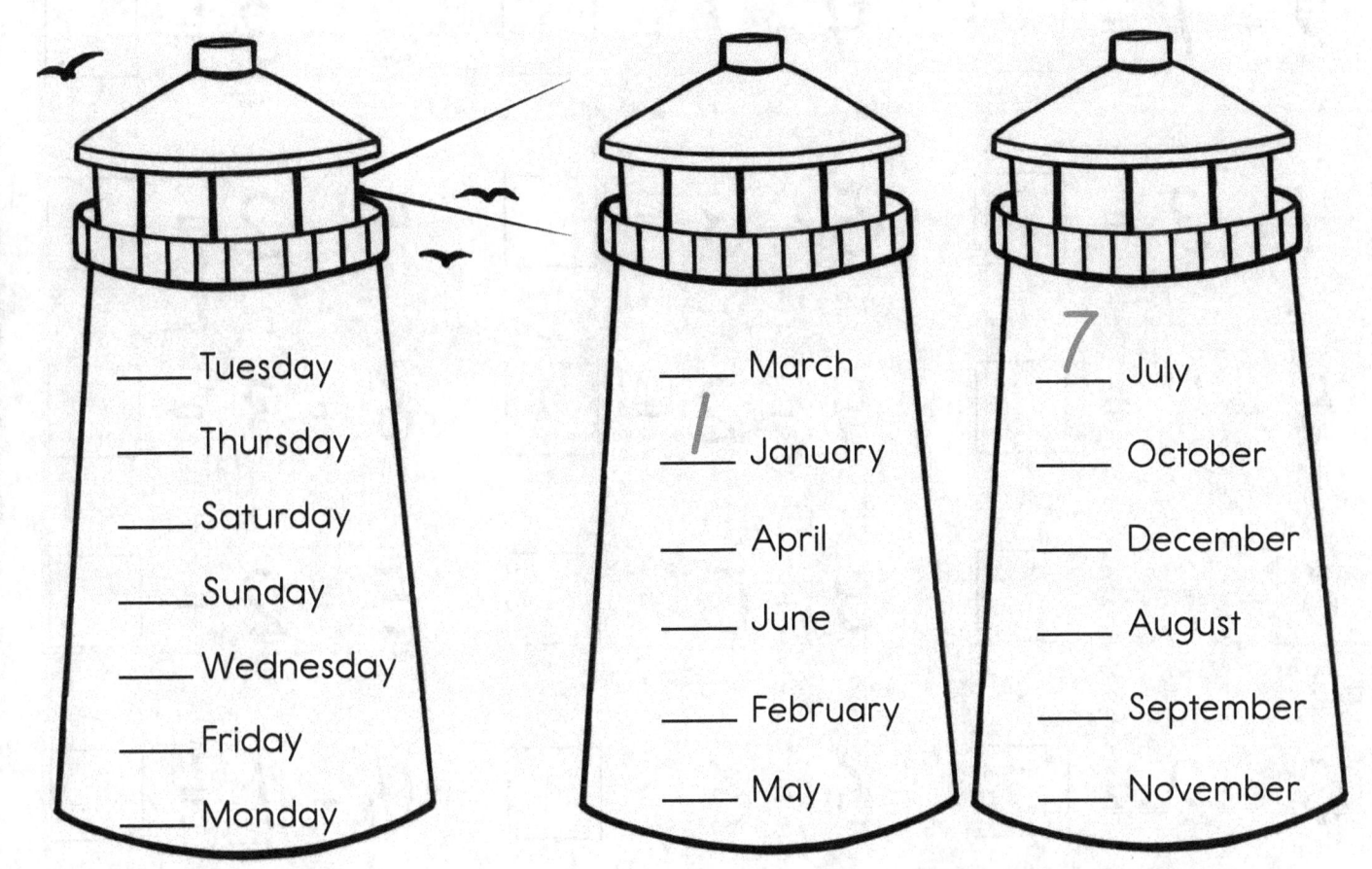

___ March
_1_ January
___ April
___ June
___ February
___ May

_7_ July
___ October
___ December
___ August
___ September
___ November

Chapter 8 • Lesson 63

Chapter 4 Review

Draw a line to match each expanded form to the correct number.

40 + 7   75

10 + 4   47

70 + 5   51

50 + 1   14

Write the expanded form for each number.

36 = ☐ + ☐

28 = ☐ + ☐

63 = ☐ + ☐

85 = ☐ + ☐

Subtraction Fact Review

Subtract.

2 − 1 = ☐   4 − 1 = ☐   3 − 3 = ☐

4 − 3 = ☐   5 − 3 = ☐   4 − 2 = ☐

5 − 4 = ☐   4 − 2 = ☐   6 − 3 = ☐

6 − 1 = ☐   3 − 0 = ☐   5 − 2 = ☐

3 − 2 = ☐   6 − 6 = ☐   4 − 1 = ☐

Chapter 8 Review

Complete each sentence.

1 hour = ☐ minutes   a half-hour = ☐ minutes

Write each time.

Look at the table. Write the time.

Job	Start	Finish
(pulling grass)	3:00	6:00
(planting a bush)	6:00	7:00

What time does pulling grass start?

What time does planting a bush finish?

Chapter 8 Review   one hundred nineteen **119**

Number the pictures in order.

[ ] [ ] [ ]

Number the days of the week in order.

[ ] Saturday   [ ] Monday   [ ] Friday
[ ] Wednesday [ ] Thursday [ ] Tuesday [1] Sunday

Color all the Fridays blue.
Find August 22. Mark an X on it.

Find the day after August 22. Mark a T on it.
Find the day before August 22. Mark a Y on it.

August

Sunday	Monday	Tuesday	Wednesday	Thursday	Friday	Saturday
			1	2	3	4
5	6	7	8	9	10	11
12	13	14	15	16	17	18
19	20	21	22	23	24	25
26	27	28	29	30	31	

Mark the correct answer.

1. How many Mondays are there in August?   4 ○   5 ○   6 ○

2. What day is August 31?   Wednesday ○   Thursday ○   Friday ○

3. What day is August 1?   Wednesday ○   Thursday ○   Friday ○

120   one hundred twenty   Math 1 Reviews

Cumulative Review

Match each number to the correct word.

0	two	6	eight
1	five	7	nine
2	zero	8	six
3	one	9	ten
4	three	10	seven
5	four		

Write the number. Circle the single pencils to make pairs.
Circle *even* if the number is even.
Circle *odd* if the number is odd.

even   odd

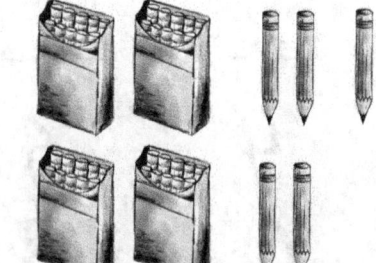

even   odd

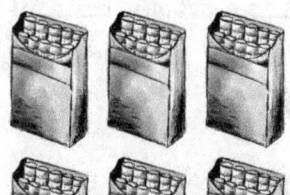

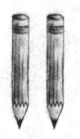

even   odd

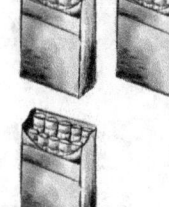

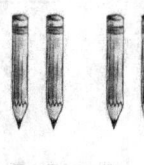

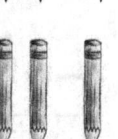

even   odd

Chapter 8 Cumulative Review — one hundred twenty-one — 121

Subtraction Fact Review

Subtract.

3	5	6	3	2	3
-1	-4	-1	-3	-0	-2
☐	☐	☐	☐	☐	☐

5	4	2	5	1	6
-2	-3	-1	-3	-0	-6
☐	☐	☐	☐	☐	☐

4	6	4	5	6	5
-1	-2	-2	-1	-5	-0
☐	☐	☐	☐	☐	☐

0	4	6	5	2	3
-0	-0	-3	-5	-1	-0
☐	☐	☐	☐	☐	☐

2	6	4	6	1	4
-2	-4	-4	-0	-1	-2
☐	☐	☐	☐	☐	☐

Halves

Circle each item that has been divided into equal parts.

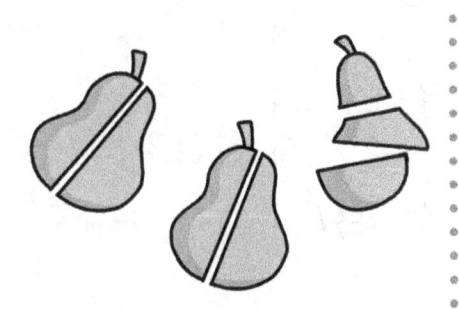

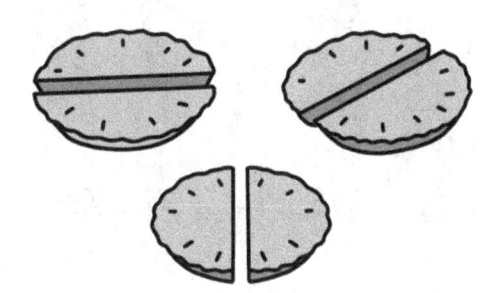

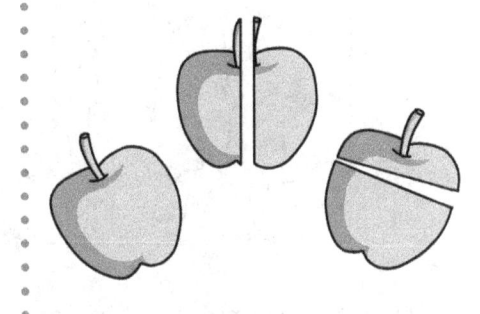

Halves
Each equal part is a *half*.

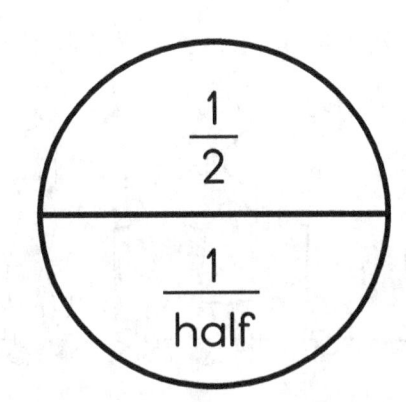

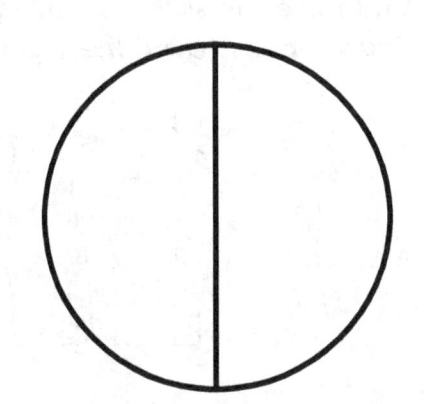

Each circle has ☐ equal parts.

Divide each shape into 2 equal parts.
Color $\frac{1}{2}$ of each shape.

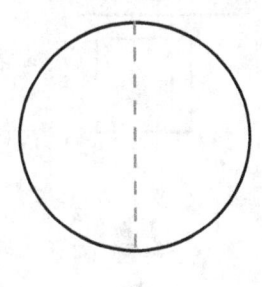

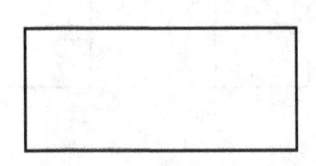

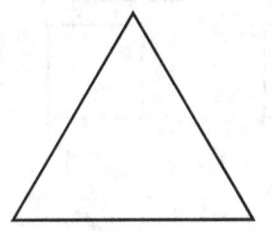

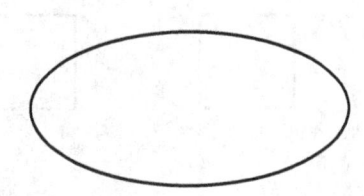

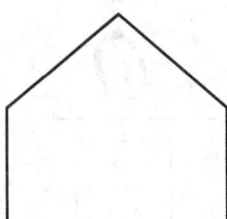

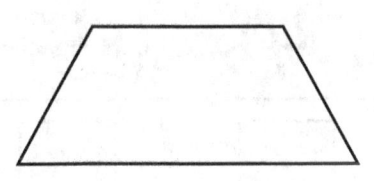

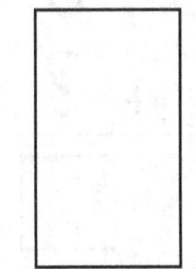

Chapter 9 • Lesson 67   one hundred twenty-three   **123**

Chapter 2 Review

Circle the first child in line.
Underline the third child in line.
Draw a box around the seventh child in line.

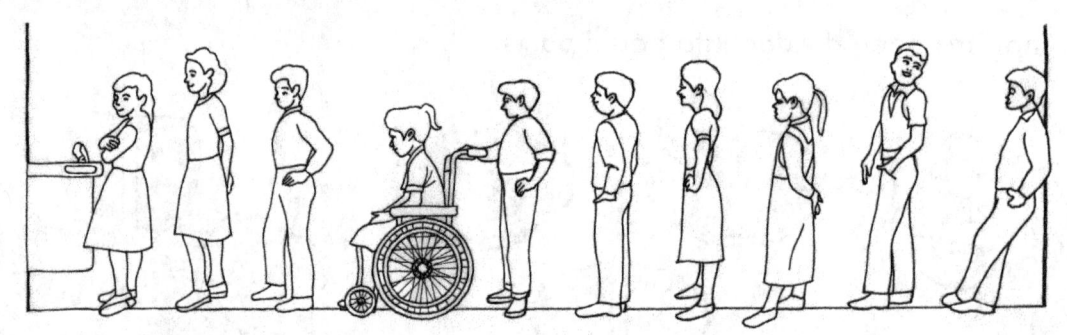

Circle the fourth child in line.
Underline the sixth child in line.
Draw a box around the eighth child in line.

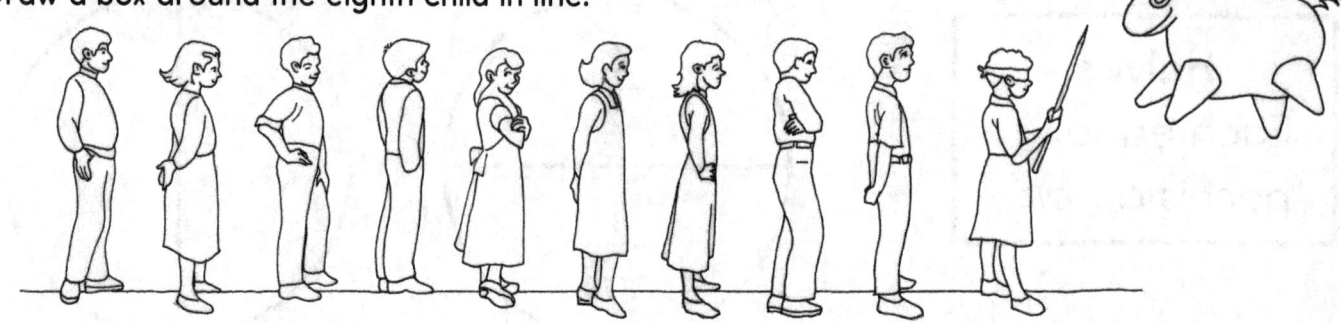

Addition Fact Review

Add.

$9 + 2 =$ ☐

$2 + 5 =$ ☐

$4 + 1 =$ ☐

$0 + 6 =$ ☐

$1 + 9 =$ ☐

$7 + 1 =$ ☐

$3 + 1 =$ ☐

$0 + 9 =$ ☐

$6 + 2 =$ ☐

$2 + 4 =$ ☐

$5 + 5 =$ ☐

$4 + 0 =$ ☐

Thirds

Thirds Each equal part is a *third*.

$\frac{1}{\text{third}}$ $\frac{1}{3}$ $\frac{1}{3}$

Each pie slice has ☐ equal parts.

Mark the correct circle to show if the object is divided into thirds.

○ yes ○ no

○ yes ○ no

○ yes ○ no

Color $\frac{1}{3}$ of each shape.

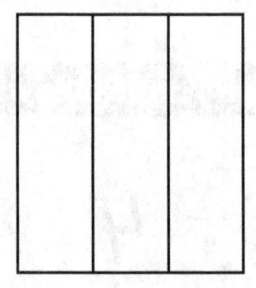

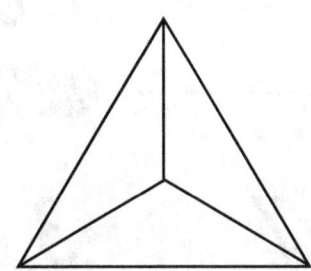

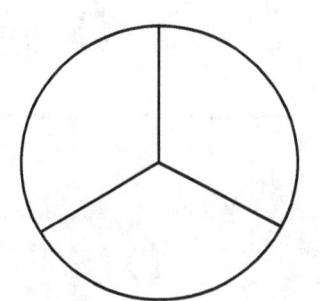

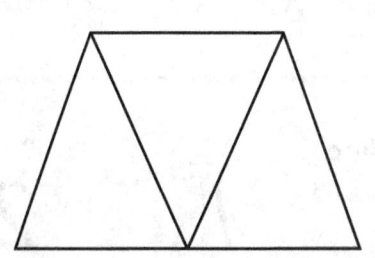

Circle the fraction that names the shaded part.

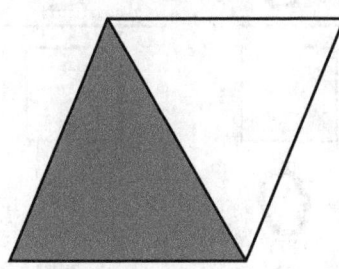

$\frac{1}{2}$ part shaded equal parts

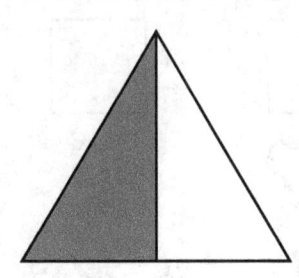

$\frac{1}{2}$ $\frac{1}{3}$ $\frac{2}{3}$

$\frac{1}{2}$ $\frac{1}{3}$ $\frac{2}{3}$

$\frac{1}{2}$ $\frac{1}{3}$ $\frac{2}{3}$

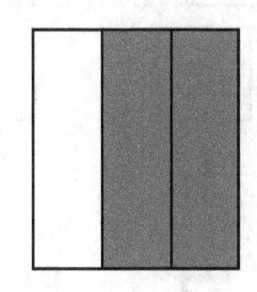

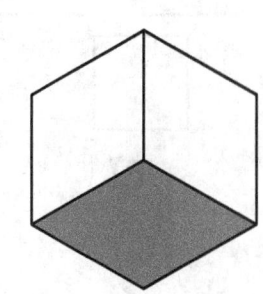

Chapter 9 • Lesson 68

Chapter 5 Review

Circle the shape that comes next in the pattern.

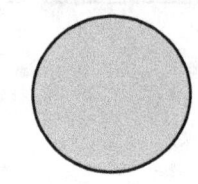

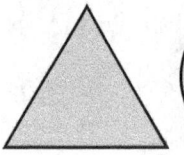

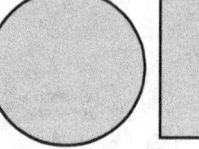

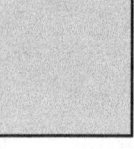

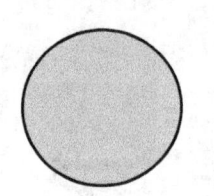

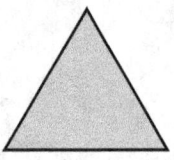

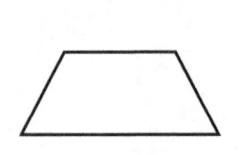

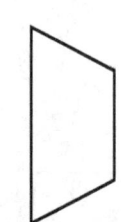

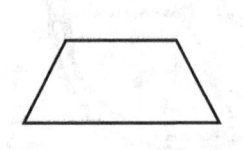

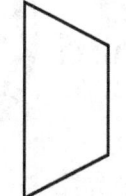

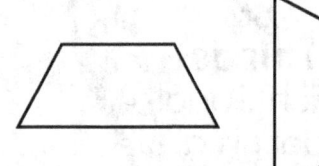

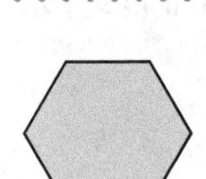

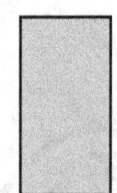

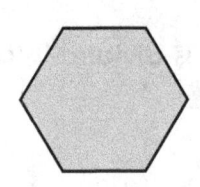

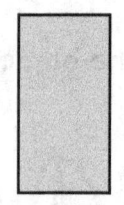

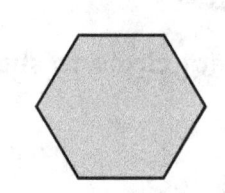

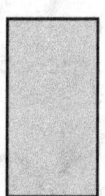

Circle the letter or number that comes next in the pattern.

X Y Z X Y Z X

5 6 6 5 6 6 5 6

X Y

5 6

Addition Fact Review

Add.

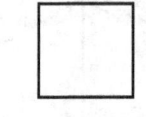

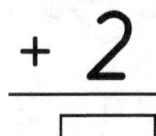

$\begin{array}{r}1\\+3\\\hline\end{array}$ $\begin{array}{r}2\\+0\\\hline\end{array}$ $\begin{array}{r}6\\+6\\\hline\end{array}$ $\begin{array}{r}7\\+2\\\hline\end{array}$ $\begin{array}{r}3\\+3\\\hline\end{array}$ $\begin{array}{r}4\\+2\\\hline\end{array}$

$\begin{array}{r}9\\+2\\\hline\end{array}$ $\begin{array}{r}4\\+4\\\hline\end{array}$ $\begin{array}{r}2\\+1\\\hline\end{array}$ $\begin{array}{r}2\\+2\\\hline\end{array}$ $\begin{array}{r}0\\+8\\\hline\end{array}$ $\begin{array}{r}1\\+7\\\hline\end{array}$

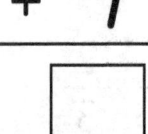

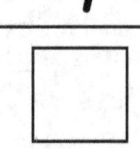

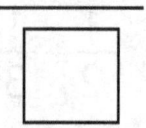

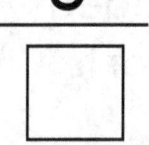

Fourths

Fourths
Each equal part is a *fourth*.

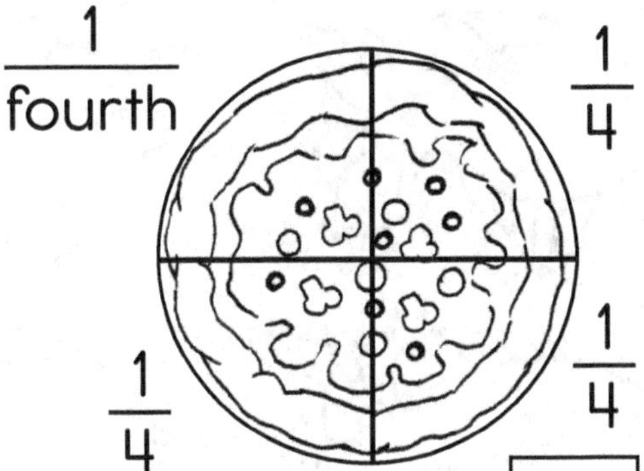

$\frac{1}{\text{fourth}}$ $\frac{1}{4}$

$\frac{1}{4}$ $\frac{1}{4}$

The pizza has ☐ equal parts.

Color $\frac{1}{4}$ of each shape.

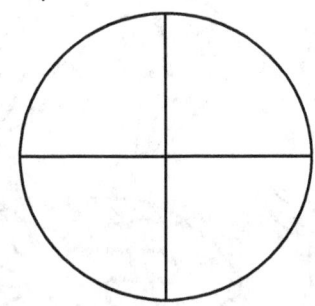

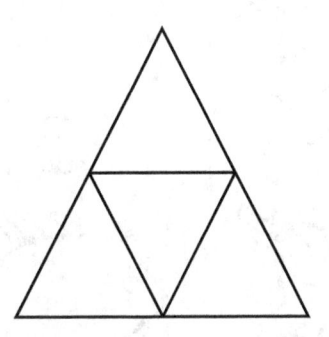

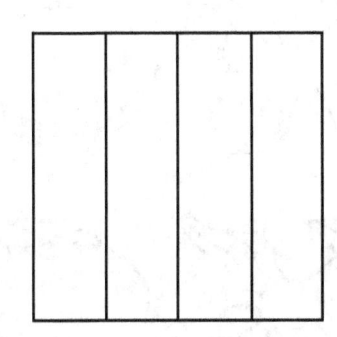

Color the shape to match the fraction.

$\frac{1}{2}$ $\frac{3}{4}$ $\frac{2}{4}$ $\frac{2}{3}$

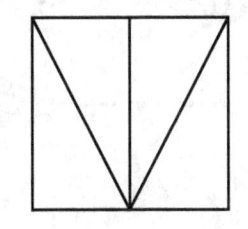

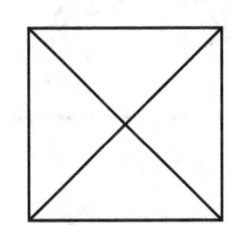

Mark the fraction that names the shaded part.

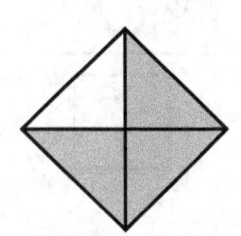

○ $\frac{3}{4}$
○ $\frac{2}{3}$

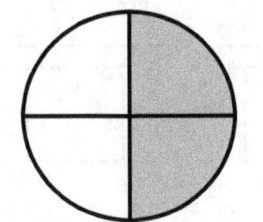

○ $\frac{2}{3}$
○ $\frac{2}{4}$

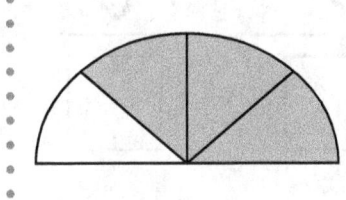

○ $\frac{2}{3}$
○ $\frac{3}{4}$

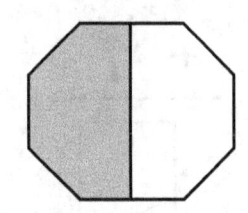

○ $\frac{1}{2}$
○ $\frac{1}{3}$

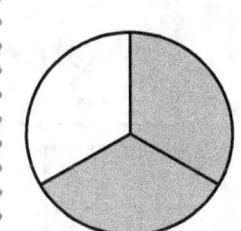

○ $\frac{1}{3}$
○ $\frac{2}{3}$

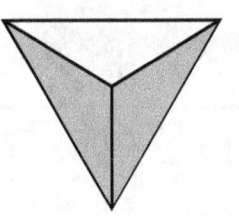

○ $\frac{2}{3}$
○ $\frac{3}{4}$

Chapter 9 • Lesson 69

Chapter 4 Review

Write the number that comes *after*.

71, ___
38, ___
89, ___
56, ___

Write the number that comes *before*.

___, 78
___, 60
___, 13
___, 95

Addition Fact Review

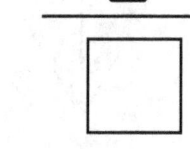

Add.

$$9 + 2 = \square \quad 2 + 6 = \square \quad 3 + 2 = \square \quad 1 + 9 = \square \quad 2 + 1 = \square \quad 7 + 0 = \square$$

$$6 + 6 = \square \quad 8 + 2 = \square \quad 4 + 0 = \square \quad 3 + 3 = \square \quad 6 + 1 = \square \quad 4 + 4 = \square$$

one hundred twenty-eight

Math 1 Reviews

Part of a Set

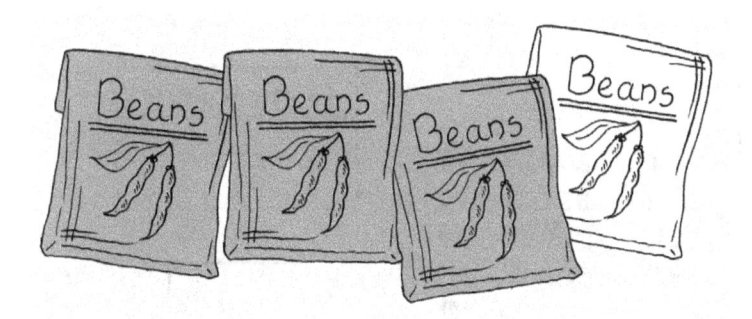

There are 4 seed packets.
3 out of 4 seed packets are shaded.
$\frac{3}{4}$ of the packets are shaded.

Circle the fraction to name the shaded part of each set.

$\frac{2}{3}$
$\frac{2}{4}$

$\frac{1}{2}$
$\frac{1}{3}$

$\frac{2}{3}$
$\frac{3}{4}$

$\frac{2}{4}$
$\frac{2}{3}$

Color to show each fraction of a set.

$\frac{2}{3}$

$\frac{1}{2}$

$\frac{3}{4}$

$\frac{1}{3}$

Chapter 9 • Lesson 70

Chapter 4 Review

Write the number of tens and ones.
Write each number.

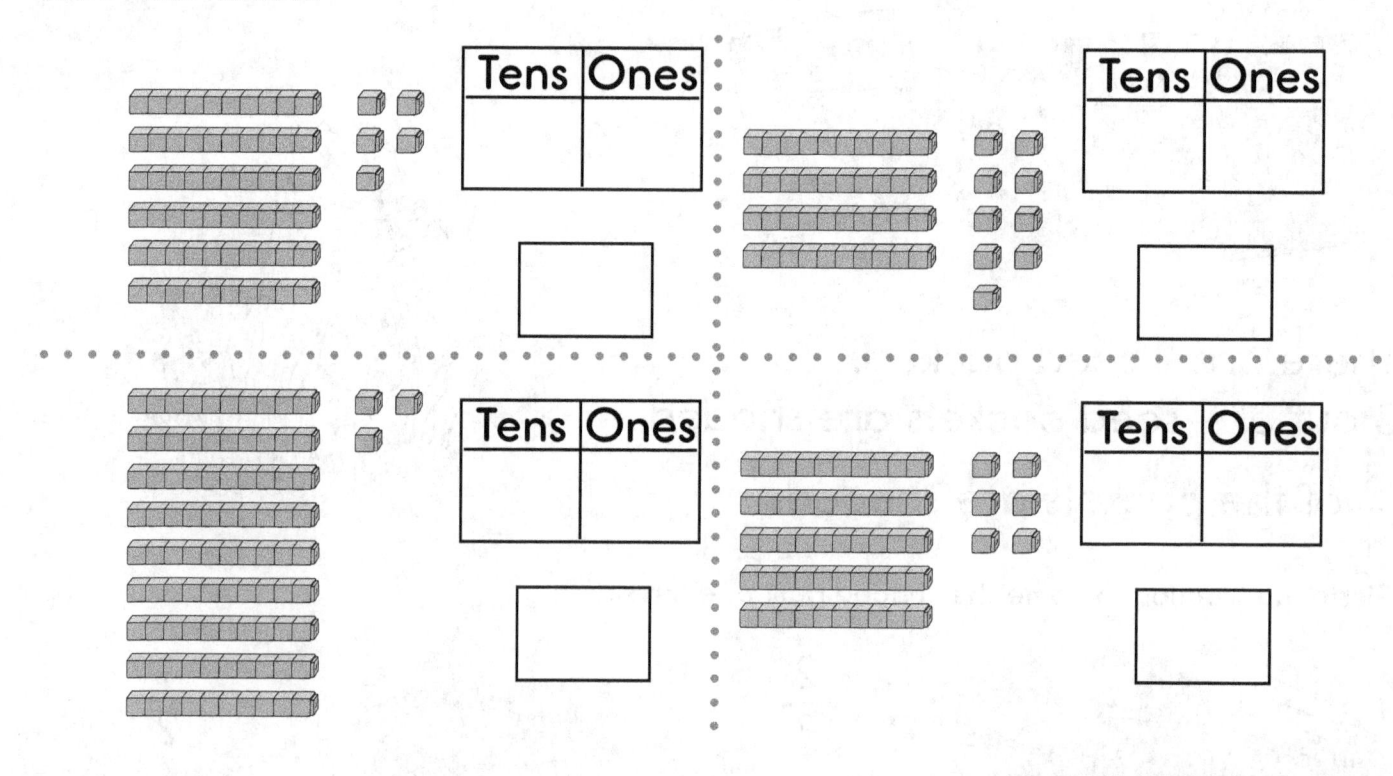

Tens	Ones

Tens	Ones

Tens	Ones

Tens	Ones

Addition Fact Review

Add.

$$2 + 2 = \square$$

$$2 + 3 = \square$$

$$5 + 5 = \square$$

$$6 + 2 = \square$$

$$0 + 3 = \square$$

$$1 + 5 = \square$$

$$1 + 7 = \square$$

$$2 + 6 = \square$$

$$2 + 1 = \square$$

$$1 + 4 = \square$$

$$4 + 4 = \square$$

$$6 + 6 = \square$$

130 one hundred thirty

Math 1 Reviews

Fair Shares

Each boy has a **fair share**. Each has the same amount.

Circle the fair share for each child. Write the number each child gets.

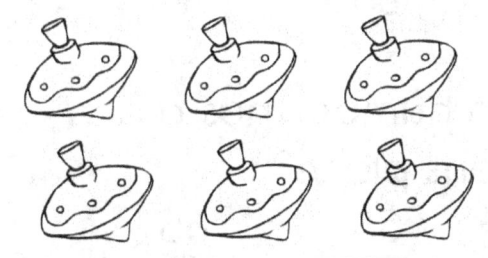

3 children get ☐ each.

2 children get ☐ each.

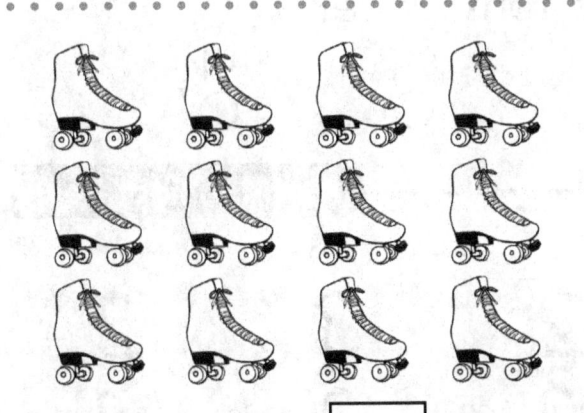

4 children get ☐ each.

3 children get ☐ each.

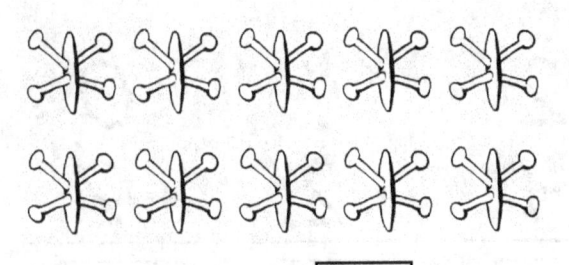

2 children get ☐ each.

4 children get ☐ each.

Chapter 9 • Lesson 71   one hundred thirty-one   **131**

Chapter 4 Review

Write the expanded form for each number.

51 = ☐ + ☐

78 = ☐ + ☐

39 = ☐ + ☐

88 = ☐ + ☐

Mark the correct answer for each question.

Which number has a 7 in the Ones place?
- 72 ○
- 47 ○

Which number has a 3 in the Ones place?
- 83 ○
- 34 ○

Which number has a 5 in the Tens place?
- 56 ○
- 65 ○

Which number has a 4 in the Tens place?
- 94 ○
- 41 ○

Addition Fact Review

Add.

7	2	9	2	3	2
+2	+6	+2	+2	+2	+8
☐	☐	☐	☐	☐	☐

5	2	8	2	1	0
+2	+3	+2	+4	+2	+2
☐	☐	☐	☐	☐	☐

Probability & Tally Marks

- - - - - - - - - - - -

Color 2 parts red. Color 6 parts blue.
Use a paper clip and pencil for a spinner. Spin 10 times.
Make a tally mark to show each spin. Write the total.

Before you spin, guess what color the paper clip is more likely to stop on. Circle the color.

red   **blue**

Color	Tallies	Total
red		
blue		

Circle the color the paper clip stopped on the most.

red   **blue**

Chapter 9 • Lesson 72   one hundred thirty-three   **133**

Chapter 7 Review

Circle the correct equation for each word problem.

Alex has 5 nuts.
He gives 2 nuts to Sam.
How many nuts does Alex have now?

$5 + 2 = 7$ nuts
$5 - 2 = 3$ nuts

Kim painted 4 pictures for her grandmother. She painted 2 more pictures for her mom. How many pictures did Kim paint in all?

$4 + 2 = 6$ pictures
$4 - 2 = 2$ pictures

Addition Fact Review

Add.

| 4 | 5 | 3 | 2 | 6 | 9 |
|+ 1 |+ 5 |+ 2 |+ 9 |+ 6 |+ 0 |

| 8 | 6 | 2 | 3 | 1 | 2 |
|+ 0 |+ 1 |+ 7 |+ 1 |+ 5 |+ 3 |

| 6 | 1 | 0 | 5 | 3 | 3 |
|+ 2 |+ 7 |+ 5 |+ 1 |+ 0 |+ 3 |

Chapter 9 Review

Circle each shape that has equal parts.
Write how many equal parts.

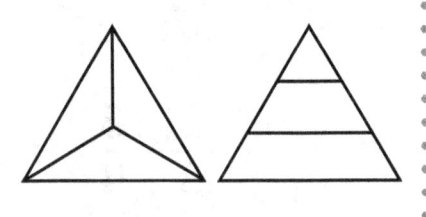

☐ equal parts

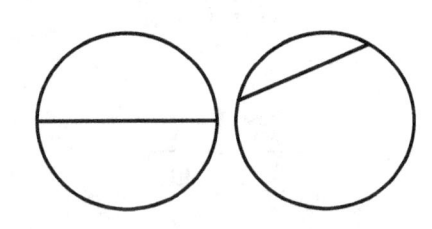

☐ equal parts

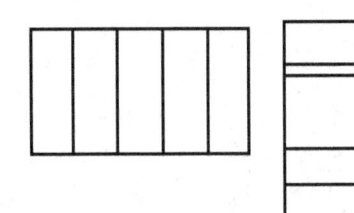

☐ equal parts

Mark the fraction that names the shaded part.

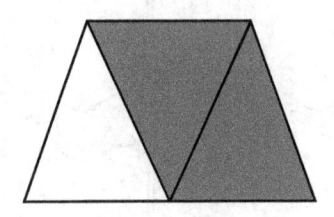

$\frac{1}{3}$ $\frac{2}{3}$
○ ○

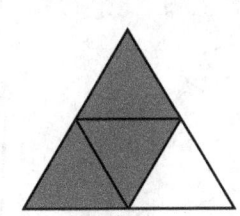

$\frac{2}{3}$ $\frac{3}{4}$
○ ○

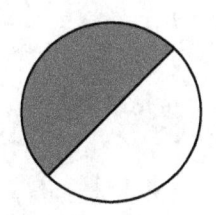

$\frac{1}{2}$ $\frac{1}{3}$
○ ○

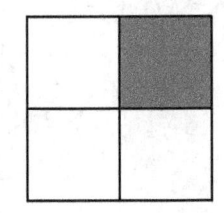

$\frac{1}{3}$ $\frac{1}{4}$
○ ○

Color the shape to match the fraction.

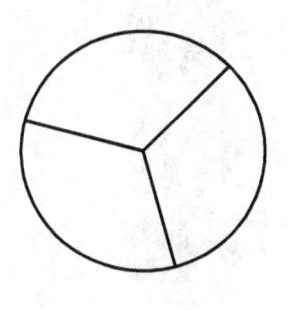

$\frac{1}{3}$

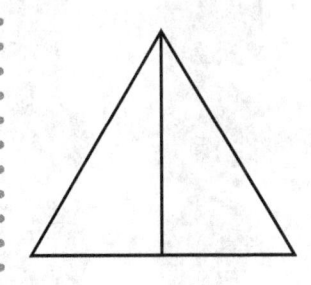

$\frac{1}{2}$

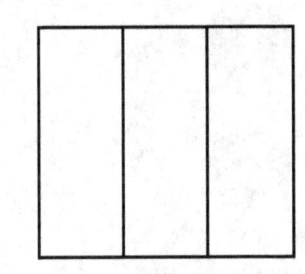

$\frac{2}{3}$

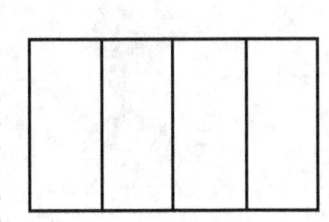

$\frac{3}{4}$

Chapter 9 Review

Circle the fraction to name the shaded part of each set.

$\frac{2}{3}$ $\frac{2}{4}$

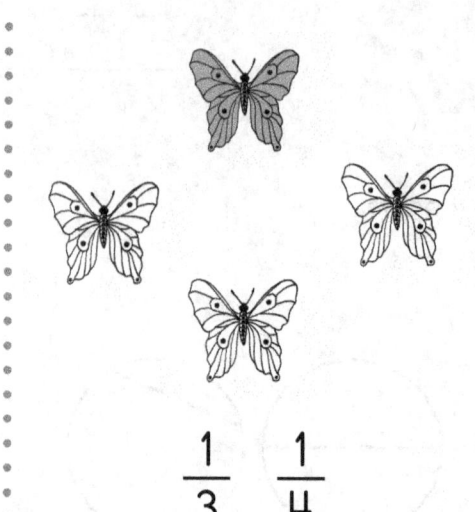

$\frac{1}{3}$ $\frac{1}{4}$

$\frac{2}{3}$ $\frac{3}{4}$

Color to show each fraction of a set.

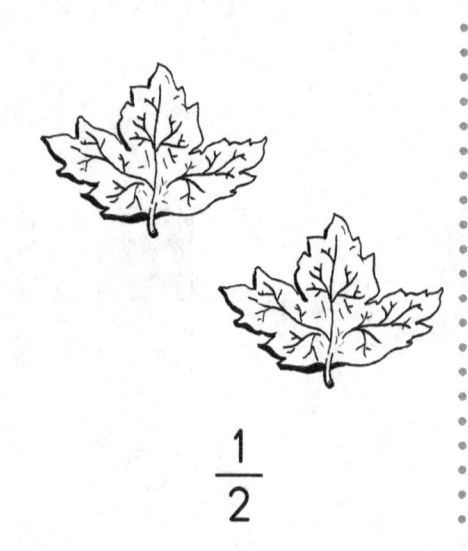

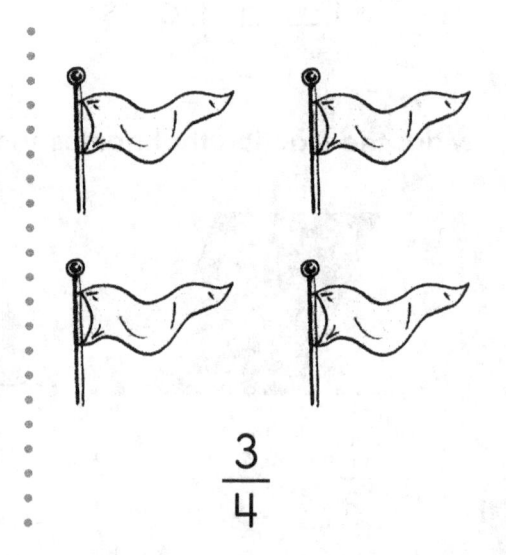

$\frac{1}{2}$

$\frac{2}{3}$

$\frac{3}{4}$

Circle the fair share for each child. Write the number each child gets.

2 children get ☐ each.

3 children get ☐ each.

136   one hundred thirty-six

Math 1 Reviews

Cumulative Review

Number the days of the week in order.

☐ Wednesday ☐ Saturday ☐ Monday
☐ Thursday   ☐ Tuesday   ☐ Friday   ☐ Sunday

Day of the Week	Kids in the Park
Monday	👤
Tuesday	👤👤👤
Wednesday	
Thursday	👤👤
Friday	👤👤👤👤👤
Saturday	👤👤👤👤👤👤

Look at the graph. Solve each problem.

How many more kids were in the park on Saturday than on Thursday?

☐ ○ ☐ ○ ☐ kids

How many kids played together in the park on Tuesday and Thursday?

☐ ○ ☐ ○ ☐ kids

How many kids went to the park on Friday and Wednesday?

☐ ○ ☐ ○ ☐ kids

How many more kids came to the park on Friday than on Monday?

☐ ○ ☐ ○ ☐ kids

Chapter 9 Cumulative Review

Addition Fact Review

Add.

5	3	6	2	8
+ 2	+ 3	+ 1	+ 7	+ 1
☐	☐	☐	☐	☐

6	3	1	3	9
+ 6	+ 2	+ 7	+ 6	+ 2
☐	☐	☐	☐	☐

9	6	1	7	0
+ 1	+ 2	+ 8	+ 2	+ 8
☐	☐	☐	☐	☐

8	1	2	1	2
+ 0	+ 6	+ 5	+ 9	+ 8
☐	☐	☐	☐	☐

5	0	8	9	2
+ 5	+ 7	+ 2	+ 0	+ 3
☐	☐	☐	☐	☐

Count Back 1 or 2; Subtract 0

Use the number line to subtract.
Draw to find each answer.

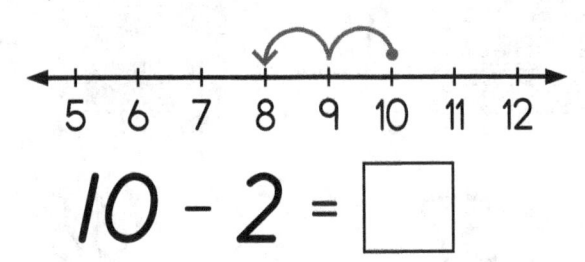

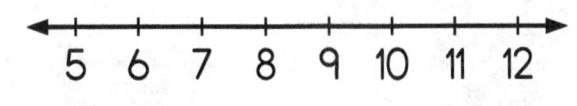

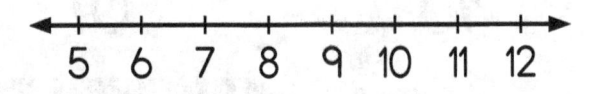

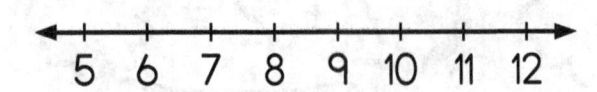

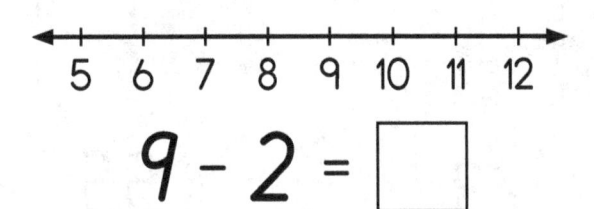

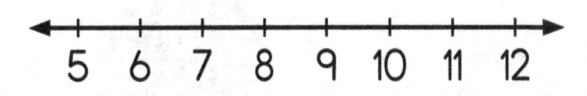

10 − 2 = ☐

8 − 1 = ☐

9 − 1 = ☐

8 − 2 = ☐

9 − 2 = ☐

11 − 2 = ☐

7 − 2 = ☐

10 − 1 = ☐

Subtract 0.

5 − 0 = ☐

9 − 0 = ☐

7 − 0 = ☐

6 − 0 = ☐

8 − 0 = ☐

Chapter 10 • Lesson 75

Chapter 4 Review

Circle the correct sign.

> is greater than   < is less than

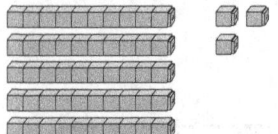

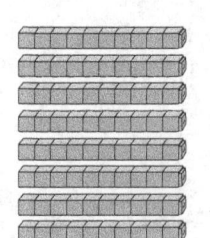

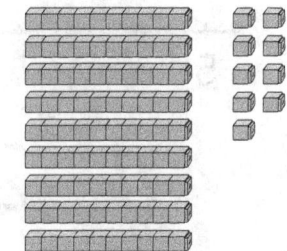

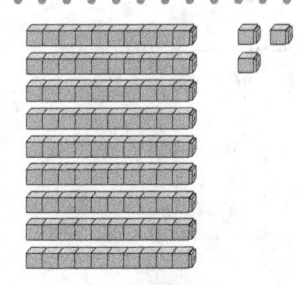

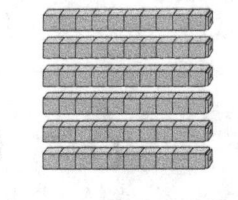

73 >/< 55   57 >/< 68

80 >/< 99   93 >/< 61

Chapter 7 Review

Choose 5 crayons. Make your own key.

Colors
8 = ○
9 = ○
10 = ○
11 = ○
12 = ○

Add.

5 + 3 =
8 + 3 =
4 + 4 =
7 + 3 =

7 + 2 =
9 + 2 =
8 + 1 =
6 + 6 =

140   one hundred forty   Math 1 Reviews

Subtract *All*;
Subtract *Nearly All*

Subtract *all*.

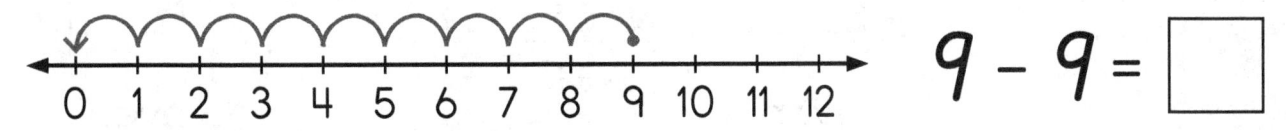

$9 - 9 = \square$

Subtract *nearly all*.

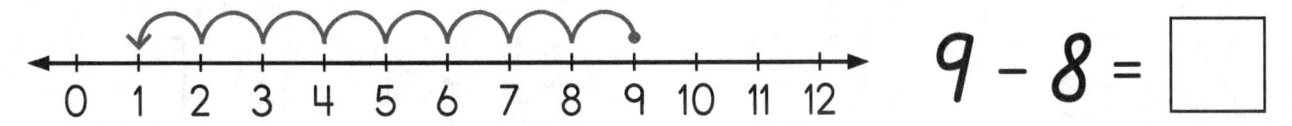

$9 - 8 = \square$

Use the number line to subtract. Draw ⤹ to find each answer.

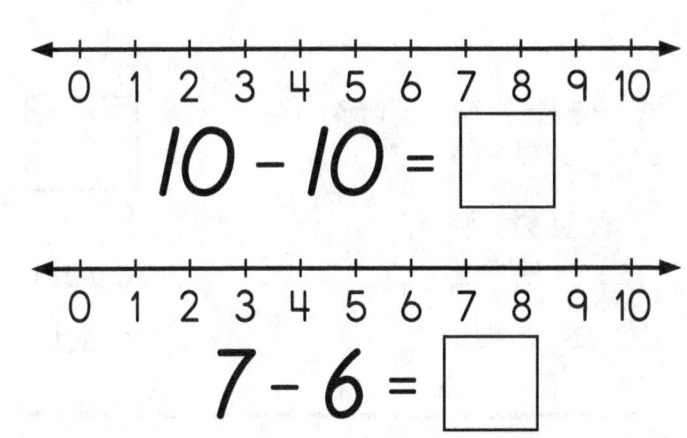

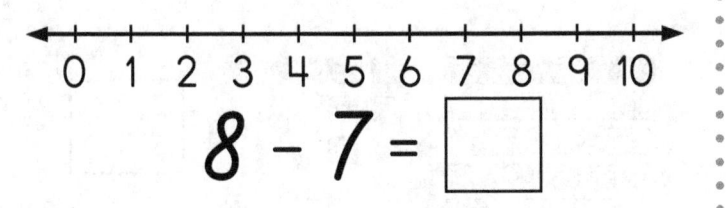

$8 - 7 = \square$

$10 - 10 = \square$

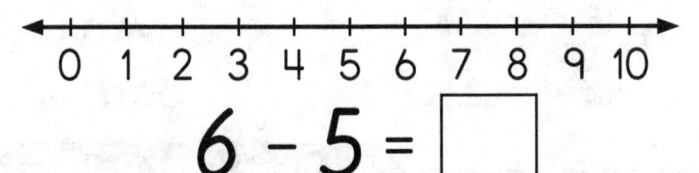

$6 - 5 = \square$

$7 - 6 = \square$

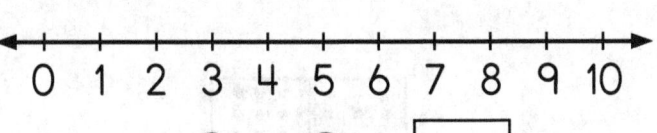

$8 - 8 = \square$

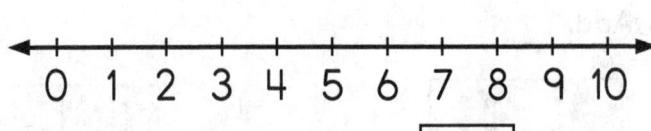

$9 - 8 = \square$

Write an equation for each word problem. Solve.

Dad had 4 cups of milk. He drank 3 of them. How many cups does he have left?

$\square \bigcirc \square \bigcirc \square$ cup(s)

Jess had 3 glasses of water. He gave the 3 glasses of water to his friends. How many glasses does he have now?

$\square \bigcirc \square \bigcirc \square$ glasses

Chapter 10 • Lesson 76

Chapter 4 Review

Write the number. Circle the single cubes to form pairs.
Circle *even* if the number is even.
Circle *odd* if the number is odd.

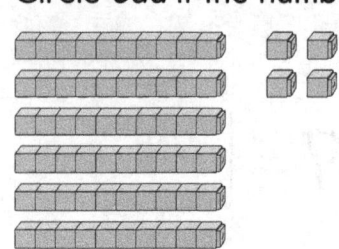

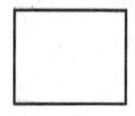

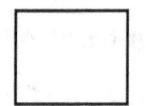

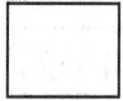

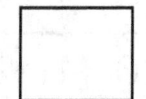

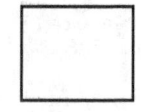

even
odd

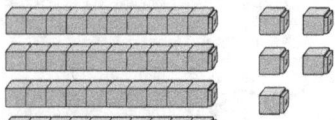

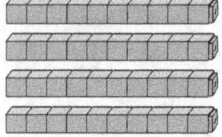

even
odd

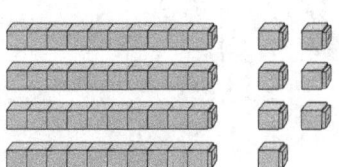

even
odd

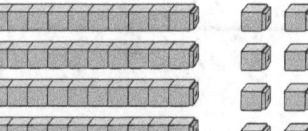

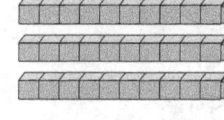

even
odd

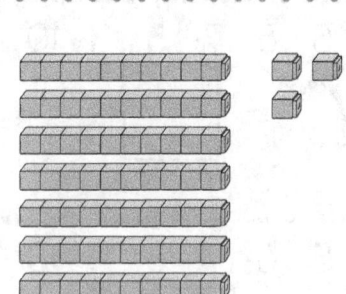

even
odd

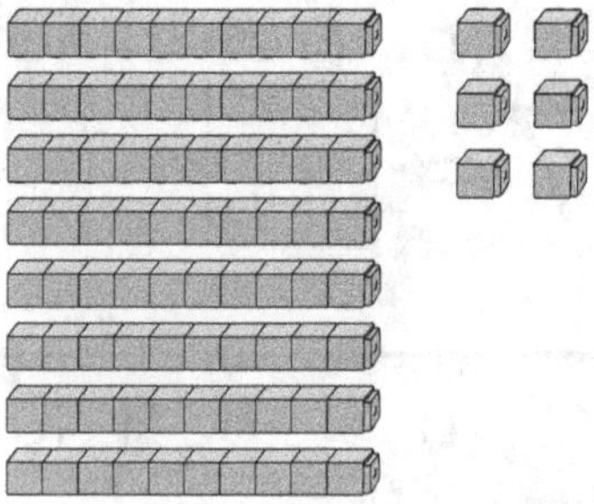

even
odd

Addition Fact Review

Add.

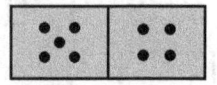

$5 + 4 =$ ☐

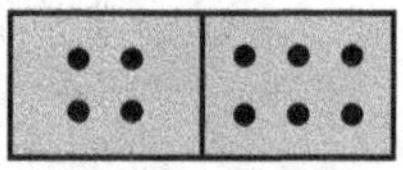

$4 + 6 =$ ☐

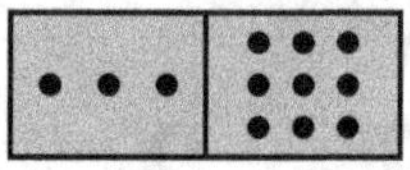

$3 + 9 =$ ☐

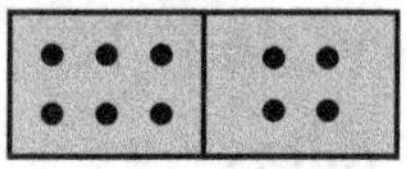

$6 + 4 =$ ☐

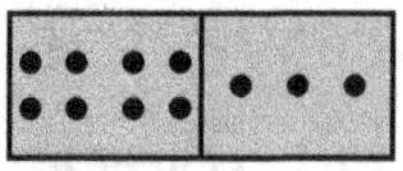

$8 + 3 =$ ☐

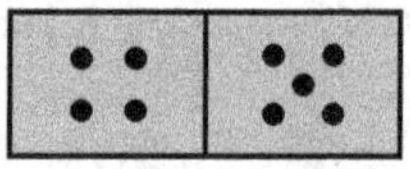

$4 + 5 =$ ☐

Compare to Subtract; Double Facts

Compare the bugs to find the difference. Write how many more.

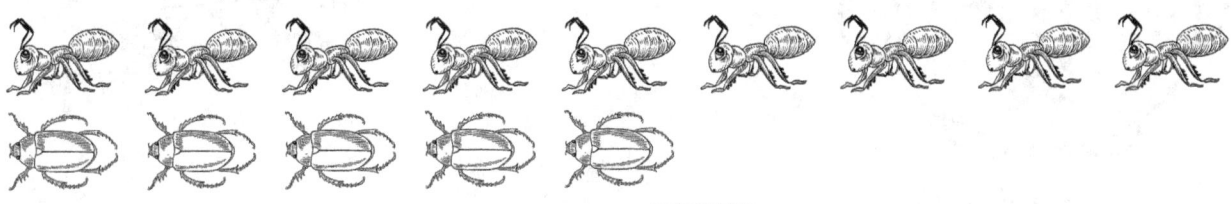

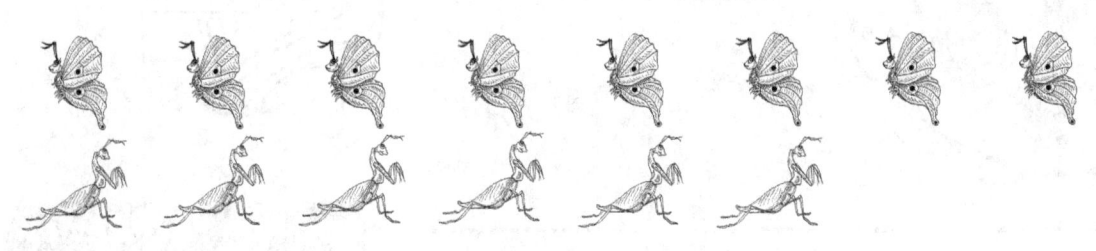

9 − 5 = ☐

7 − 4 = ☐

8 − 6 = ☐

Write an equation for each word problem. Draw pictures to solve.

Emma got 5 bugs one day. The next day she got 2 butterflies. How many more bugs than butterflies did Emma get?

Ava sees 6 red ants and 3 black ants. How many more red ants than black ants does she see?

☐ ○ ☐ ○ ☐ bugs

☐ ○ ☐ ○ ☐ red ants

Chapter 10 • Lesson 77

Chapter 6 Review

Draw a line to match each set of coins with the correct total value.

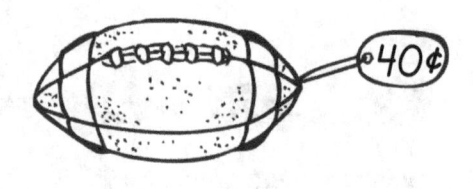

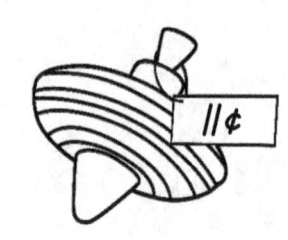

Addition Fact Review

Add.

$$\begin{array}{r} 6 \\ +1 \\ \hline \end{array}$$ $$\begin{array}{r} 7 \\ +2 \\ \hline \end{array}$$ $$\begin{array}{r} 3 \\ +4 \\ \hline \end{array}$$ $$\begin{array}{r} 6 \\ +6 \\ \hline \end{array}$$ $$\begin{array}{r} 8 \\ +3 \\ \hline \end{array}$$ $$\begin{array}{r} 3 \\ +9 \\ \hline \end{array}$$

$$\begin{array}{r} 1 \\ +8 \\ \hline \end{array}$$ $$\begin{array}{r} 4 \\ +4 \\ \hline \end{array}$$ $$\begin{array}{r} 7 \\ +3 \\ \hline \end{array}$$ $$\begin{array}{r} 9 \\ +2 \\ \hline \end{array}$$ $$\begin{array}{r} 2 \\ +5 \\ \hline \end{array}$$ $$\begin{array}{r} 6 \\ +3 \\ \hline \end{array}$$

one hundred forty-four

Math 1 Reviews

Missing Addend; Subtract from 10

```
←+--+--+--+--+--+--+--+--+--+--+--+--→
  0  1  2  3  4  5  6  7  8  9 10 11 12
```

Add or subtract. Use the number line if needed.

```
   10              3           10              7
 -  3           + ☐          -  7           + ☐
 ----           ----          ----           ----
   ☐             10            ☐             10
```

```
   10              8           10              4
 -  8           + ☐          -  4           + ☐
 ----           ----          ----           ----
   ☐             10            ☐             10
```

Complete the equation for the word problem.

Miss Cash needs 8 Bibles for her Sunday school class. She has 4 Bibles in her bag. How many more Bibles does she need?

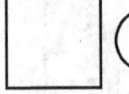

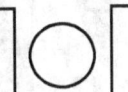

☐ ○ ☐ ○ ☐ Bibles

Draw a line to match each pair of related equations.

7 + 3 = 10   10 − 6 = 4

6 + 4 = 10   10 − 8 = 2

9 + 1 = 10   10 − 3 = 7

8 + 2 = 10   10 − 1 = 9

Chapter 8 Review

If the activity takes more than an hour, color the clock yellow.
If the activity takes less than an hour, color the clock red.

Addition Fact Review

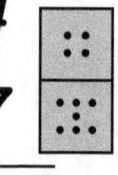

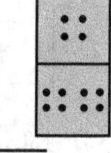

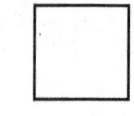

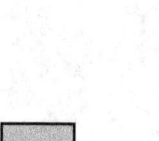

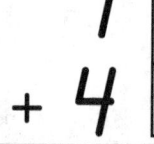

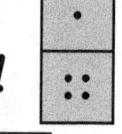

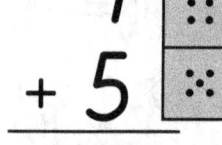

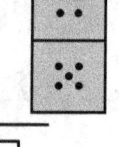

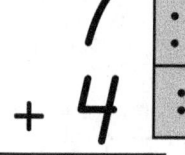

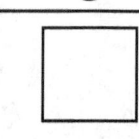

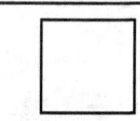

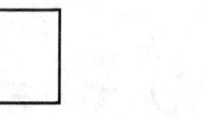

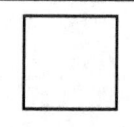

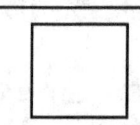

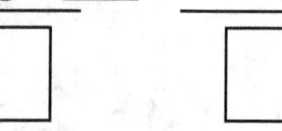

Add.

$5 + 4 =$ ☐

$4 + 7 =$ ☐

$4 + 0 =$ ☐

$6 + 4 =$ ☐

$4 + 8 =$ ☐

$4 + 6 =$ ☐

$8 + 4 =$ ☐

$1 + 4 =$ ☐

$4 + 5 =$ ☐

$7 + 4 =$ ☐

146 one hundred forty-six

Math 1 Reviews

Fact Families

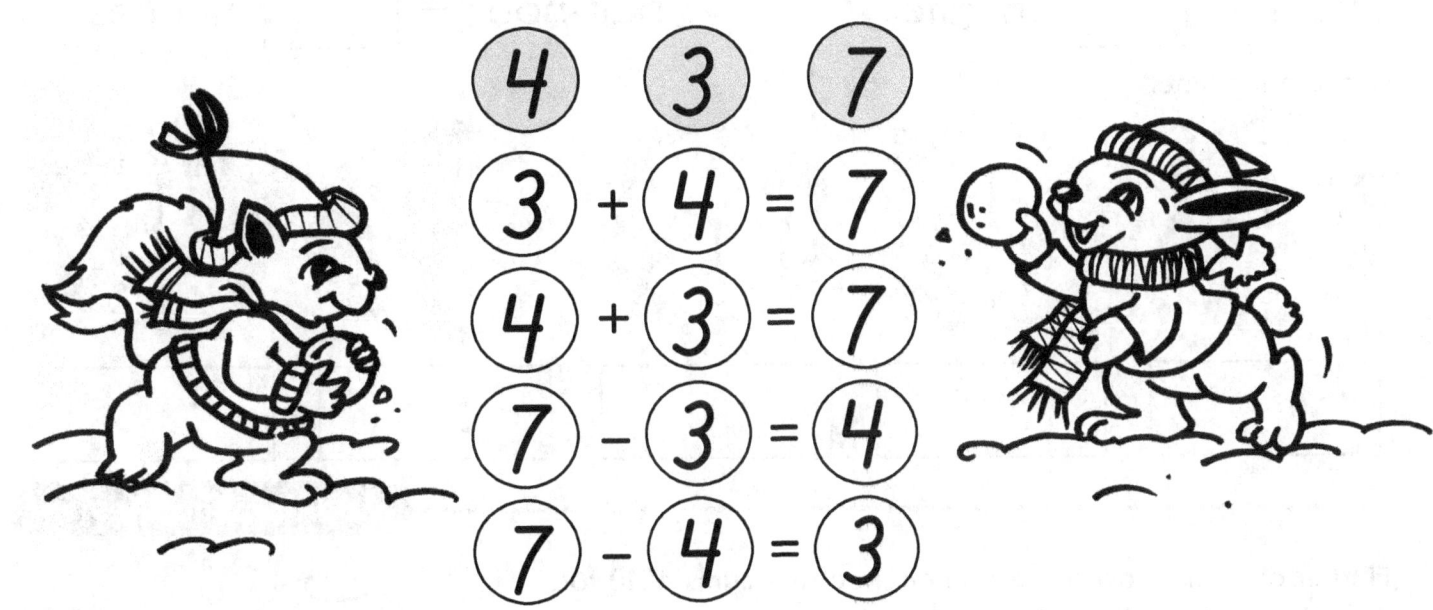

4 3 7

3 + 4 = 7
4 + 3 = 7
7 − 3 = 4
7 − 4 = 3

Add or subtract.
Write the numbers for each fact family.

○ ○ ○   ○ ○ ○   ○ ○ ○

3 + 5 = ☐   4 + 6 = ☐   3 + 6 = ☐
5 + 3 = ☐   6 + 4 = ☐   6 + 3 = ☐
8 − 3 = ☐   10 − 4 = ☐   9 − 3 = ☐
8 − 5 = ☐   10 − 6 = ☐   9 − 6 = ☐

Write the fact family.

3 7 10

☐ + ☐ = 10   10 − ☐ = ☐
☐ + ☐ = 10   10 − ☐ = ☐

Chapter 10 • Lesson 79

Chapter 8 Review

Complete each sentence.

1 hour = ☐ minutes   a half-hour = ☐ minutes

Write each time.

☐ : ☐   ☐ : ☐   ☐ : ☐   ☐ : ☐   ☐ : ☐

Addition Fact Review

Find each sum. Draw a line and connect the sums of 10 to help the snail carrier get the mail delivered.

5 + 5 = 10

4 + 3 = ☐

7 + 3 = ☐

8 + 3 = ☐

5 + 2 = ☐

2 + 8 = ☐

6 + 2 = ☐

9 + 1 = ☐

148 one hundred forty-eight   Math 1 Reviews

Fact Families for 11; Missing Addend

Complete each equation.

11 − 5 = ☐ monkeys

11 − 8 = ☐ monkeys

Write each fact family.

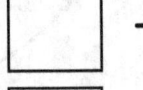

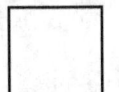

② ⑨ ⑪

☐ + ☐ = ☐
☐ + ☐ = ☐
☐ − ☐ = ☐
☐ − ☐ = ☐

④ ⑦ ⑪

☐ + ☐ = ☐
☐ + ☐ = ☐
☐ − ☐ = ☐
☐ − ☐ = ☐

Write each missing addend.

☐
+ 3
———
 4

☐
+ 3
———
 5

☐
+ 3
———
 6

Complete the equation for the word problem.

Clay wants to plant 6 trees.
He planted 3 trees.
How many more trees does he want to plant?

3 + ☐ = 6
   trees

Chapter 10 • Lesson 80   one hundred forty-nine   149

Chapter 8 Review

Circle the time passed.

Job	Start	Finish	Time Passed
(boy mowing)	3:00	4:00	30 minutes or 1 hour
(boy washing car)	10:00	10:30	30 minutes or 1 hour

Addition Fact Review

Make your own color key.
Add and color the fish.

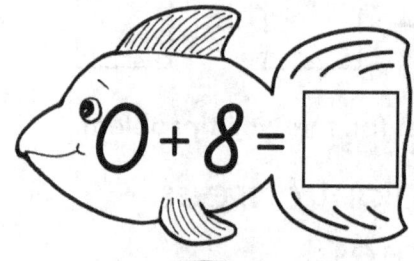

7 = ○
8 = ○
9 = ○

3 + 6 = ☐
6 + 2 = ☐
0 + 9 = ☐
4 + 3 = ☐
0 + 8 = ☐
5 + 2 = ☐
2 + 7 = ☐
1 + 8 = ☐
4 + 4 = ☐
5 + 3 = ☐

Fact Families for 12; Missing Addend

Complete each missing addend equation.

4 + ☐ = 7   8 + ☐ = 10

6 + ☐ = 12   7 + ☐ = 9

Add or subtract.
Write each number for the fact family.

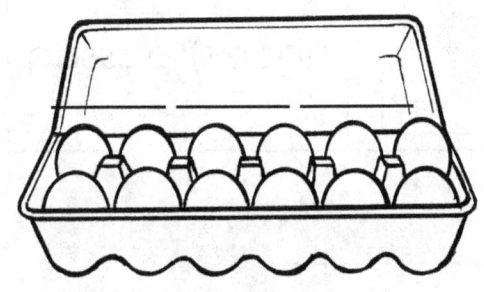

4 + 8 = ☐   12 − 4 = ☐
8 + 4 = ☐   12 − 8 = ☐

Write each fact family.

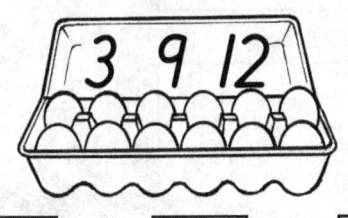

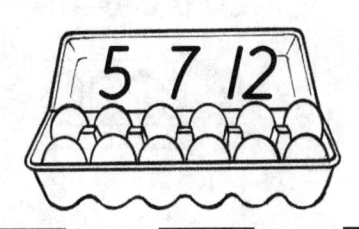

3 9 12   5 7 12

☐ + ☐ = ☐   ☐ + ☐ = ☐
☐ + ☐ = ☐   ☐ + ☐ = ☐
☐ − ☐ = ☐   ☐ − ☐ = ☐
☐ − ☐ = ☐   ☐ − ☐ = ☐

Write an addition or subtraction equation for the word problem. Solve.

Eloise had 6 eggs.
She broke 4 eggs.
How many of the eggs are not broken?

☐ ○ ☐ ○ ☐ eggs

Chapter 10 • Lesson 81

Chapter 8 Review

Write the time under each clock.
Fill in the time passed for each job.

Job	Start	Finish	Time Passed
Store / Farm Fresh Eggs truck	10:00	11:00	☐ hour(s) passed
Woman gathering eggs	12:00	6:00	☐ hour(s) passed

Look at the table and write the time.

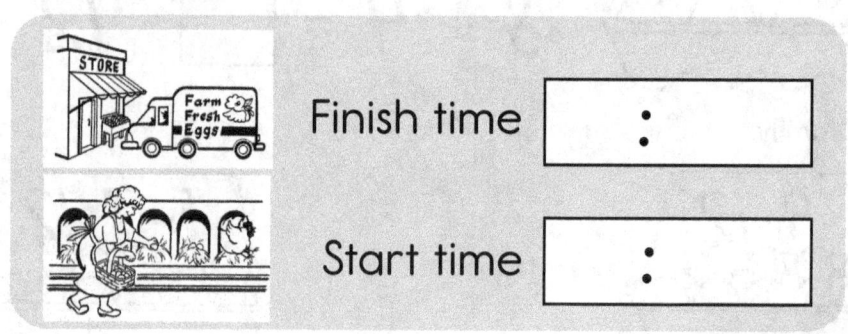

Finish time ☐ : ☐

Start time ☐ : ☐

Addition Fact Review

Add.

$5 + 4 =$ ☐ $6 + 5 =$ ☐ $8 + 4 =$ ☐ $4 + 6 =$ ☐ $7 + 4 =$ ☐

$4 + 8 =$ ☐ $7 + 5 =$ ☐ $6 + 4 =$ ☐ $4 + 5 =$ ☐ $5 + 6 =$ ☐

Chapter 10 Review

Use the number line to subtract.
Draw ⌒ to find each answer.

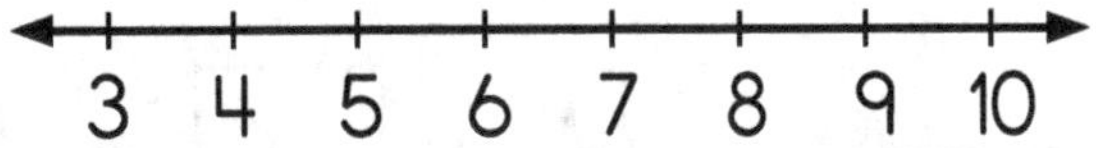

3 4 5 6 7 8 9 10
10 − 1 = ☐

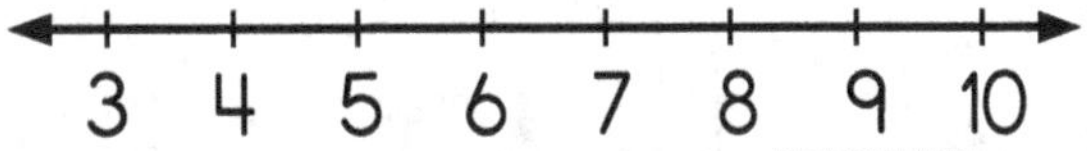

3 4 5 6 7 8 9 10
8 − 2 = ☐

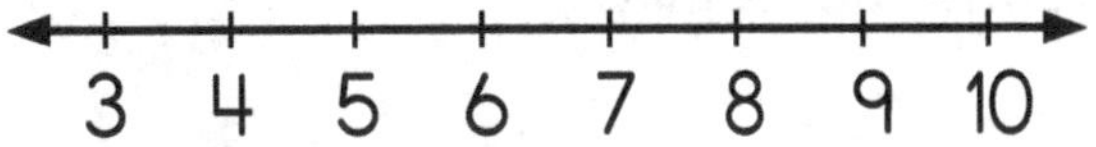

3 4 5 6 7 8 9 10
9 − 2 = ☐

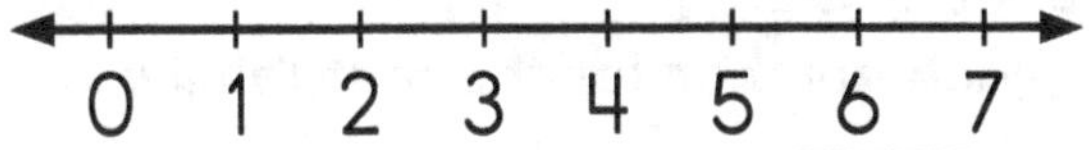

0 1 2 3 4 5 6 7
7 − 6 = ☐

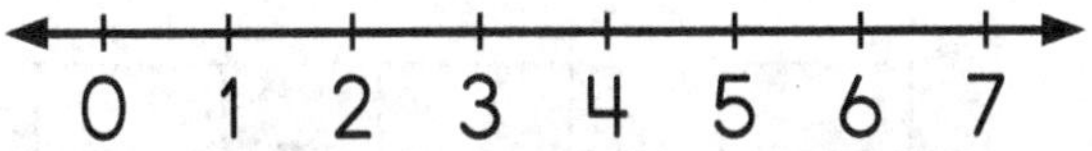

0 1 2 3 4 5 6 7
7 − 7 = ☐

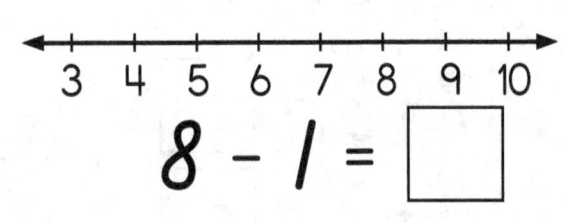

3 4 5 6 7 8 9 10
8 − 1 = ☐

Compare the flowers to find the difference.
Write how many more.

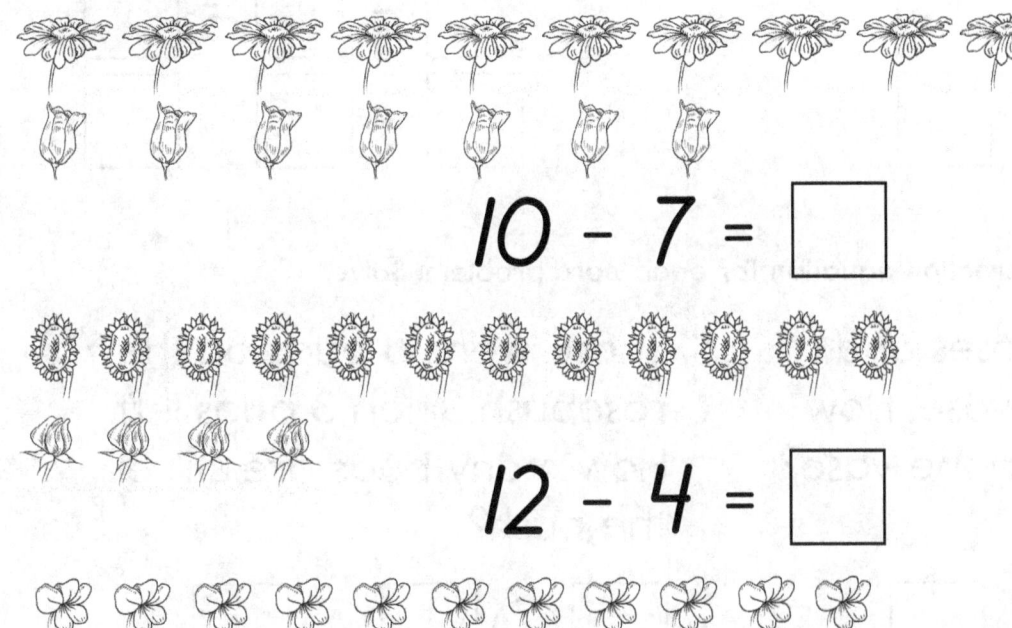

10 − 7 = ☐

12 − 4 = ☐

11 − 6 = ☐

Chapter 10 Review one hundred fifty-three 153

0 1 2 3 4 5 6 7 8 9 10 11 12

Add or subtract. Use the number line if needed.

10 − 4 = ☐   11 − 5 = ☐   12 − 4 = ☐

4 + ☐ = 10   5 + ☐ = 11   4 + ☐ = 12

Add or subtract.
Write each number for the fact family.

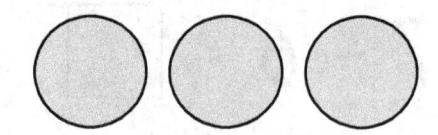

◯ ◯ ◯

7 + 3 = ☐
3 + 7 = ☐
10 − 3 = ☐
10 − 7 = ☐

Write the fact family.

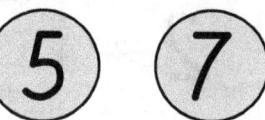

5   7   12

☐ + ☐ = ☐
☐ + ☐ = ☐
☐ − ☐ = ☐
☐ − ☐ = ☐

Write an addition or subtraction equation for each word problem. Solve.

Mom put 4 red roses and 4 pink roses in a vase. How many roses are in the vase?

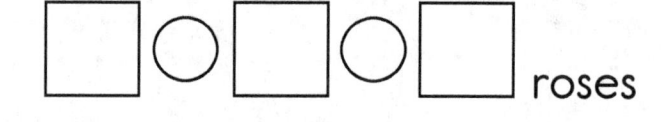

☐ ◯ ☐ ◯ ☐ roses

There were 6 bugs on the rosebush. Then 3 bugs left. How many bugs are still on the bush?

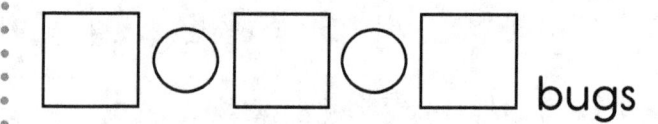

☐ ◯ ☐ ◯ ☐ bugs

154   one hundred fifty-four   Math 1 Reviews

Cumulative Review

Circle the longest object. Mark an *X* on the shortest object.

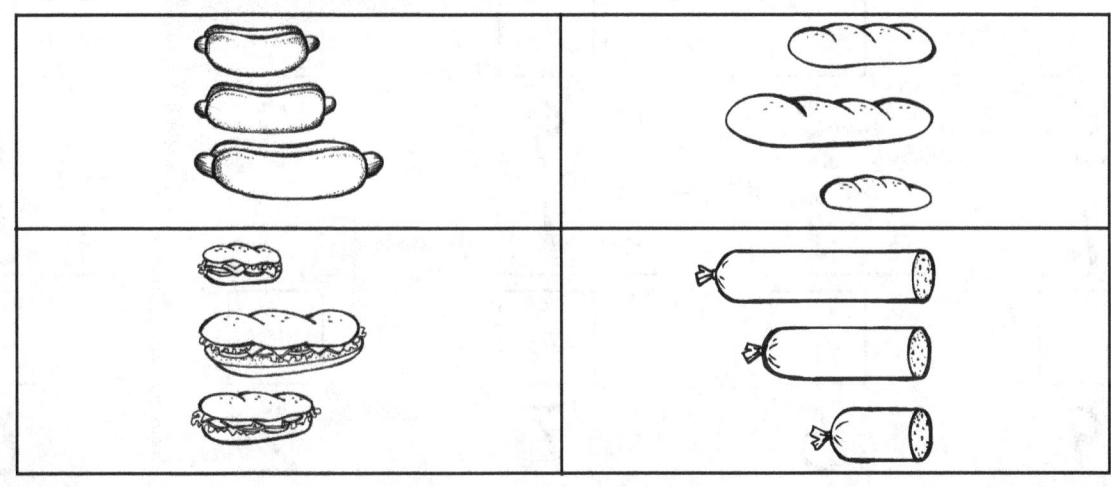

Draw a line from each number to the correct number of tens.

50	1 ten
20	5 tens
10	2 tens

9 tens	40
4 tens	10
1 ten	90

30	7 tens
70	10 tens
100	3 tens

10 tens	80
6 tens	60
8 tens	100

Fill in the boxes.

10	ten	1 ten	60 sixty	6	tens
	twenty	2 tens	70 seventy		tens
	thirty	3 tens	80 eighty		tens
	forty	4 tens	90 ninety		tens
	fifty	5 tens	100 one hundred		tens

Chapter 10 Cumulative Review   one hundred fifty-five   **155**

Add.

3	8	2	7	2
+9	+2	+6	+1	+5
☐	☐	☐	☐	☐

6	2	9	3	1
+1	+7	+1	+4	+8
☐	☐	☐	☐	☐

9	5	7	2	6
+2	+5	+3	+9	+0
☐	☐	☐	☐	☐

4	5	3	6	5
+4	+3	+7	+6	+2
☐	☐	☐	☐	☐

1	8	4	6	0
+7	+3	+3	+2	+5
☐	☐	☐	☐	☐

9	7	2	1	7
+3	+0	+8	+6	+2
☐	☐	☐	☐	☐

Math 1 Reviews

Measure in Paper Clips & Inches

Count the paper clips to find each length.

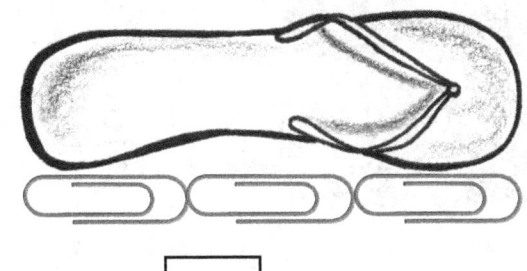

☐ 🖇

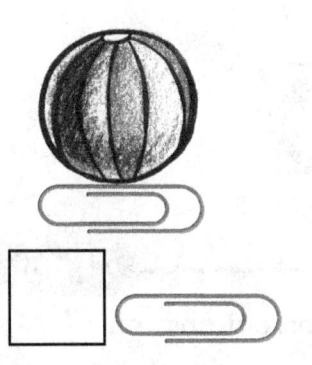

☐ 🖇

Write the number of inches.

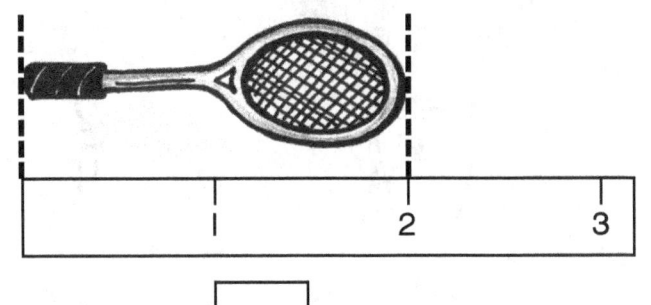

☐ inches

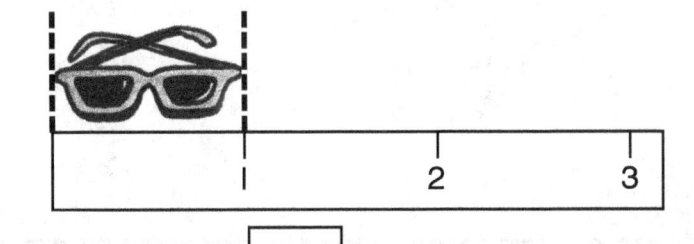

☐ inches

Use an inch ruler to measure.

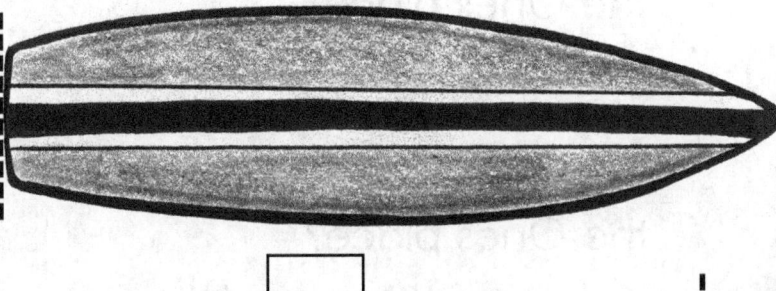

☐ inches

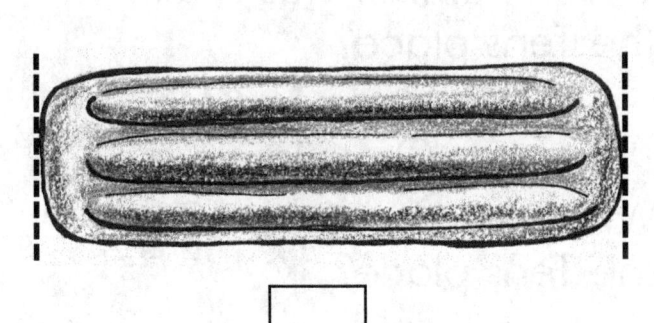

☐ inches

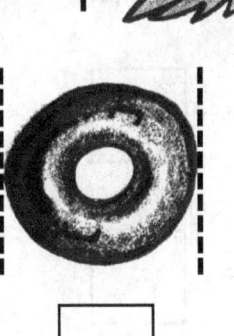

☐ inches

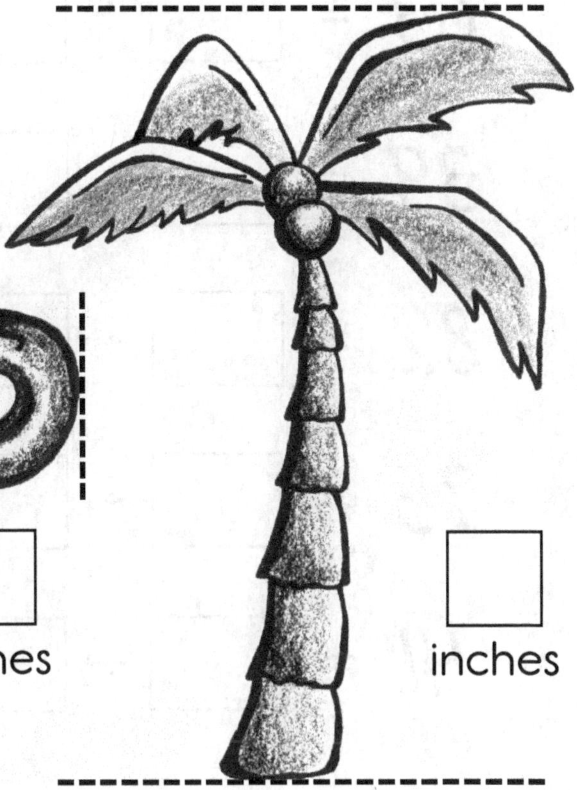

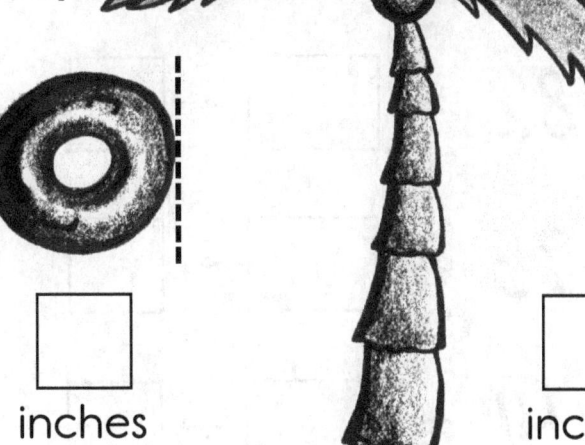

☐ inches

Chapter 11 • Lesson 84   one hundred fifty-seven   **157**

Chapter 9 Review

Circle the fraction to name the shaded part of each set.

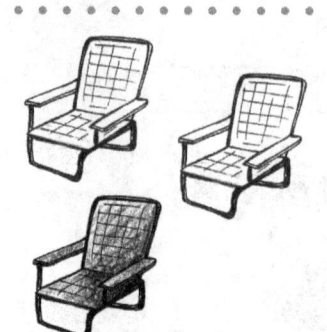

$\frac{2}{3}$

$\frac{2}{4}$

$\frac{2}{3}$

$\frac{3}{4}$

$\frac{1}{2}$

$\frac{1}{3}$

$\frac{1}{4}$

$\frac{1}{3}$

Chapter 4 Review

Write the expanded form for each number.

63 = ☐ + ☐

39 = ☐ + ☐

82 = ☐ + ☐

75 = ☐ + ☐

41 = ☐ + ☐

Mark the correct answer.

Which number has an 8 in the Ones place?

48 ○   86 ○

Which number has a 4 in the Ones place?

49 ○   64 ○

Which number has a 6 in the Tens place?

62 ○   86 ○

Which number has a 2 in the Tens place?

52 ○   26 ○

Estimate & Measure in Inches

Guess the length. Use an inch ruler to measure.

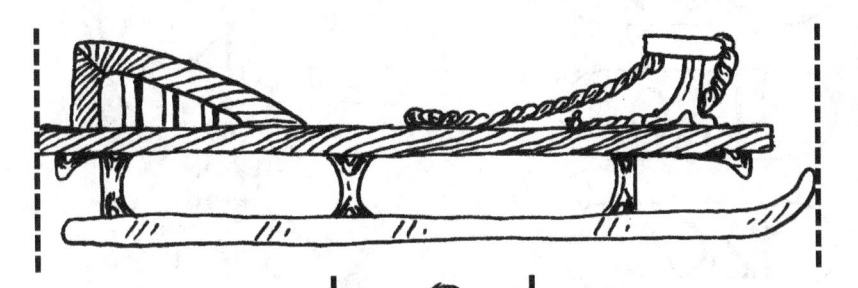

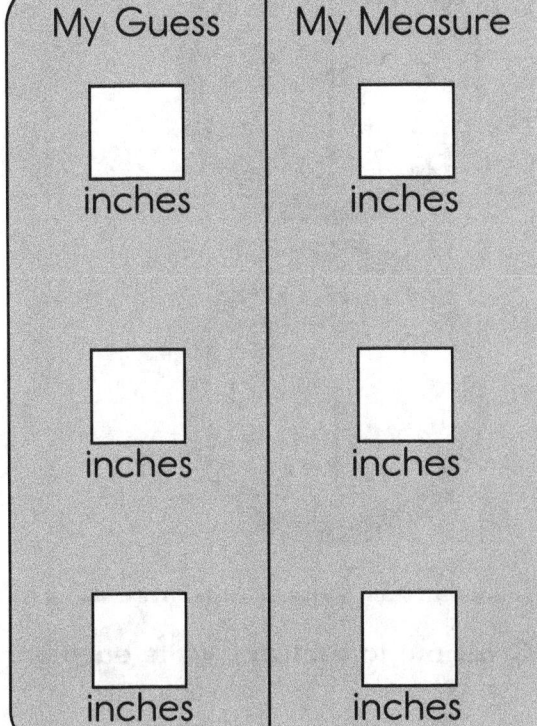

	My Guess	My Measure
	☐ inches	☐ inches
	☐ inches	☐ inches
	☐ inches	☐ inches

Use an inch ruler to draw each length. Start at the star.

(4 inches ✱)

(3 inches ✱)

(5 inches ✱)

Mark the correct equation.

There were 2 inches of snow on the grass. It snowed 8 more inches. How many inches of snow are there now?

○ 8 − 2 = 6 inches
○ 2 + 8 = 10 inches

Mason made 5 snowmen. His 5 snowmen melted. How many snowmen does Mason have left?

○ 5 − 5 = 0 snowmen
○ 5 + 5 = 10 snowmen

Chapter 11 • Lesson 85

Chapter 8 Review

Draw a line from each clock to the correct time.

6:30

4:30

12:00

9:30

Subtraction Fact Review

Cross out to subtract. Write each answer.

9 − 0 = ☐

7 − 7 = ☐

10 − 9 = ☐

6 − 3 = ☐

9 − 8 = ☐

7 − 0 = ☐

8 − 8 = ☐

7 − 6 = ☐

8 − 0 = ☐

5 − 4 = ☐

9 − 9 = ☐

8 − 7 = ☐

Cups & Pints

Color 2 cups for each pint. Fill in each box.

☐ pint = ☐ cups

☐ pints = ☐ cups

☐ pints = ☐ cups

☐ pints = ☐ cups

Circle which set holds *less*.

Circle which set holds *more*.

Chapter 11 • Lesson 86   one hundred sixty-one   **161**

Chapter 3 Review

Color each number name.

- 2 red
- 3 yellow
- 4 green
- 5 blue
- 6 purple
- 7 orange

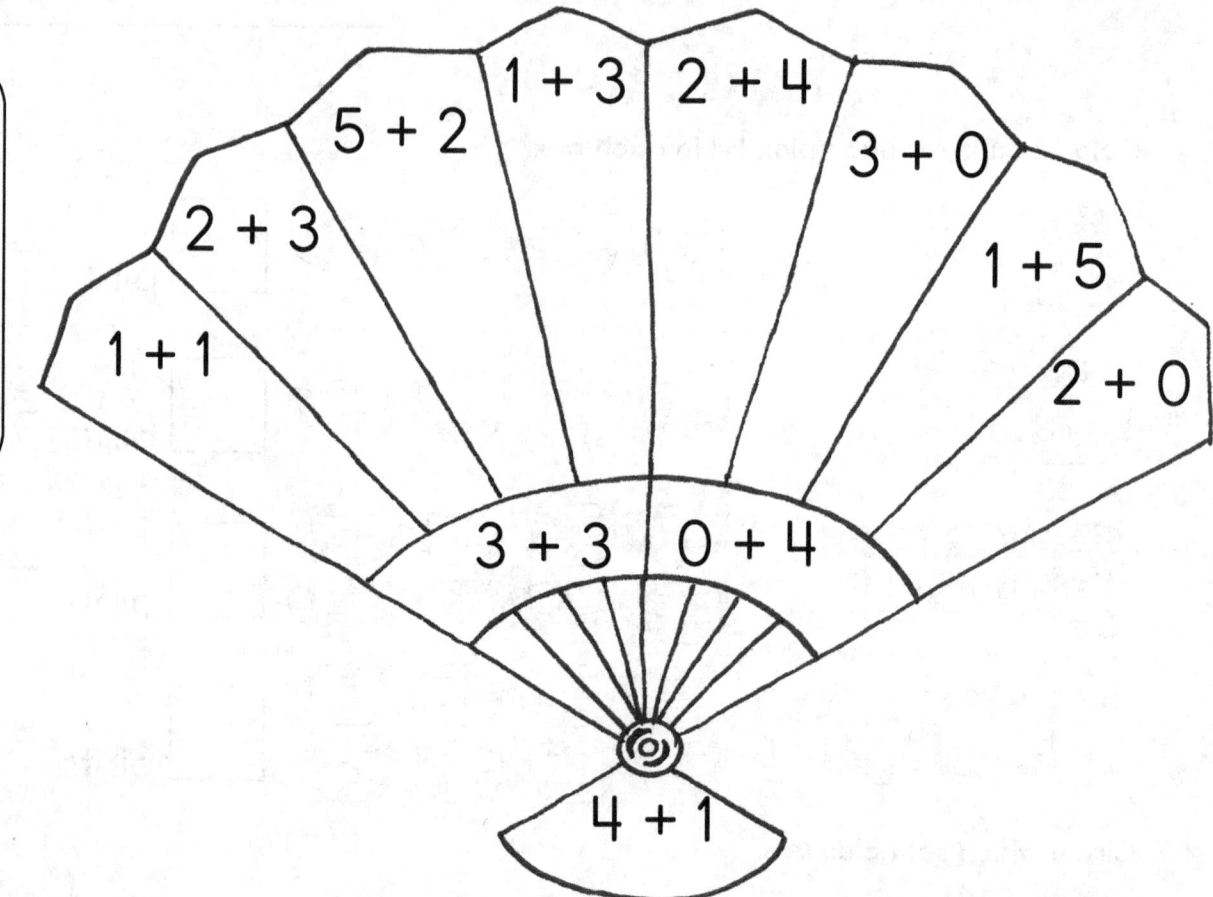

Fan sections:
1 + 1, 2 + 3, 5 + 2, 1 + 3, 2 + 4, 3 + 0, 1 + 5, 2 + 0, 3 + 3, 0 + 4, 4 + 1

Chapter 7 Review

Write an addition or subtraction equation for each word problem.

Grandma has 4 cans of yams and 4 cans of carrots. How many cans does she have in all?

☐ ○ ☐ ○ ☐ cans

Mother has 6 fans. She gives 4 fans to a friend. How many fans does Mother have left?

☐ ○ ☐ ○ ☐ fans

Jack had 5 goats in a pen. The gate was left open, and 3 goats ran away. How many goats are left in the pen?

☐ ○ ☐ ○ ☐ goats

Emma made 3 cakes to sell. Charlotte made 7 cakes to sell. How many cakes did they make in all?

☐ ○ ☐ ○ ☐ cakes

162   one hundred sixty-two   Math 1 Reviews

Pints, Quarts & Gallons

1 quart = 2 pints

1 gallon = 4 quarts

Circle 2 pints for each quart.

Circle 4 quarts for the gallon.

Circle the container that holds *less*.

Chapter 11 • Lesson 87   one hundred sixty-three   **163**

Chapter 9 Review

Circle the fair share for each child.
Write the number each child gets.

3 children get ☐ each.

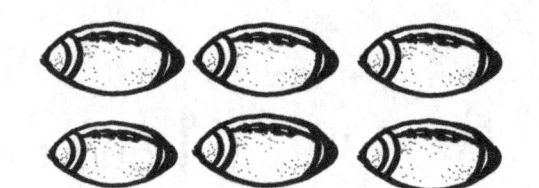

2 children get ☐ each.

4 children get ☐ each.

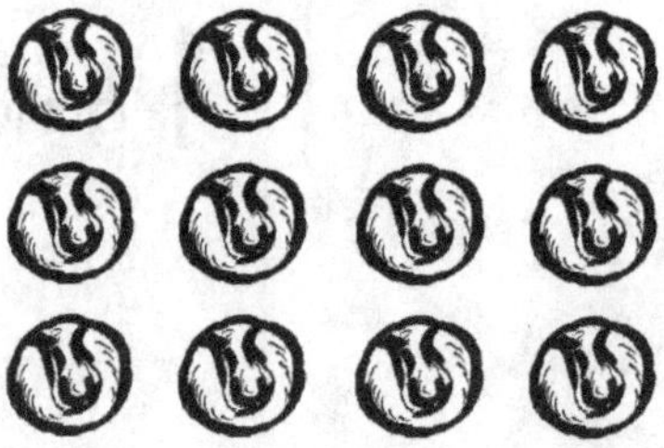

3 children get ☐ each.

2 children get ☐ each.

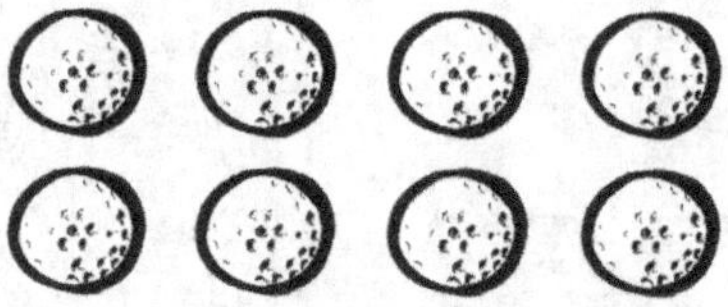

4 children get ☐ each.

Subtraction Fact Review

Cross out to subtract. Write each answer.

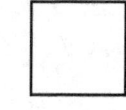

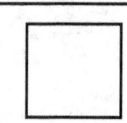

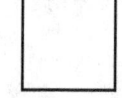

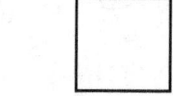

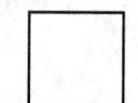

7 − 1 = ☐

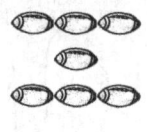

10 − 2 = ☐

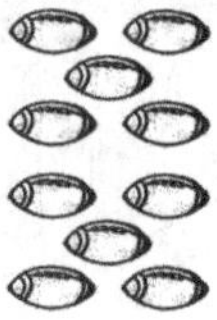

9 − 1 = ☐

11 − 2 = ☐

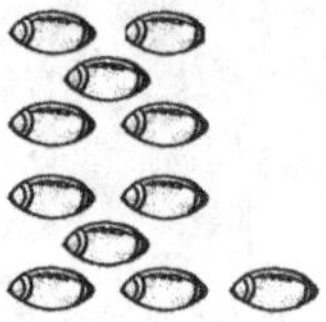

9 − 2 = ☐

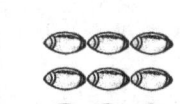

8 − 1 = ☐

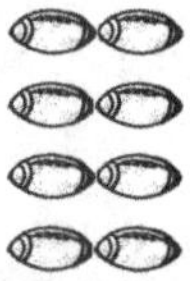

7 − 2 = ☐

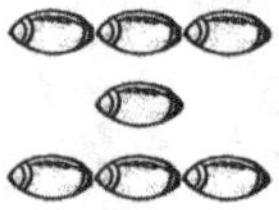

10 − 1 = ☐

More Than or Less Than a Pound

1 pound   1 pound

Mark the correct answer.

○ more   ○ less
than 1 pound

○ more   ○ less
than 1 pound

Write the number by each object in the correct circle.

1. (ruler)   2. (backpack)

1. (easel)   2. (tube)

1. (briefcase)   2. (apple)

Circle each object that is *less* than 1 pound.

Chapter 11 • Lesson 88   one hundred sixty-five   **165**

Chapter 10 Review

0 1 2 3 4 5 6 7 8 9 10 11 12

Add or subtract.
Use the number line if needed.

11 − 2 = ☐

2 + ☐ = 11

7 − 6 = ☐

6 + ☐ = 7

10 − 9 = ☐

9 + ☐ = 10

12 − 3 = ☐

3 + ☐ = 12

8 − 1 = ☐

1 + ☐ = 8

9 − 0 = ☐

0 + ☐ = 9

Complete the equation for each word problem.

Mom needs 12 stamps.
She has 6 stamps.
How many more stamps does Mom need?

6 + ☐ = 12
stamps

Milo needs 6 insects.
He has 3 insects.
How many more insects does he need to find?

3 + ☐ = 6
insects

Read a Thermometer

Draw a line from each thermometer to the correct picture.

Circle the correct temperature.

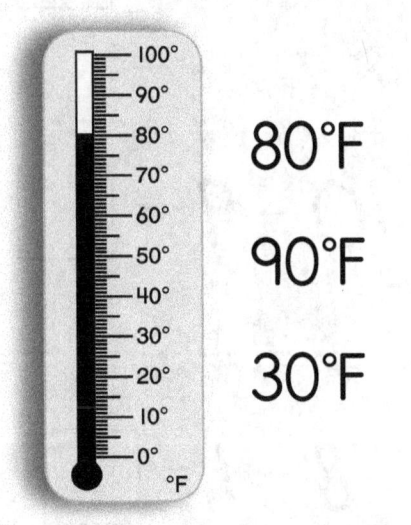

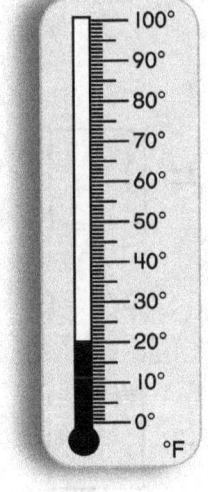

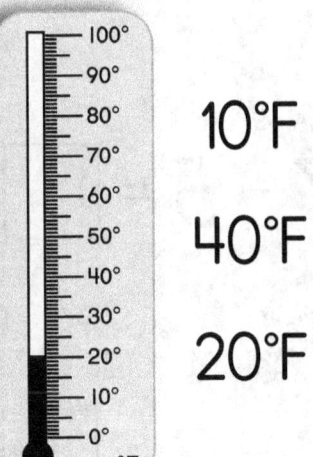

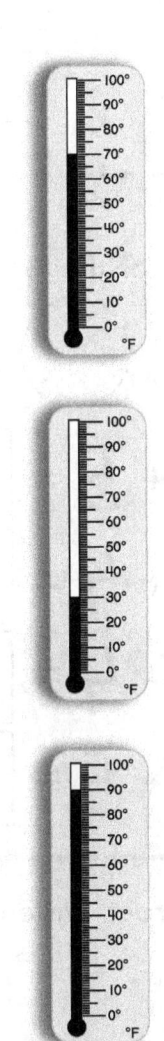

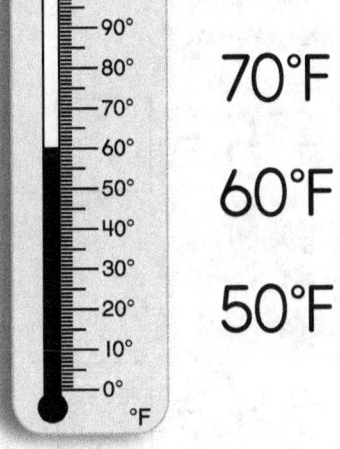

80°F
90°F
30°F

10°F
40°F
20°F

70°F
60°F
50°F

Chapter 11 • Lesson 89   one hundred sixty-seven   **167**

Chapter 8 Review

Number the days of the week in order.

☐ Tuesday ☐ Saturday ☐ Wednesday
☐ Sunday ☐ Thursday ☐ Monday ☐ Friday

Chapter 10 Review

Add or subtract.
Write the numbers for the fact family.

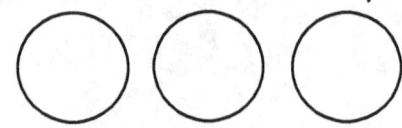

○ ○ ○

4 + 5 = ☐
5 + 4 = ☐
9 − 4 = ☐
9 − 5 = ☐

Write the fact family.

5   6   11

☐ + ☐ = ☐
☐ + ☐ = ☐
☐ − ☐ = ☐
☐ − ☐ = ☐

Subtraction Fact Review

Cross out to subtract. Write each answer.

7 − 3 = ☐   8 − 3 = ☐   9 − 2 = ☐

8 − 5 = ☐   7 − 5 = ☐   10 − 9 = ☐

7 − 4 = ☐   10 − 2 = ☐   8 − 1 = ☐

168   one hundred sixty-eight   Math 1 Reviews

Chapter 11 Review

Circle the correct measuring tool.

How hot is it outside?

How long is this rope?

Guess the length. Use an inch ruler to measure.

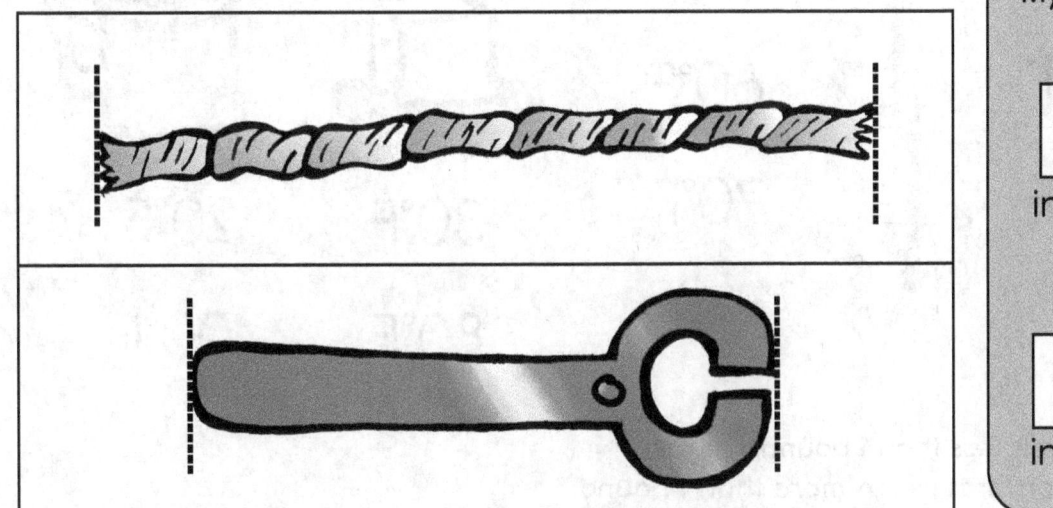

My Guess	My Measure
☐ inches	☐ inches
☐ inches	☐ inches

Use an inch ruler to draw each length. Start at the star.

3 inches ✶

5 inches ✶

6 inches ✶

Chapter 11 Review — one hundred sixty-nine 169

Chapter 9 Review

Color 2 pints for each quart.

Circle which set holds *more*.

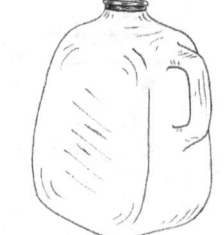

Circle which set holds *less*.

Circle the correct temperature.

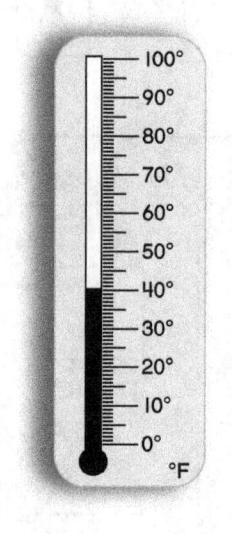

20°F
30°F
40°F

50°F
60°F
70°F

30°F
80°F

20°F
100°F

Circle each object that is *less* than 1 pound.
Mark an *X* on the objects that weigh *more* than 1 pound.

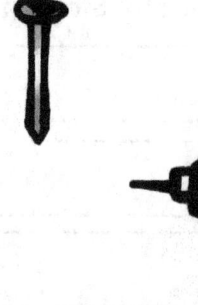

170   one hundred seventy

Math 1 Reviews

Cumulative Review

Write an equation for each word problem. Solve.

Grandma fixed 5 eggs in a pan. She fixed 4 eggs in another pan. How many eggs did she fix in all?

☐ ○ ☐ ○ ☐ eggs

Luna has 5 sticks of gum. While she was playing, she lost 3 sticks of gum. How many sticks of gum does Luna have left?

☐ ○ ☐ ○ ☐ sticks of gum

Color the shape to match the fraction.

$\frac{2}{3}$ 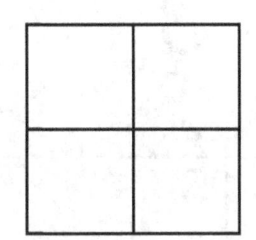 $\frac{3}{4}$  $\frac{1}{2}$

Mark the fraction that names the shaded part.

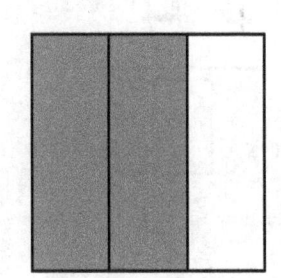

○ $\frac{1}{3}$
○ $\frac{2}{3}$

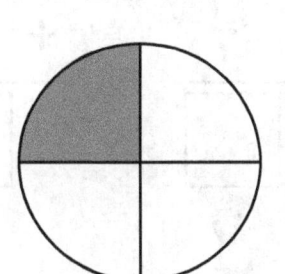

○ $\frac{1}{4}$
○ $\frac{2}{4}$

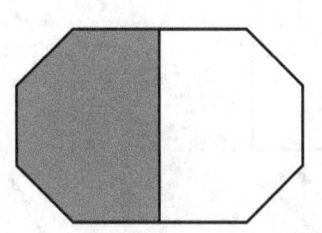

○ $\frac{1}{2}$
○ $\frac{1}{3}$

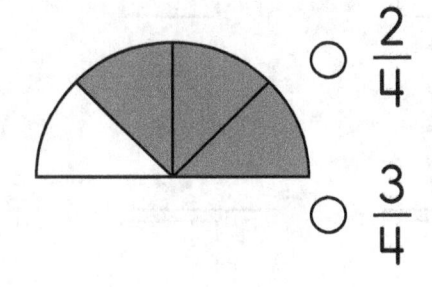

○ $\frac{2}{4}$
○ $\frac{3}{4}$

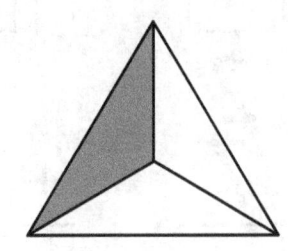

○ $\frac{1}{3}$
○ $\frac{2}{3}$

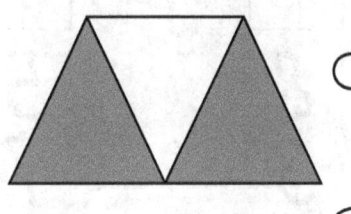

○ $\frac{2}{3}$
○ $\frac{2}{4}$

Chapter 11 Cumulative Review

Addition Fact Review

Add.

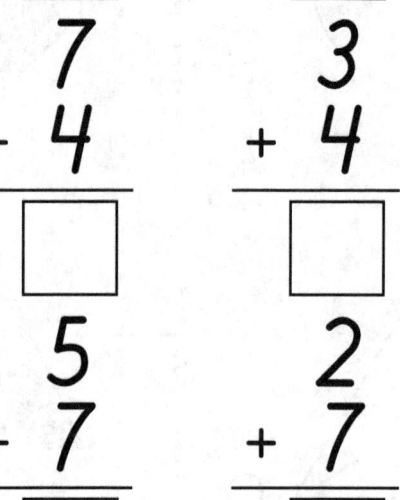

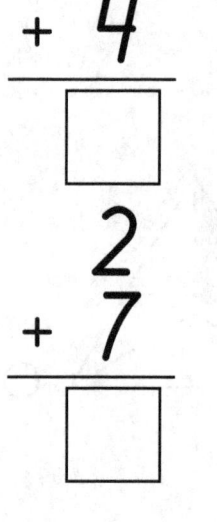

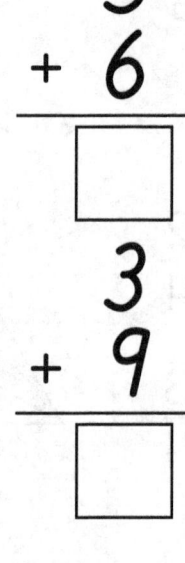

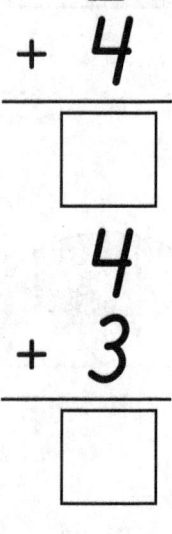

8 + 0 = ☐	5 + 2 = ☐	3 + 8 = ☐	7 + 1 = ☐	6 + 6 = ☐
3 + 3 = ☐	4 + 6 = ☐	9 + 0 = ☐	6 + 3 = ☐	8 + 2 = ☐
4 + 2 = ☐	5 + 5 = ☐	3 + 2 = ☐	7 + 5 = ☐	6 + 4 = ☐
9 + 1 = ☐	6 + 5 = ☐	4 + 4 = ☐	9 + 3 = ☐	8 + 4 = ☐
2 + 6 = ☐	7 + 4 = ☐	3 + 4 = ☐	5 + 6 = ☐	2 + 4 = ☐
8 + 3 = ☐	5 + 7 = ☐	2 + 7 = ☐	3 + 9 = ☐	4 + 3 = ☐

Add the Ones First

Tens	Ones
3	4
+ 2	3
5	7

Add the Ones first.

Add the Tens next.

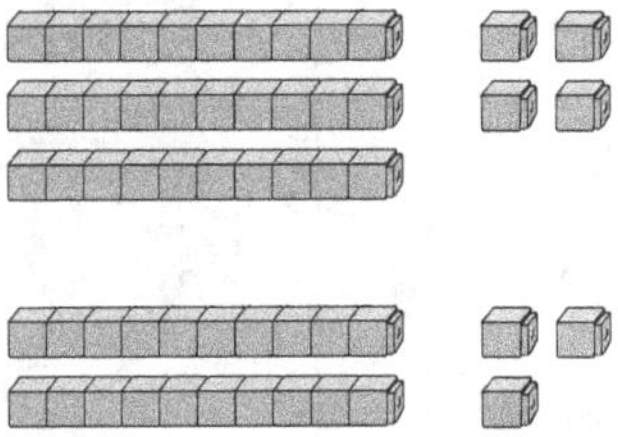

Add.

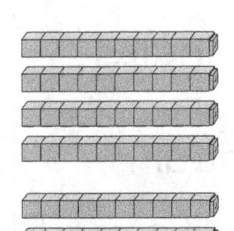

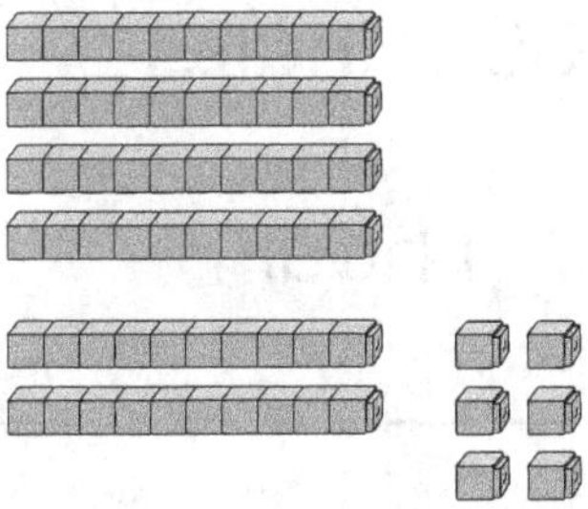

Tens	Ones
4	0
+ 2	6

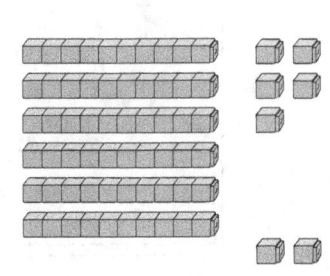

Tens	Ones
6	5
+	2

Add.

Tens	Ones
3	6
+ 4	2

Tens	Ones
5	3
+ 3	5

Tens	Ones
4	6
+ 4	3

Tens	Ones
4	4
+ 2	2

Write an equation for the word problem.
Solve using the Tens/Ones frame.

Ava picked up 23 seashells. Soren picked up 16 seashells. How many seashells did they pick up in all?

Tens	Ones
2	3
+	

23 + ☐ = ☐ seashells

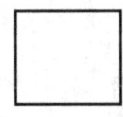

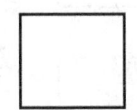

Chapter 12 • Lesson 92

Chapter 8 Review

Write the times under each clock.
Circle the time passed.

Activity	Start	Finish	Time Passed
	3:00 ⬜ : ⬜	4:00 ⬜ : ⬜	30 minutes or 1 hour
	1:00 ⬜ : ⬜	1:30 ⬜ : ⬜	30 minutes or 1 hour
	2:00 ⬜ : ⬜	3:00 ⬜ : ⬜	30 minutes or 1 hour

Addition Fact Review

Add.

$3 + 2 =$ ☐ $6 + 1 =$ ☐ $5 + 2 =$ ☐

$1 + 8 =$ ☐ $5 + 4 =$ ☐ $6 + 6 =$ ☐

$9 + 3 =$ ☐ $6 + 5 =$ ☐ $8 + 2 =$ ☐

$4 + 3 =$ ☐ $7 + 2 =$ ☐ $6 + 3 =$ ☐

$8 + 0 =$ ☐ $4 + 4 =$ ☐ $9 + 2 =$ ☐

Adding Two-Digit Numbers

Add.

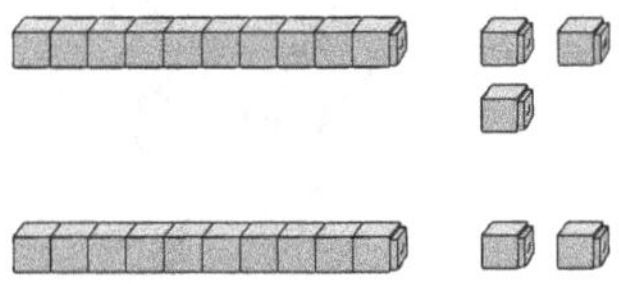

Tens	Ones
1	3
+1	2

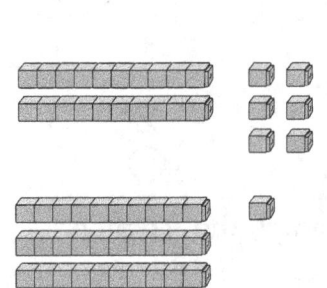

Tens	Ones
2	6
+3	1

Add.

Tens	Ones
6	4
+2	3

Tens	Ones
1	3
+5	2

Tens	Ones
2	4
+2	5

Tens	Ones
4	3
+2	0

```
  91        31        62        53
+  8      +42      +11      + 5
____      ____      ____      ____
 ☐         ☐         ☐         ☐
```

Write an equation for the word problem.
Solve using the Tens/Ones frame.

Iris sent 32 letters.
Mia sent 12 letters.
How many letters did they send in all?

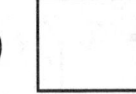

☐ ◯ ☐ ◯ ☐ letters

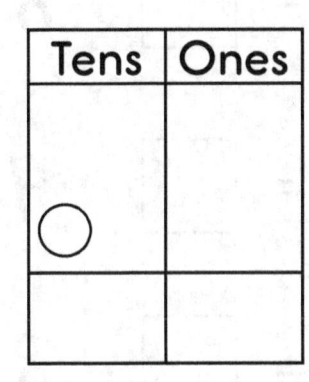

Tens	Ones
◯	

Chapter 12 • Lesson 93

Chapter 9 Review

Mark the fraction that names the shaded part.

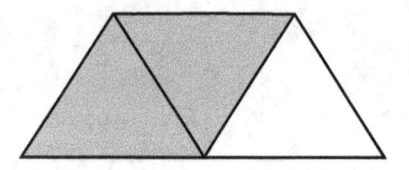

$\frac{1}{2}$ 　 $\frac{2}{3}$ 　 $\frac{2}{4}$ 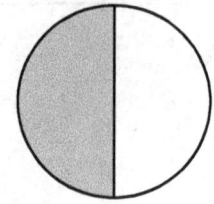 　　　 $\frac{1}{2}$ 　 $\frac{1}{3}$ 　 $\frac{1}{4}$ 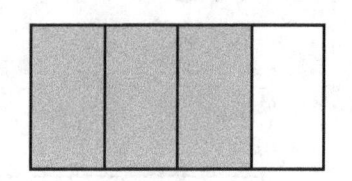 　　　 $\frac{2}{3}$ 　 $\frac{2}{4}$ 　 $\frac{3}{4}$ 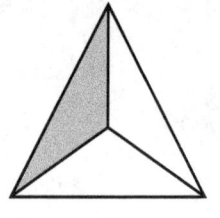 　　　 $\frac{1}{3}$ 　 $\frac{2}{3}$ 　 $\frac{1}{4}$

Color the shape to match the fraction.

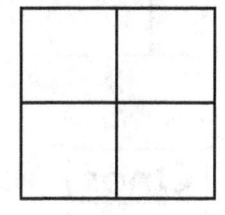

$\frac{1}{4}$ 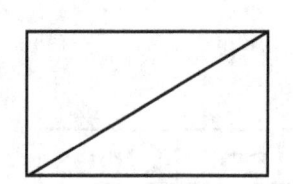 　　　 $\frac{1}{2}$ 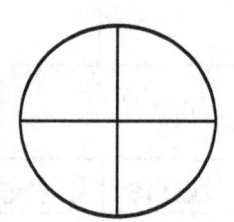 　　　 $\frac{3}{4}$ 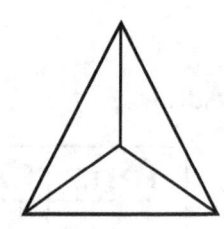 　　　 $\frac{2}{3}$

Subtraction Fact Review

Subtract.

10 − 1 = ☐   5 − 3 = ☐   6 − 2 = ☐

5 − 0 = ☐   7 − 5 = ☐   9 − 8 = ☐

6 − 6 = ☐   8 − 3 = ☐   7 − 6 = ☐

5 − 4 = ☐   9 − 0 = ☐   6 − 4 = ☐

4 − 2 = ☐   8 − 8 = ☐   5 − 2 = ☐

Adding Dimes & Pennies

Add.

Tens	Ones
2	0 ¢
+1	2 ¢
	¢

Tens	Ones
1	2 ¢
+1	2 ¢
	¢

Add.

Tens	Ones
1	3 ¢
+2	5 ¢
	¢

Tens	Ones
3	2 ¢
+4	1 ¢
	¢

Tens	Ones
5	2 ¢
+2	6 ¢
	¢

Tens	Ones
4	4 ¢
+5	1 ¢
	¢

Tens	Ones
3	2 ¢
+2	3 ¢
	¢

Tens	Ones
4	6 ¢
+2	1 ¢
	¢

Tens	Ones
8	1 ¢
+1	4 ¢
	¢

Tens	Ones
6	3 ¢
+3	6 ¢
	¢

Write an equation for the word problem. Solve.

The baker baked 23 muffins. Jill baked 16 muffins. How many muffins did they bake in all?

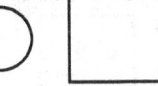

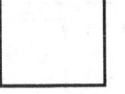

☐ ◯ ☐ ◯ ☐ muffins

Tens	Ones

Chapter 12 • Lesson 94   one hundred seventy-seven   **177**

Chapter 10 Review

Add or subtract.
Write the numbers for each fact family.

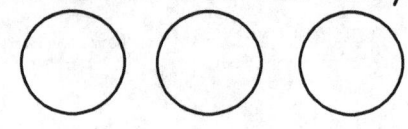

○ ○ ○   ○ ○ ○

3 + 5 = ☐   4 + 7 = ☐
5 + 3 = ☐   7 + 4 = ☐
8 − 3 = ☐   11 − 4 = ☐
8 − 5 = ☐   11 − 7 = ☐

Write each fact family.

④ ⑤ ⑨   ③ ⑦ ⑩

☐ + ☐ = ☐   ☐ + ☐ = ☐
☐ + ☐ = ☐   ☐ + ☐ = ☐
☐ − ☐ = ☐   ☐ − ☐ = ☐
☐ − ☐ = ☐   ☐ − ☐ = ☐

Subtraction Fact Review

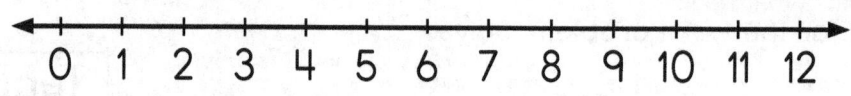

←―――――――――――――――→
0 1 2 3 4 5 6 7 8 9 10 11 12

Subtract. Use the number line if needed.

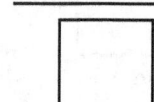

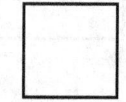

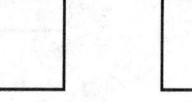

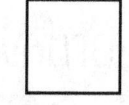

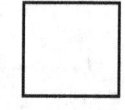

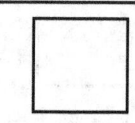

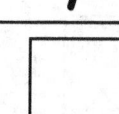

  9   9   8   9   8   9   9
− 3   − 5   − 6   − 4   − 4   − 6   − 7
 ___   ___   ___   ___   ___   ___   ___
 ☐   ☐   ☐   ☐   ☐   ☐   ☐

178   one hundred seventy-eight   Math 1 Reviews

Adding Money

Add.

Tens	Ones
2	3 ¢
+1	3 ¢
	¢

Tens	Ones
4	0 ¢
+2	2 ¢
	¢

Add.

Tens	Ones
6	5 ¢
+1	3 ¢
	¢

Tens	Ones
4	2 ¢
+2	4 ¢
	¢

Tens	Ones
2	6 ¢
+1	3 ¢
	¢

Tens	Ones
6	7 ¢
+2	1 ¢
	¢

 34 ¢
+52 ¢
☐ ¢

 14 ¢
+25 ¢
☐ ¢

 23 ¢
+34 ¢
☐ ¢

 75 ¢
+11 ¢
☐ ¢

Write an equation for the word problem. Solve.

Alice got 20 jellybeans.
John got 15 jellybeans.
How many jellybeans did they get in all?

☐ ○ ☐ ○ ☐ jellybeans

Tens	Ones

Chapter 12 • Lesson 95   one hundred seventy-nine   **179**

Chapter 11 Review

Use a ruler to draw each length. Start at the star.

3 inches *

4 inches *

5 inches *

Subtraction Fact Review

Subtract.

5	9	7	6	10
−4	−2	−4	−5	−2

7	6	3	8	9
−3	−4	−3	−2	−3

6	5	9	10	7
−2	−2	−8	−1	−1

180   one hundred eighty   Math 1 Reviews

Rename 10 Ones

Write the number of tens and ones.
Circle 10 ones and draw an arrow to show renaming.
Write the number.

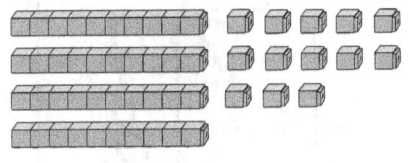

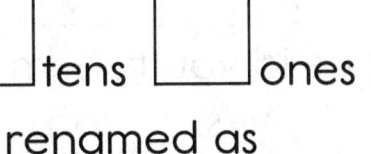

☐ tens ☐ ones
renamed as
☐ tens ☐ ones

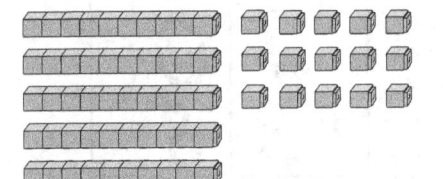

☐ tens ☐ ones
renamed as
☐ tens ☐ ones

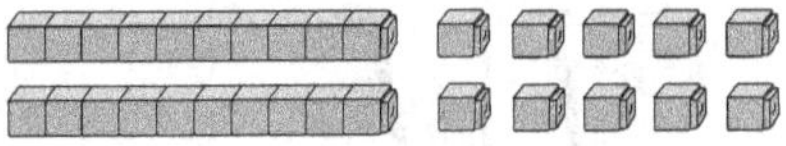

☐ tens ☐ ones
renamed as
☐ tens ☐ ones

Add. Rename if needed.
Circle *yes* if you renamed.
Circle *no* if you did not rename.

Think
1. Add the ones.
2. Do I rename?
3. Add the tens.

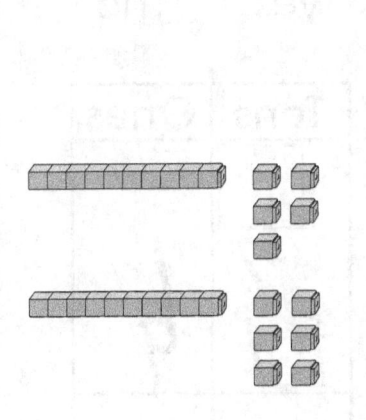

Tens	Ones
1	5
+1	6

Did you rename?
yes   no

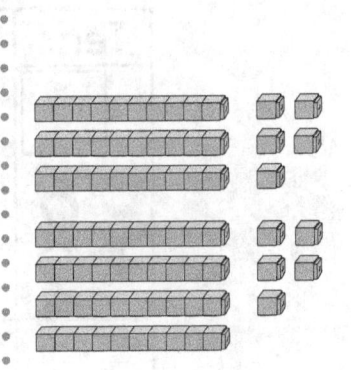

Tens	Ones
3	5
+4	5

Did you rename?
yes   no

Chapter 12 • Lessons 96–97   one hundred eighty-one   **181**

Chapter 9 Review

Add. Rename if needed.
Circle *yes* if you renamed.
Circle *no* if you did not rename.

Tens	Ones
1	1
+1	8

Did you rename?
yes   no

Tens	Ones
1	0
+	38

Did you rename?
yes   no

Tens	Ones
2	4
+4	8

Did you rename?
yes   no

Tens	Ones
1	6
+2	5

Did you rename?
yes   no

Tens	Ones
3	0
+	9

Did you rename?
yes   no

Tens	Ones
2	4
+2	6

Did you rename?
yes   no

Tens	Ones
3	3
+4	5

Did you rename?
yes   no

Tens	Ones
6	9
+2	3

Did you rename?
yes   no

Tens	Ones
7	1
+1	9

Did you rename?
yes   no

Rename 10 Pennies

Write the number of dimes and pennies.
Circle the number of pennies that can be traded for 1 dime.
Write the total value.

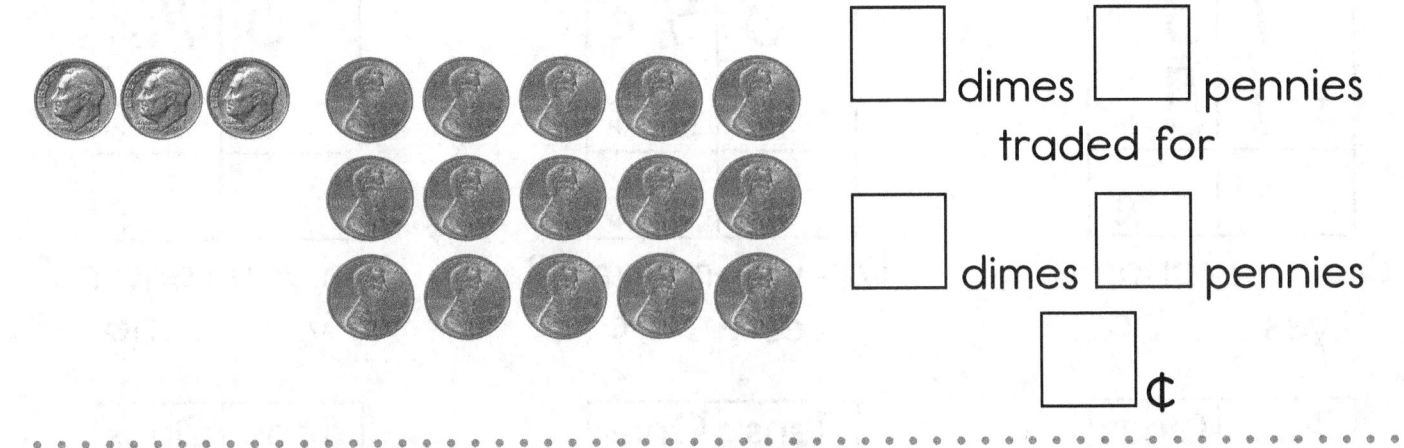

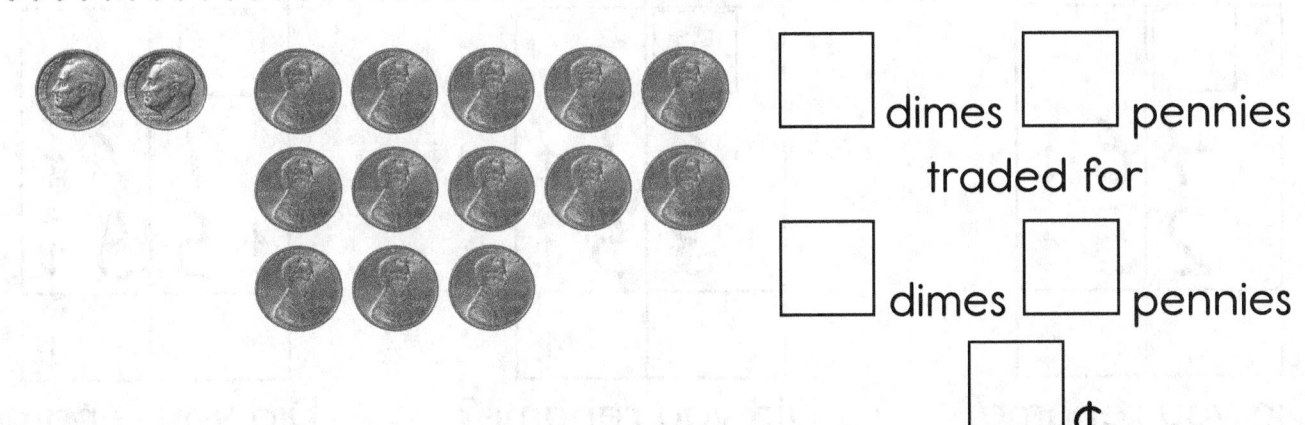

☐ dimes ☐ pennies
traded for
☐ dimes ☐ pennies
☐ ¢

☐ dimes ☐ pennies
traded for
☐ dimes ☐ pennies
☐ ¢

Write an equation for the word problem. Solve.

At the bake sale, a cupcake is 33¢ and a slice of pie is 38¢. How much will a cupcake and a slice of pie cost?

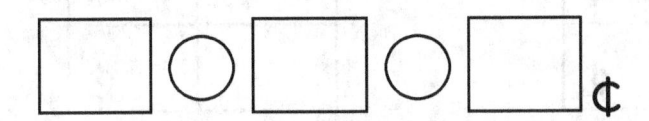

☐ ◯ ☐ ◯ ☐ ¢

Tens	Ones
☐	
	¢
◯	¢
	¢

Chapter 12 • Lesson 98

Add. Rename if needed.
Circle *yes* if you renamed.
Circle *no* if you did not rename.

Tens	Ones
	1 3 ¢
+ 1	5 ¢
	¢

Did you rename?
yes   no

Tens	Ones
	3 4 ¢
+ 1	8 ¢
	¢

Did you rename?
yes   no

Tens	Ones
	6 1 ¢
+ 1	9 ¢
	¢

Did you rename?
yes   no

Tens	Ones
	4 6 ¢
+ 2	5 ¢
	¢

Did you rename?
yes   no

Tens	Ones
	6 1 ¢
+ 3	5 ¢
	¢

Did you rename?
yes   no

Tens	Ones
	1 7 ¢
+ 5	5 ¢
	¢

Did you rename?
yes   no

Tens	Ones
	1 2 ¢
+ 1	9 ¢
	¢

Did you rename?
yes   no

Tens	Ones
	1 6 ¢
+ 2	5 ¢
	¢

Did you rename?
yes   no

Tens	Ones
	5 3 ¢
+ 1	1 ¢
	¢

Did you rename?
yes   no

Chapter 12 Review

_____

Add.

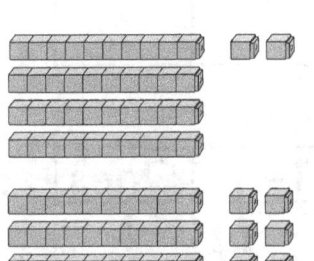

Tens	Ones
4	2
+3	6

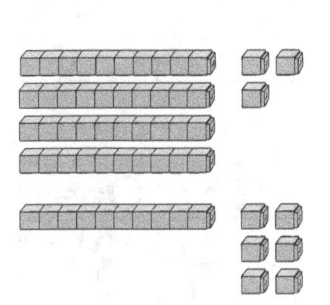

Tens	Ones
4	3
+1	6

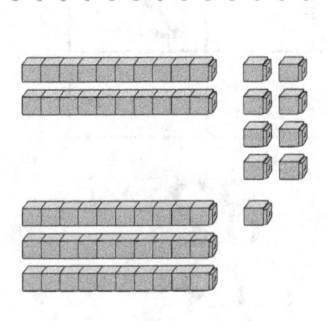

Tens	Ones
2	8
+3	1

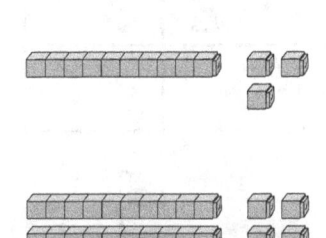

Tens	Ones
1	3
+2	5

Add.

Tens	Ones
2	0 ¢
+3	2 ¢
	¢

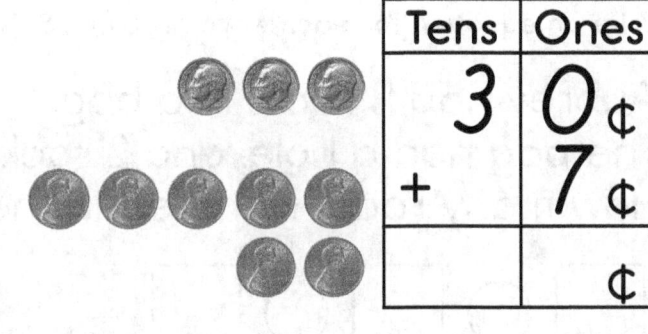

Tens	Ones
3	0 ¢
+	7 ¢
	¢

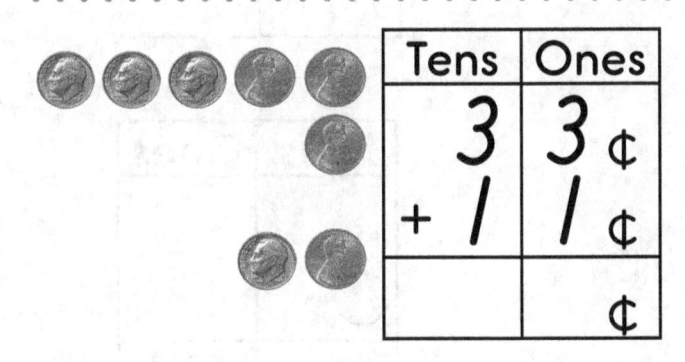

Tens	Ones
3	3 ¢
+1	1 ¢
	¢

Tens	Ones
6	1 ¢
+1	4 ¢
	¢

Chapter 12 Review   one hundred eighty-five   **185**

Add.

Tens	Ones
9	3¢
+	5¢
	¢

Tens	Ones
7	1¢
+1	4¢
	¢

Tens	Ones
6	3¢
+3	2¢
	¢

Tens	Ones
5	4¢
+3	3¢
	¢

Add.

Tens	Ones
2	3
+4	1

Tens	Ones
4	3
+2	3

Tens	Ones
5	2
+2	5

Tens	Ones
4	4
+	2

```
  33        15        25        82
 +54       + 4       +12       +13
 ___       ___       ___       ___
```

Write an equation for each word problem. Solve.

Aubrey had 9 rocks in a bag.
The bag had a hole, and 2 rocks fell out.
How many rocks were left in the bag?

☐ ○ ☐ ○ ☐ rocks

Tens	Ones
○	

Matt has 23 rocks.
Tim has 15 rocks.
How many rocks do they have in all?

☐ ○ ☐ ○ ☐ rocks

Tens	Ones
○	

Math 1 Reviews

Cumulative Review

Write the number. Circle the single cubes to make pairs.
Circle *even* if the number is even.
Circle *odd* if the number is odd.

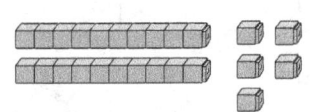

even   odd

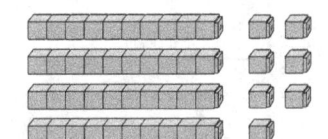

even   odd

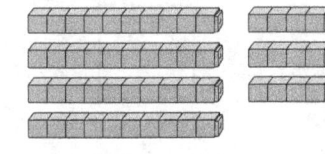

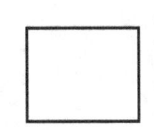

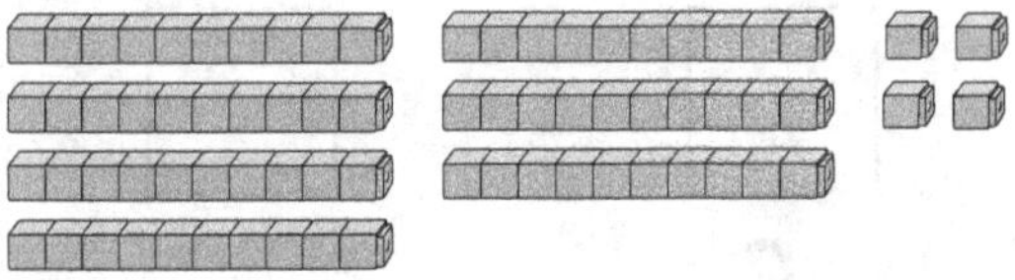

even   odd

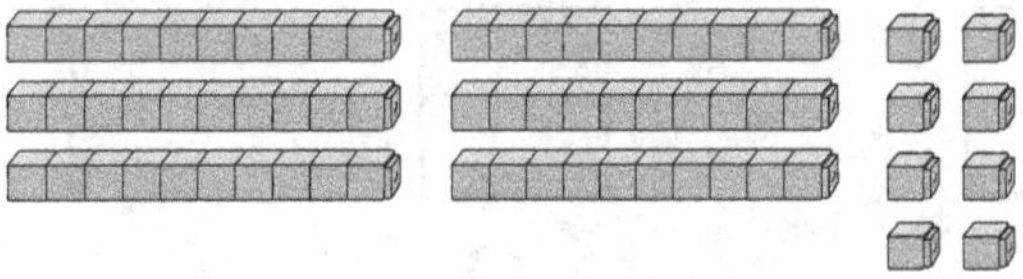

even   odd

Write the time.

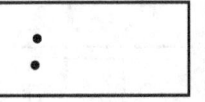

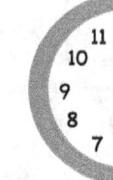

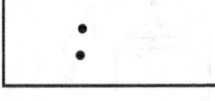

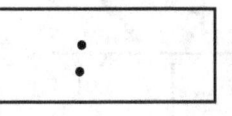

Write each fact family.

4 7 11

☐ + ☐ = ☐
☐ + ☐ = ☐
☐ − ☐ = ☐
☐ − ☐ = ☐

3 9 12

☐ + ☐ = ☐
☐ + ☐ = ☐
☐ − ☐ = ☐
☐ − ☐ = ☐

Chapter 12 Cumulative Review   one hundred eighty-seven   **187**

Addition Fact Review

Add.

2 + 4 ☐	1 + 1 ☐	9 + 1 ☐	1 + 5 ☐	3 + 8 ☐
3 + 3 ☐	2 + 5 ☐	4 + 1 ☐	6 + 3 ☐	2 + 9 ☐
6 + 0 ☐	2 + 3 ☐	7 + 2 ☐	5 + 4 ☐	6 + 6 ☐
3 + 1 ☐	2 + 2 ☐	6 + 1 ☐	7 + 3 ☐	6 + 5 ☐
1 + 2 ☐	4 + 2 ☐	6 + 2 ☐	8 + 0 ☐	5 + 5 ☐
4 + 4 ☐	3 + 2 ☐	2 + 8 ☐	4 + 3 ☐	7 + 5 ☐

Subtracting the Ones First

Cross out the cubes to subtract.
Write the answer.

First, subtract the Ones.
Next, subtract the Tens.

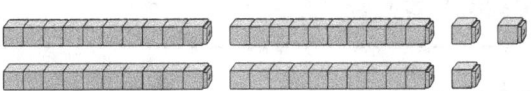

Tens	Ones
4	3
-2	1

Subtract.

Tens	Ones
5	9
-2	8

Tens	Ones
4	8
-2	2

Tens	Ones
8	3
-5	2

Tens	Ones
3	7
-1	7

Tens	Ones
6	5
-4	0

Tens	Ones
8	3
-	2

Tens	Ones
4	3
-3	1

Tens	Ones
7	5
-3	1

Write an equation for the word problem.
Solve using the Tens/Ones frame.

Jude had 35 bags of popcorn.
He gave 22 bags away.
How many bags of popcorn does Jude have left?

☐ ◯ ☐ ◯ ☐ bags

Tens	Ones
3	5
-2	2

Chapter 13 • Lesson 102

Chapter 4 Review

Mark the number that matches the description.

3 in the Tens place	34 ○	43 ○
5 in the Ones place	56 ○	65 ○
7 in the Tens place	72 ○	27 ○
9 in the Tens place	19 ○	90 ○
2 tens, 5 ones	25 ○	52 ○
3 tens, 6 ones	63 ○	36 ○
7 tens, 4 ones	74 ○	47 ○

Subtraction Fact Review

Subtract.

5 9 8 10 4
-3 -2 -7 -9 -4

6 4 9 6 9
-1 -3 -9 -3 -6

8 7 3 8 10
-0 -5 -2 -3 -2

Subtracting 2-Digit Numbers

Cross out the cubes to subtract. Write the answer.

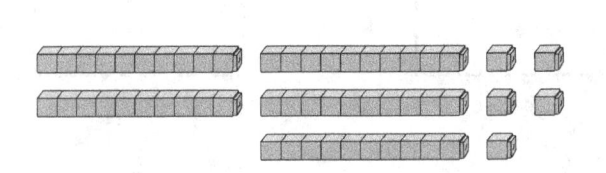

Tens	Ones
5	5
-2	2

Subtract.

```
  65        44        26        67
- 43      - 32      - 13      - 47
 ☐         ☐         ☐         ☐

  56        45        13        68
- 42      - 30      -  2      - 42
 ☐         ☐         ☐         ☐
```

Write an equation for the word problem. Solve.

There were 36 trucks in the parking lot.
The boss sent 22 trucks out on jobs.
How many trucks are left in the parking lot?

☐ ○ ☐ ○ ☐ trucks

Tens	Ones
○	

Chapter 13 • Lesson 103   one hundred ninety-one   191

Chapter 8 Review

Write each time.

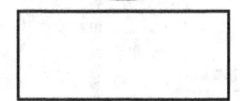

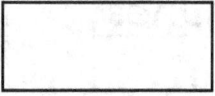

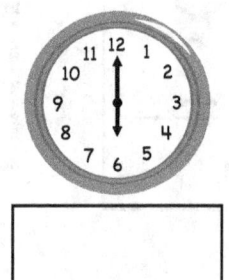

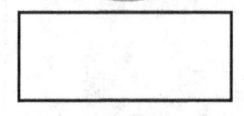

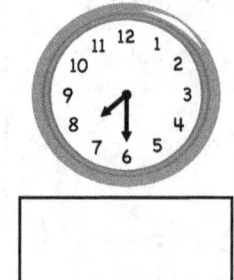

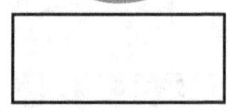

[ ] [ ] [ ] [ ]

[ ] [ ] [ ] [ ]

Complete each sentence.

1 hour = [ ] minutes   a half hour = [ ] minutes

Addition Fact Review

Add.

```
  6      1      6      5      9
+ 2    + 5    + 4    + 5    + 3
 [ ]    [ ]    [ ]    [ ]    [ ]

  8      5      7      5      8
+ 1    + 0    + 5    + 2    + 4
 [ ]    [ ]    [ ]    [ ]    [ ]

  2      3      6      3      7
+ 9    + 3    + 5    + 4    + 3
 [ ]    [ ]    [ ]    [ ]    [ ]
```

one hundred ninety-two — Math 1 Reviews

Subtracting Dimes & Pennies

Cross out the coins to subtract.
Write the answer.

Tens	Ones
5	4 ¢
- 3	2 ¢
	¢

Subtract.

Tens	Ones
1	7 ¢
-	6 ¢
	¢

Tens	Ones
9	3 ¢
- 3	1 ¢
	¢

Tens	Ones
8	2 ¢
- 4	0 ¢
	¢

Tens	Ones
7	4 ¢
- 5	3 ¢
	¢

Tens	Ones
6	6 ¢
- 4	5 ¢
	¢

Tens	Ones
9	8 ¢
- 7	6 ¢
	¢

Tens	Ones
4	6 ¢
- 2	2 ¢
	¢

Tens	Ones
2	7 ¢
- 1	4 ¢
	¢

Write an equation for the word problem. Solve.

Hugo and Miles are going camping.
Hugo got a jug of water for 92¢.
Miles got a bag of ice for 71¢.
How much more did Hugo spend than Miles?

☐ ○ ☐ ○ ☐ ¢

Chapter 13 • Lesson 104

Chapter 11 Review

Circle cups for each pint. Fill in each box.

☐ pints = ☐ cups

Circle pints for each quart. Fill in each box.

☐ quarts = ☐ pints

Circle quarts for each gallon. Fill in each box.

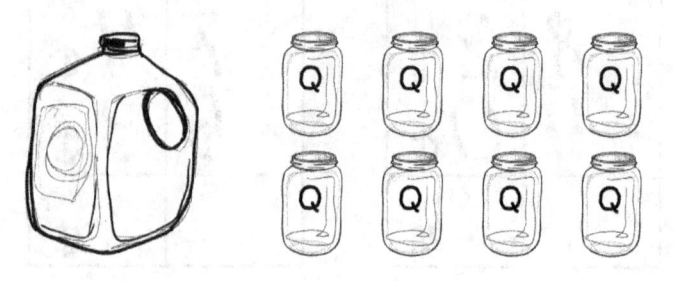

☐ gallon = ☐ quarts

Subtraction Fact Review

Subtract.

8 − 6 = ☐   9 − 6 = ☐   7 − 7 = ☐

7 − 1 = ☐   7 − 2 = ☐   9 − 4 = ☐

5 − 4 = ☐   9 − 3 = ☐   6 − 3 = ☐

9 − 7 = ☐   8 − 2 = ☐   9 − 0 = ☐

8 − 4 = ☐   9 − 5 = ☐   11 − 2 = ☐

Subtracting Money

Cross out the coins to subtract. Write the answer.

Tens	Ones
4	2 ¢
−1	1 ¢
	¢

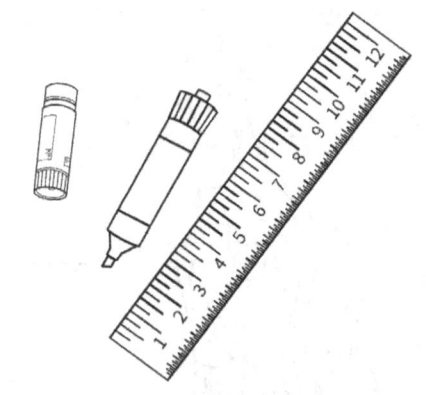

Subtract.

87 ¢
− 32 ¢
☐ ¢

96 ¢
− 54 ¢
☐ ¢

18 ¢
− 6 ¢
☐ ¢

88 ¢
− 77 ¢
☐ ¢

46 ¢
− 23 ¢
☐ ¢

59 ¢
− 37 ¢
☐ ¢

63 ¢
− 12 ¢
☐ ¢

80 ¢
− 40 ¢
☐ ¢

Write an equation for each word problem. Solve.

Lucy got a glue stick for 42¢ and a set of markers for 40¢. How much did Lucy spend in all?

☐ ◯ ☐ ◯ ☐ ¢

Finn had 75¢. He spent 43¢ on a ruler. How much money does Finn have left?

☐ ◯ ☐ ◯ ☐ ¢

Chapter 13 • Lesson 105

Chapter 11 Review

Use an inch ruler to measure.

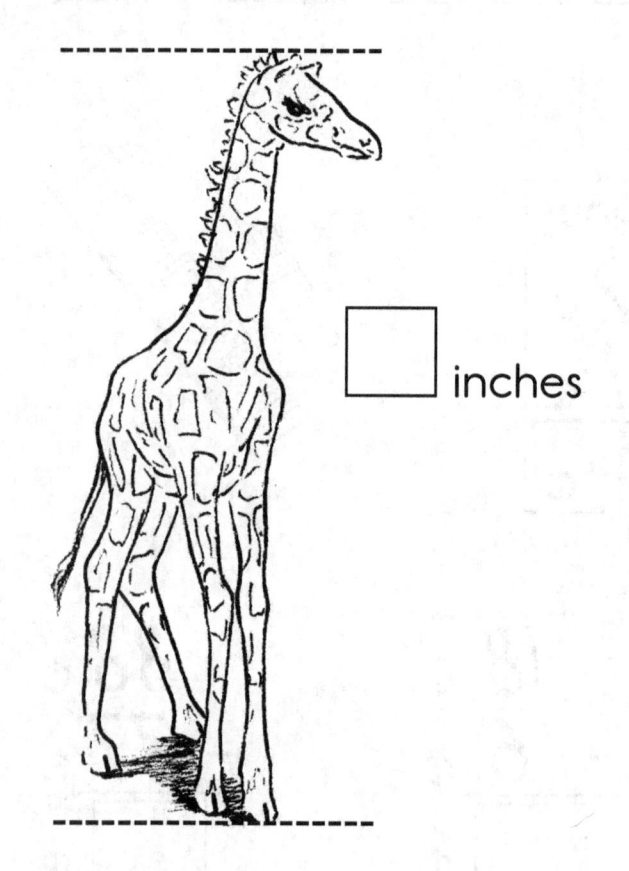

☐ inches

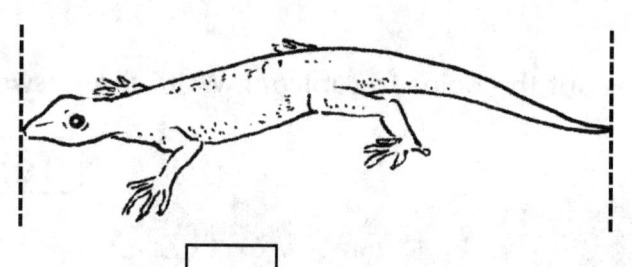

☐ inches

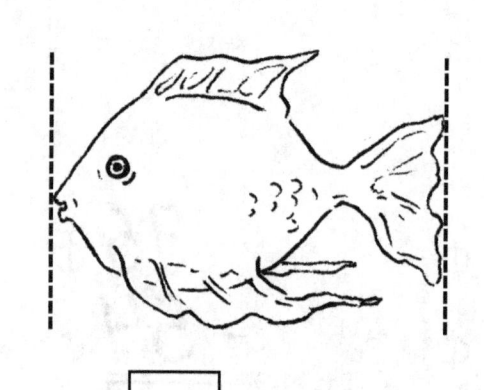

☐ inches

Addition Fact Review

Add.

9 + 1 = ☐   4 + 2 = ☐   9 + 2 = ☐
7 + 0 = ☐   3 + 3 = ☐   3 + 2 = ☐
6 + 3 = ☐   8 + 4 = ☐   3 + 9 = ☐
5 + 4 = ☐   5 + 5 = ☐   7 + 2 = ☐
5 + 6 = ☐   5 + 2 = ☐   7 + 5 = ☐

Chapter 13 Review

_____

Cross out the cubes to subtract. Write the answer.

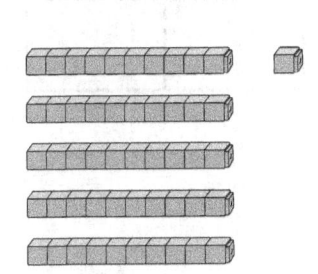

Tens	Ones
5	1
-4	0

Subtract.

Tens	Ones
4	7 ¢
-1	1 ¢
	¢

Tens	Ones
2	5 ¢
-1	3 ¢
	¢

Subtract.

```
  49        76        64        93
- 36      - 33      - 52      - 41
 ___       ___       ___       ___
```

Write an equation for each word problem. Solve.

The train had 43 cars. Then 22 cars were taken off. How many cars does the train have now?

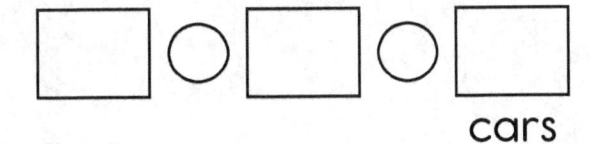

□ ○ □ ○ □ cars

There are 60 little boxes and 35 big boxes. How many boxes are there in all?

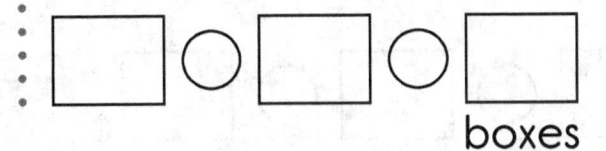

□ ○ □ ○ □ boxes

Chapter 13 Review

Cross out the coins to subtract.
Write the answer.

Tens	Ones
3	5 ¢
-1	4 ¢
	¢

Subtract.

Tens	Ones
5	6 ¢
-4	2 ¢
	¢

Tens	Ones
6	4 ¢
-5	3 ¢
	¢

Subtract.

72 ¢
- 50 ¢
☐ ¢

83 ¢
- 21 ¢
☐ ¢

96 ¢
- 43 ¢
☐ ¢

26 ¢
- 4 ¢
☐ ¢

46 ¢
- 22 ¢
☐ ¢

42 ¢
- 31 ¢
☐ ¢

94 ¢
- 71 ¢
☐ ¢

83 ¢
- 10 ¢
☐ ¢

Write an equation for the word problem. Solve.

Ella got a ride on the train for 95¢. Ruby got a ride for 70¢. How much more did Ella spend than Ruby?

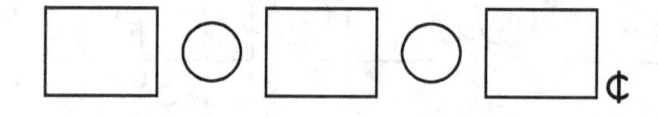

☐ ◯ ☐ ◯ ☐ ¢

work space

198   one hundred ninety-eight   Math 1 Reviews

Cumulative Review

Add.

$$\begin{array}{r}4\\1\\+\ 3\\\hline\square\end{array}\qquad\begin{array}{r}8\\2\\+\ 2\\\hline\square\end{array}\qquad\begin{array}{r}5\\1\\+\ 6\\\hline\square\end{array}\qquad\begin{array}{r}2\\2\\+\ 2\\\hline\square\end{array}\qquad\begin{array}{r}5\\4\\+\ 3\\\hline\square\end{array}$$

Write each missing number.

60 ☐ 62   43 44 ☐   ☐ 87 88

78 ☐ 80   ☐ 99 100   64 65 ☐

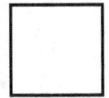

Circle the third child in line.
Draw a line under the ninth child in line.
Draw a box around the fifth child in line.

Chapter 13 Cumulative Review — one hundred ninety-nine

Subtraction Fact Review

Subtract.

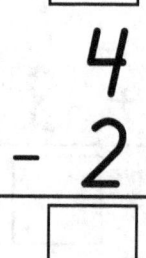

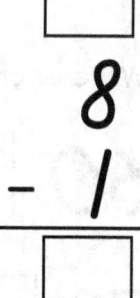

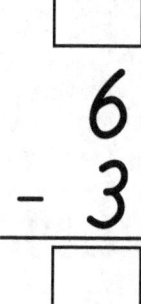

4 − 0 = ☐	7 − 5 = ☐	5 − 2 = ☐	7 − 1 = ☐	3 − 1 = ☐
7 − 3 = ☐	6 − 0 = ☐	5 − 4 = ☐	6 − 5 = ☐	4 − 2 = ☐
5 − 1 = ☐	4 − 4 = ☐	9 − 0 = ☐	2 − 2 = ☐	8 − 1 = ☐
3 − 2 = ☐	5 − 3 = ☐	8 − 5 = ☐	4 − 1 = ☐	8 − 8 = ☐
6 − 4 = ☐	4 − 3 = ☐	2 − 0 = ☐	5 − 5 = ☐	6 − 3 = ☐
2 − 1 = ☐	6 − 6 = ☐	9 − 2 = ☐	10 − 9 = ☐	11 − 2 = ☐

Counting Dimes, Nickels & Pennies

Color green the apples with the value of a penny.
Color red the apples with the value of a nickel.
Color yellow the apples with the value of a dime.

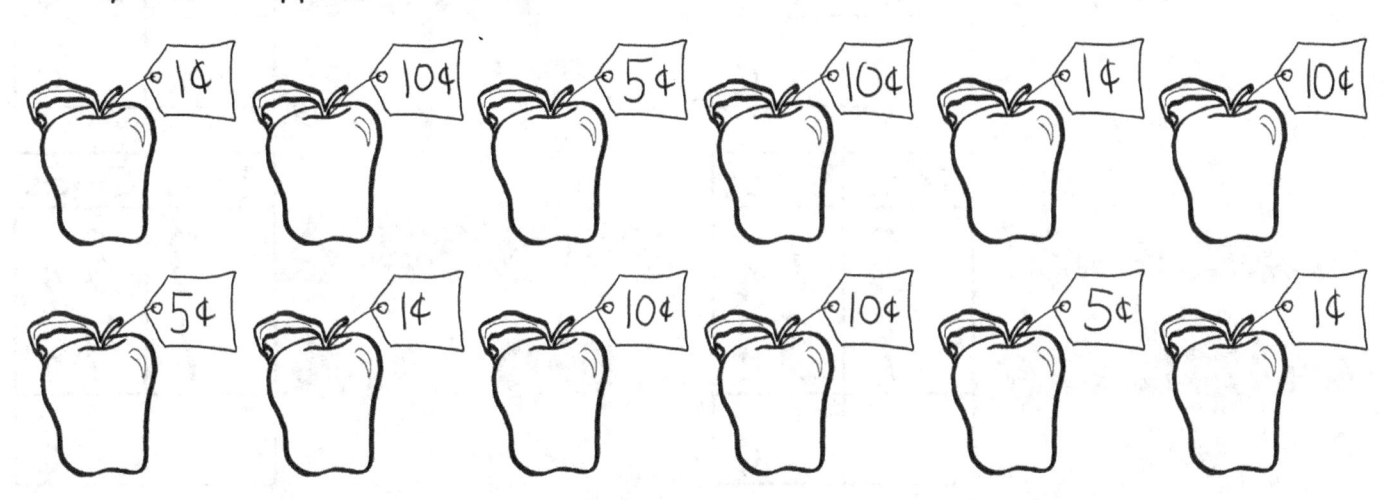

1¢   10¢   5¢   10¢   10¢   1¢   10¢

5¢   1¢   10¢   10¢   5¢   1¢

Write the value of Jason's coins as you *count on*. Does he have enough money to buy each fruit? Circle *yes* or *no*.

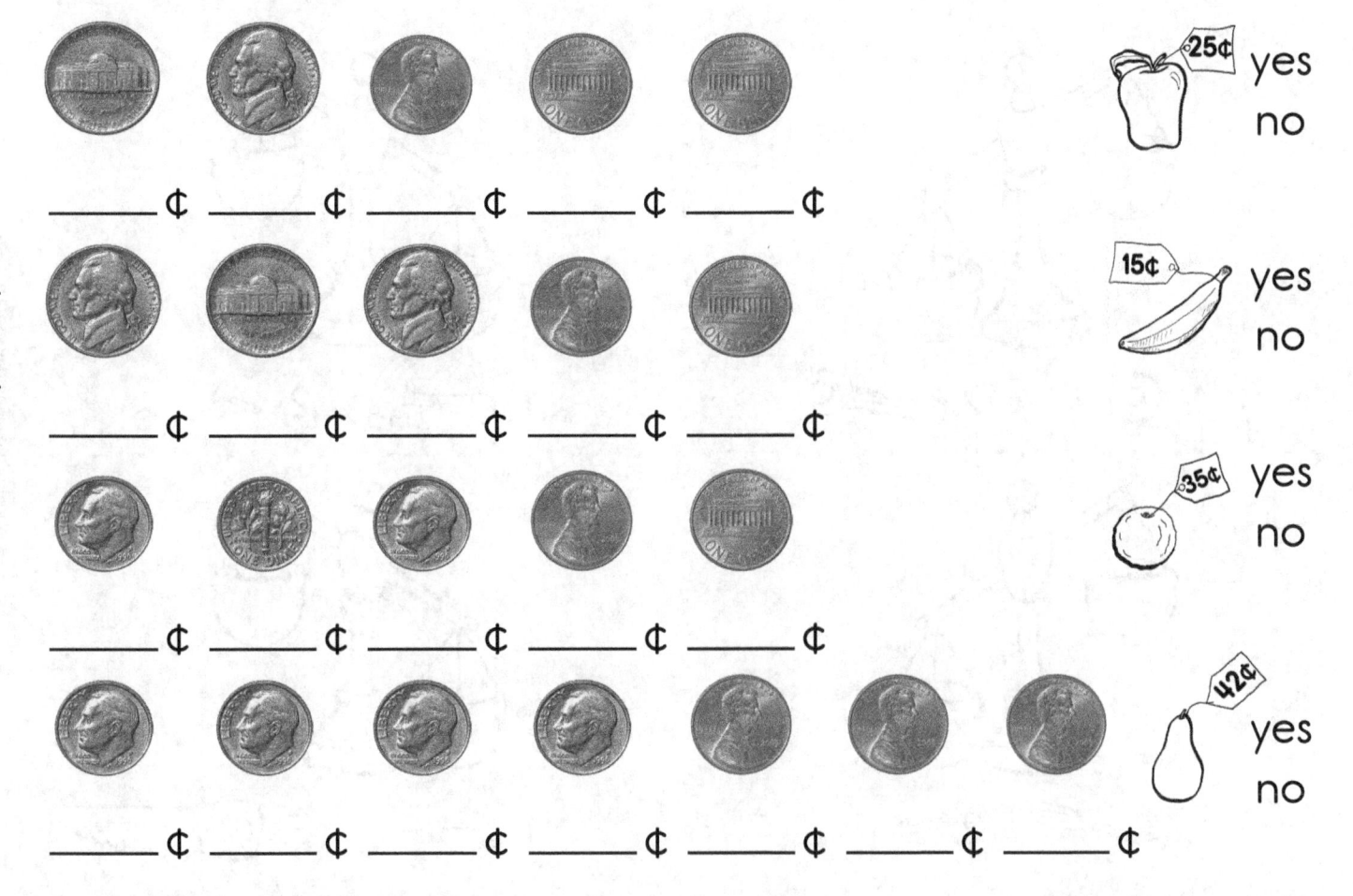

25¢ yes no

____¢ ____¢ ____¢ ____¢ ____¢

15¢ yes no

____¢ ____¢ ____¢ ____¢ ____¢

35¢ yes no

____¢ ____¢ ____¢ ____¢ ____¢

42¢ yes no

____¢ ____¢ ____¢ ____¢ ____¢ ____¢ ____¢

Chapter 14 • Lesson 108

Chapter 12 Review

Add.

Tens	Ones
3	3 ¢
+ 3	2 ¢
	¢

Tens	Ones
4	2 ¢
+ 2	2 ¢
	¢

Tens	Ones
5	1 ¢
+ 1	5 ¢
	¢

Tens	Ones
3	3 ¢
+ 1	4 ¢
	¢

Chapter 11 Review

Write the number by each object in the correct cup.

1.  ⬡   2.  🔧

1. 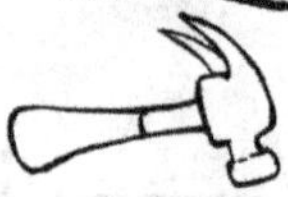 🔨   2.  📌

1.  🪛   2.  🔩

1.  ✂   2.  🪚

202 two hundred two   Math 1 Reviews

Solving Money Word Problems

Write the value as you *count on*. Write the total.

___¢ ___¢ ___¢ ___¢ ___¢ ___¢ ___¢   ___¢

___¢ ___¢ ___¢ ___¢ ___¢ ___¢ ___¢   ___¢

Write the total value. Do you have enough money to buy the item?
Circle *yes* or *no*.

___¢   yes / no

___¢   yes / no

Write an equation for the word problem. Solve.

James has 3 dimes in his bank.
He puts in 1 dime and 4 pennies.
How much money is in the bank?

☐ ¢ ◯ ☐ ¢ ◯ ☐ ¢

Chapter 14 • Lesson 109

Chapter 12 Review

Write an equation for each word problem. Solve.

Farmer Bob picked 42 baskets of corn.
His helper picked 25 baskets of corn.
How many baskets of corn did they pick in all?

☐ ◯ ☐ ◯ ☐ baskets

work space

Grandma had 67 pea plants.
In the storm, 43 pea plants died.
How many pea plants does Grandma have left?

☐ ◯ ☐ ◯ ☐ pea plants

work space

Mom canned 53 jars of string beans on Friday and 36 jars on Saturday.
How many jars of beans did Mom can in all?

☐ ◯ ☐ ◯ ☐ jars

work space

Subtraction Fact Review

Cross out to subtract. Write the answer.

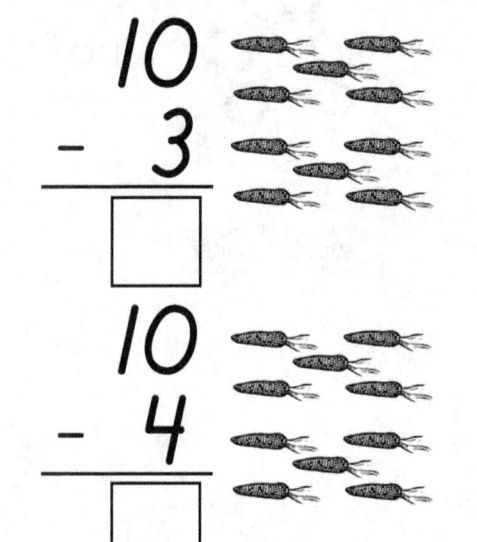

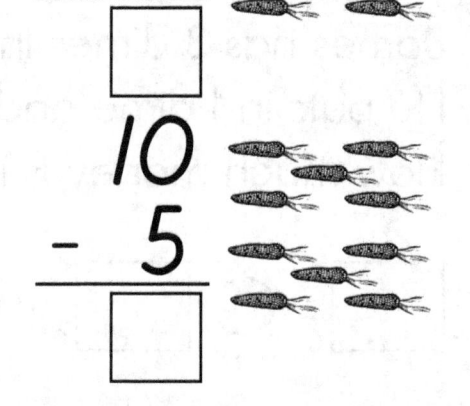

10 − 3 = ☐

10 − 7 = ☐

10 − 8 = ☐

10 − 4 = ☐

10 − 6 = ☐

10 − 5 = ☐

Money & Probability

Write the total value in each box.
Draw a line from each present to the child with the same amount of money.

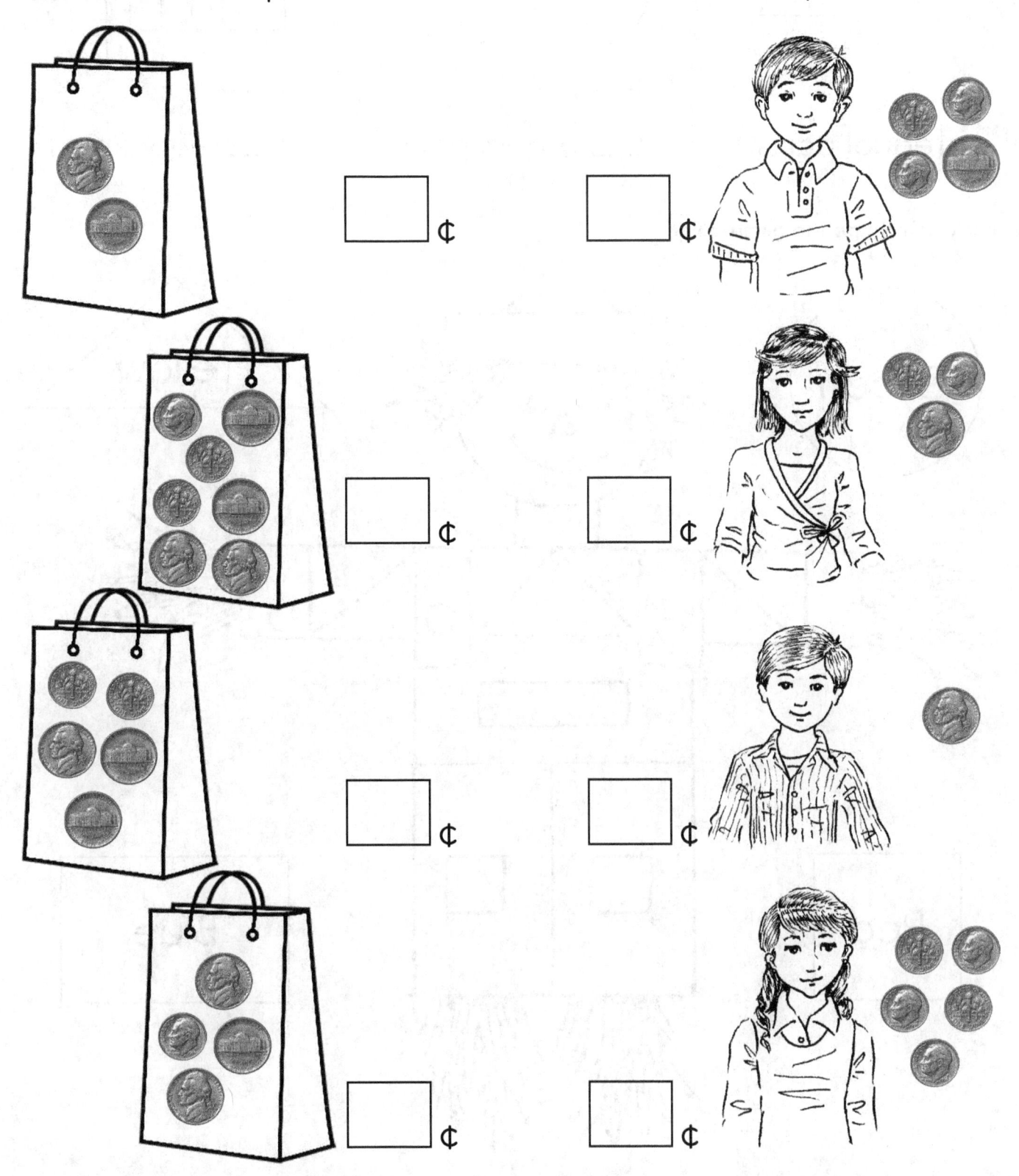

Chapter 14 • Lesson 110 two hundred five 205

Chapter 9 Review

Circle each shape that has equal parts.
Write how many equal parts.

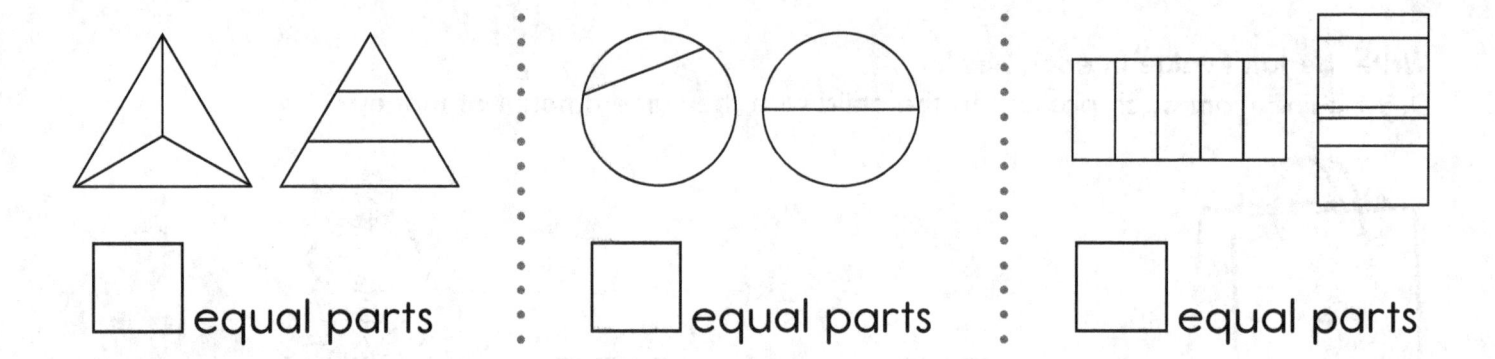

☐ equal parts   ☐ equal parts   ☐ equal parts

Color each shape the given color.

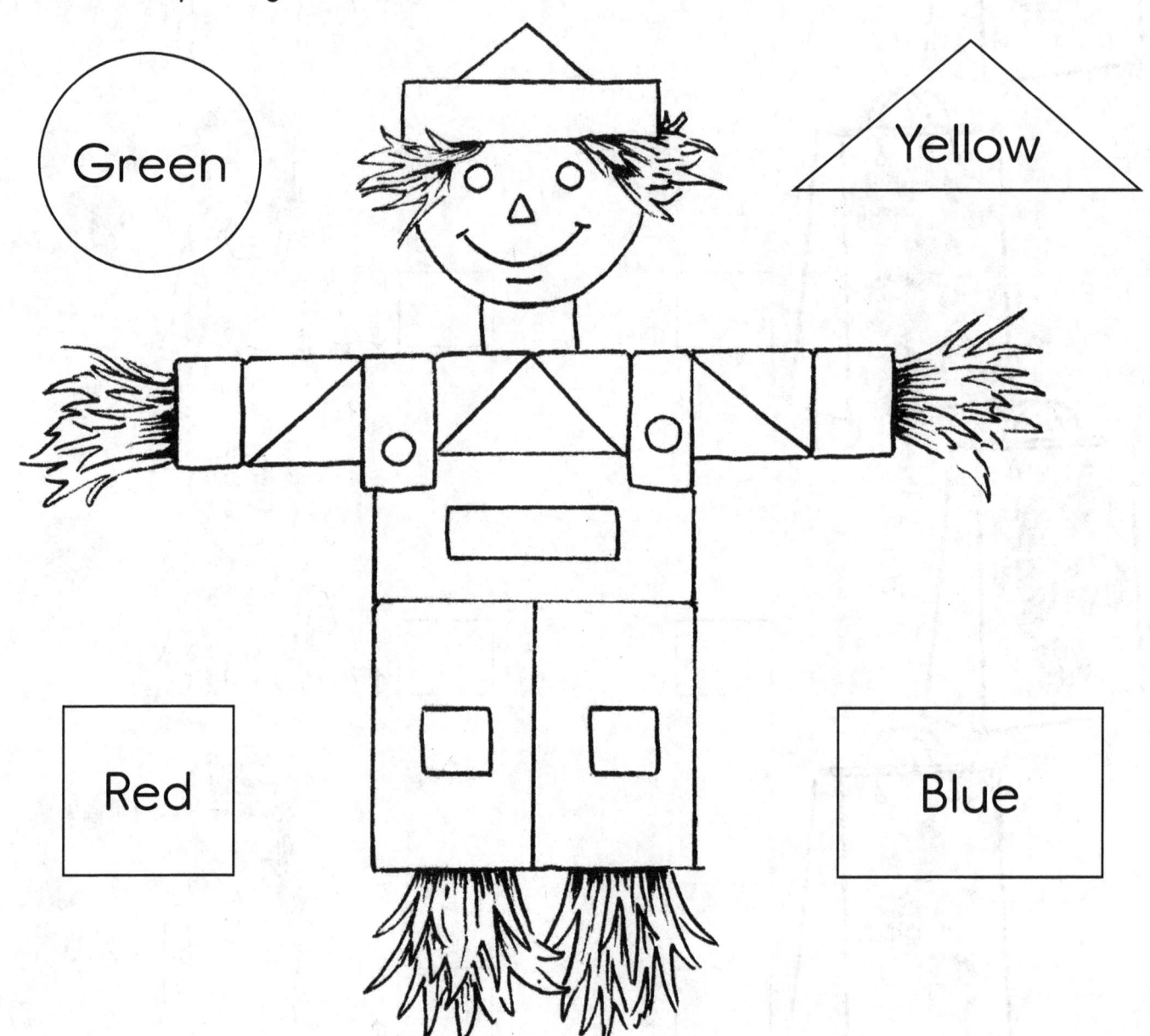

Green   Yellow

Red   Blue

206 two hundred six   Math 1 Reviews

Quarters

Help Mother find her way to the cashier.
Write the value of each set.
Draw lines to join all the coin sets that equal a quarter.

Chapter 14 • Lesson 111
two hundred seven **207**

Chapter 8 Review

If the activity takes more than an hour, color the clock yellow.

If the activity takes less than an hour, color the clock red.

Subtraction Fact Review

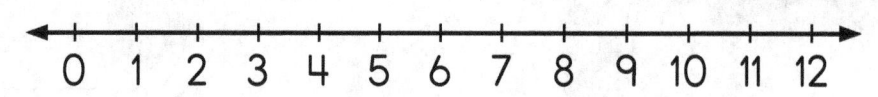

Subtract.
Use the number line if needed.

$\begin{array}{r}11\\-4\\\hline\square\end{array}$ $\begin{array}{r}11\\-7\\\hline\square\end{array}$ $\begin{array}{r}11\\-9\\\hline\square\end{array}$ $\begin{array}{r}11\\-8\\\hline\square\end{array}$ $\begin{array}{r}11\\-3\\\hline\square\end{array}$

two hundred eight   Math 1 Reviews

Chapter 14 Review

Write the value of each coin.

____¢ penny    ____¢ dime    ____¢ nickel    ____¢ quarter

Write the total value. Do you have enough money to buy the item? Circle *yes* or *no*.

☐ ¢   25¢   yes / no

☐ ¢    20¢   yes / no

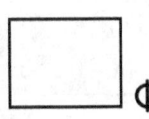

☐ ¢    45¢   yes / no

Mark an *X* on the coins needed to make the same value as a quarter.

Chapter 14 Review   two hundred nine   **209**

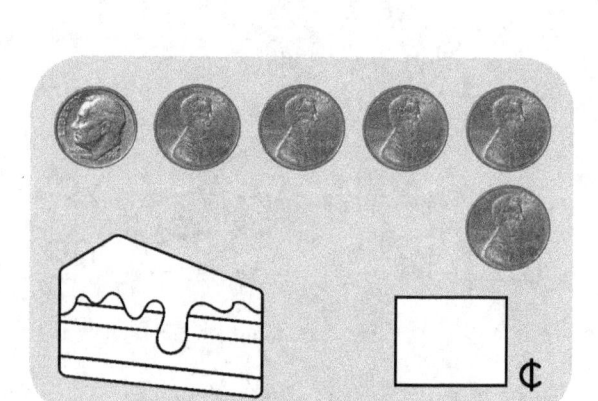

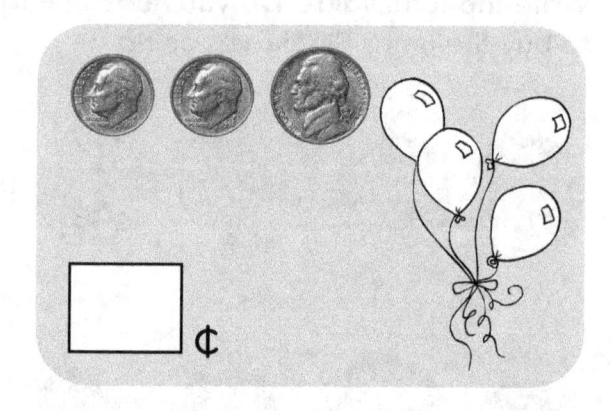

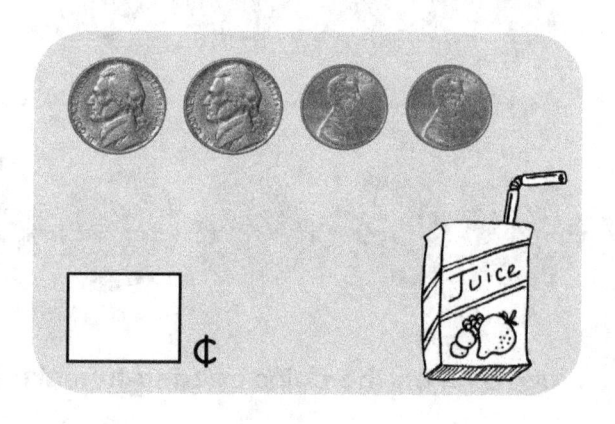

Write the total value in each box.
Draw a line to match the items with the same price.

Cumulative Review

Subtract.

Tens	Ones
5	8
-2	7

Tens	Ones
4	9
-3	2

Tens	Ones
8	3
-5	2

Tens	Ones
3	7
-1	4

$$67 - 40 = \square$$

$$95 - 3 = \square$$

$$26 - 13 = \square$$

$$75 - 33 = \square$$

Draw a line to match each picture to the correct thermometer.

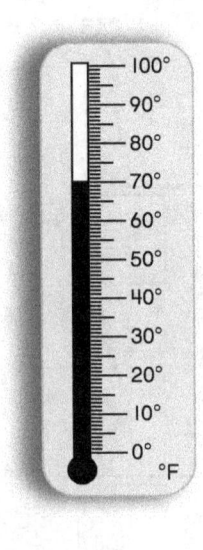

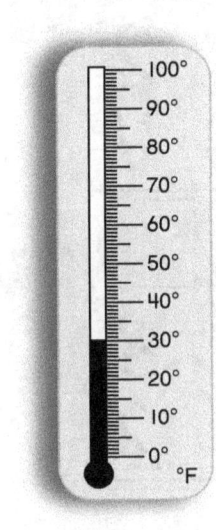

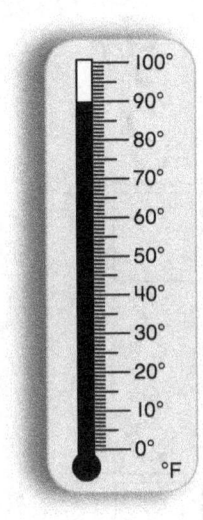

Chapter 14 Cumulative Review — two hundred eleven — 211

Addition Fact Review

Add.

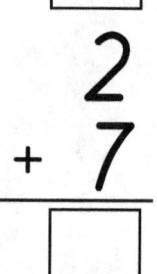

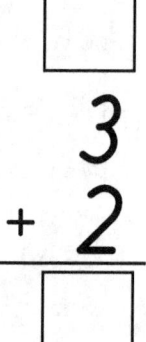

7 + 2 = ☐	5 + 4 = ☐	9 + 2 = ☐	8 + 1 = ☐	3 + 3 = ☐
8 + 0 = ☐	6 + 4 = ☐	5 + 5 = ☐	7 + 3 = ☐	8 + 4 = ☐
3 + 6 = ☐	6 + 6 = ☐	5 + 3 = ☐	5 + 6 = ☐	7 + 0 = ☐
7 + 5 = ☐	9 + 3 = ☐	2 + 4 = ☐	7 + 4 = ☐	3 + 4 = ☐
4 + 4 = ☐	6 + 5 = ☐	8 + 2 = ☐	1 + 9 = ☐	2 + 7 = ☐
0 + 9 = ☐	4 + 7 = ☐	6 + 3 = ☐	5 + 7 = ☐	3 + 2 = ☐

Solid Figures

- - - - - - - - - - -

Draw a line to match each solid figure with its name.

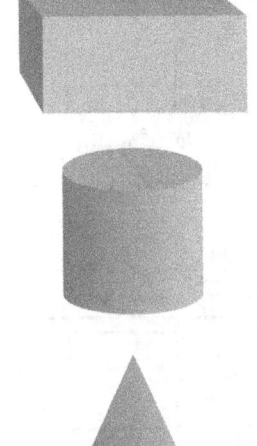

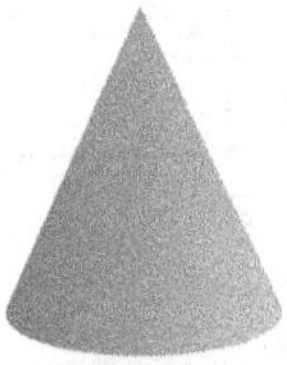

cylinder
(can shaped)

sphere
(ball shaped)

rectangular prism
(box shaped)

cone

Does each solid figure have flat sides, curved sides, or both?
Mark the correct answer.

○ flat sides
○ curved sides
○ both

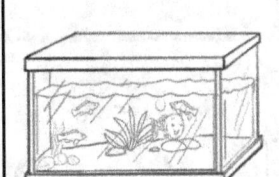

○ flat sides
○ curved sides
○ both

○ flat sides
○ curved sides
○ both

○ flat sides
○ curved sides
○ both

Chapter 15 • Lesson 114

Chapter 13 Review

Subtract.

```
  56        74        89        48
- 42      - 43      - 37      - 38
 ☐         ☐         ☐         ☐

  65        83        58        77
- 34      - 51      - 12      - 26
 ☐         ☐         ☐         ☐
```

Write an equation for the word problem. Solve.

The boy picked up 36 rocks on the riverbank. He lost 12 on his walk back to camp. How many rocks does the boy have left?

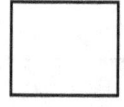

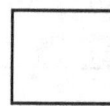

☐ ◯ ☐ ◯ ☐ rocks

work space

Addition Fact Review

Add.

2 + 8 = ☐   4 + 2 = ☐   5 + 6 = ☐

5 + 3 = ☐   9 + 2 = ☐   4 + 3 = ☐

7 + 4 = ☐   8 + 3 = ☐   4 + 6 = ☐

More Solid Figures

blue   red   yellow   green   purple   orange

Use the color key to color the objects.

Chapter 15 • Lesson 115 two hundred fifteen **215**

Chapter 12 Review

Add.

```
  43        59        83        35
 +52       +30       +14       +24
 ☐         ☐         ☐         ☐

  65        76        24        56
 +31       +22       +14       +12
 ☐         ☐         ☐         ☐
```

Write an equation for the word problem. Solve.

In the morning 12 hummingbirds came to the feeder. In the afternoon 10 hummingbirds came to the feeder. How many hummingbirds came in all?

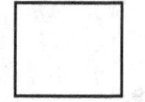

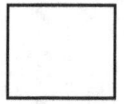

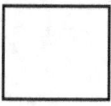

☐ ○ ☐ ○ ☐ hummingbirds

work space

Subtraction Fact Review

Subtract.

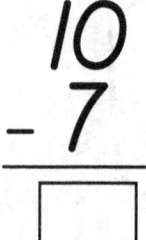

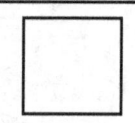

```
  10         8        11         7        11
 - 3       - 4       - 8       - 4       - 2
 ☐         ☐         ☐         ☐         ☐

  10         9        10         8        10
 - 7       - 5       - 6       - 2       - 4
 ☐         ☐         ☐         ☐         ☐
```

Faces, Curves & Corners

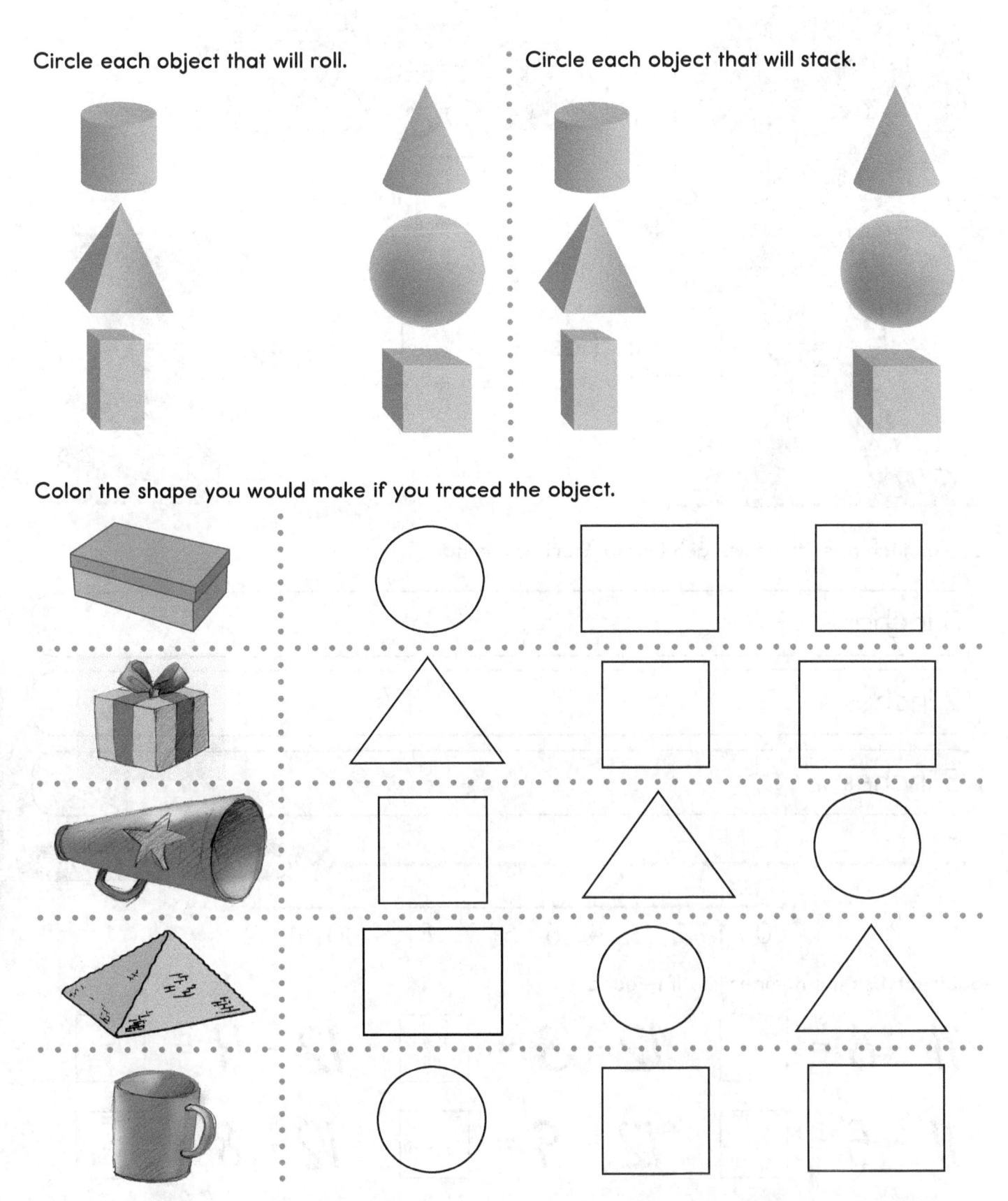

Circle each object that will roll.

Circle each object that will stack.

Color the shape you would make if you traced the object.

Chapter 15 • Lesson 116

two hundred seventeen **217**

Chapter 11 Review

Use an inch ruler to measure.

☐ inches

☐ inches

☐ inches

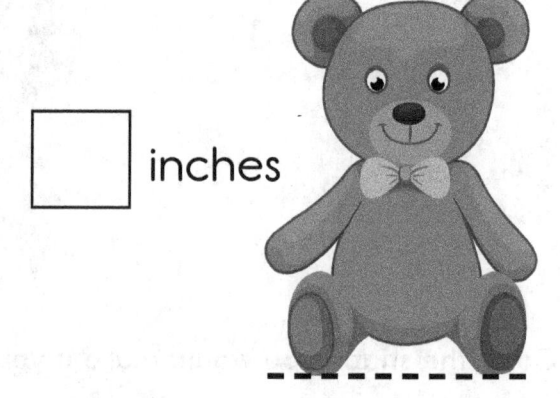

Use an inch ruler to draw each length. Start at the star.

3 inches ★

2 inches ★

5 inches ★

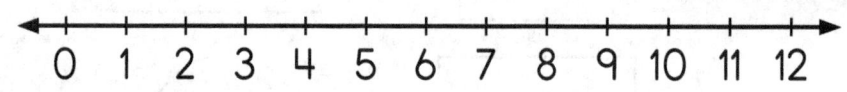

Subtraction Fact Review

0 1 2 3 4 5 6 7 8 9 10 11 12

Subtract. Use the number line if needed.

11 − 6 = ☐   12 − 3 = ☐   12 − 4 = ☐

11 − 5 = ☐   12 − 9 = ☐   12 − 8 = ☐

Plane Figures

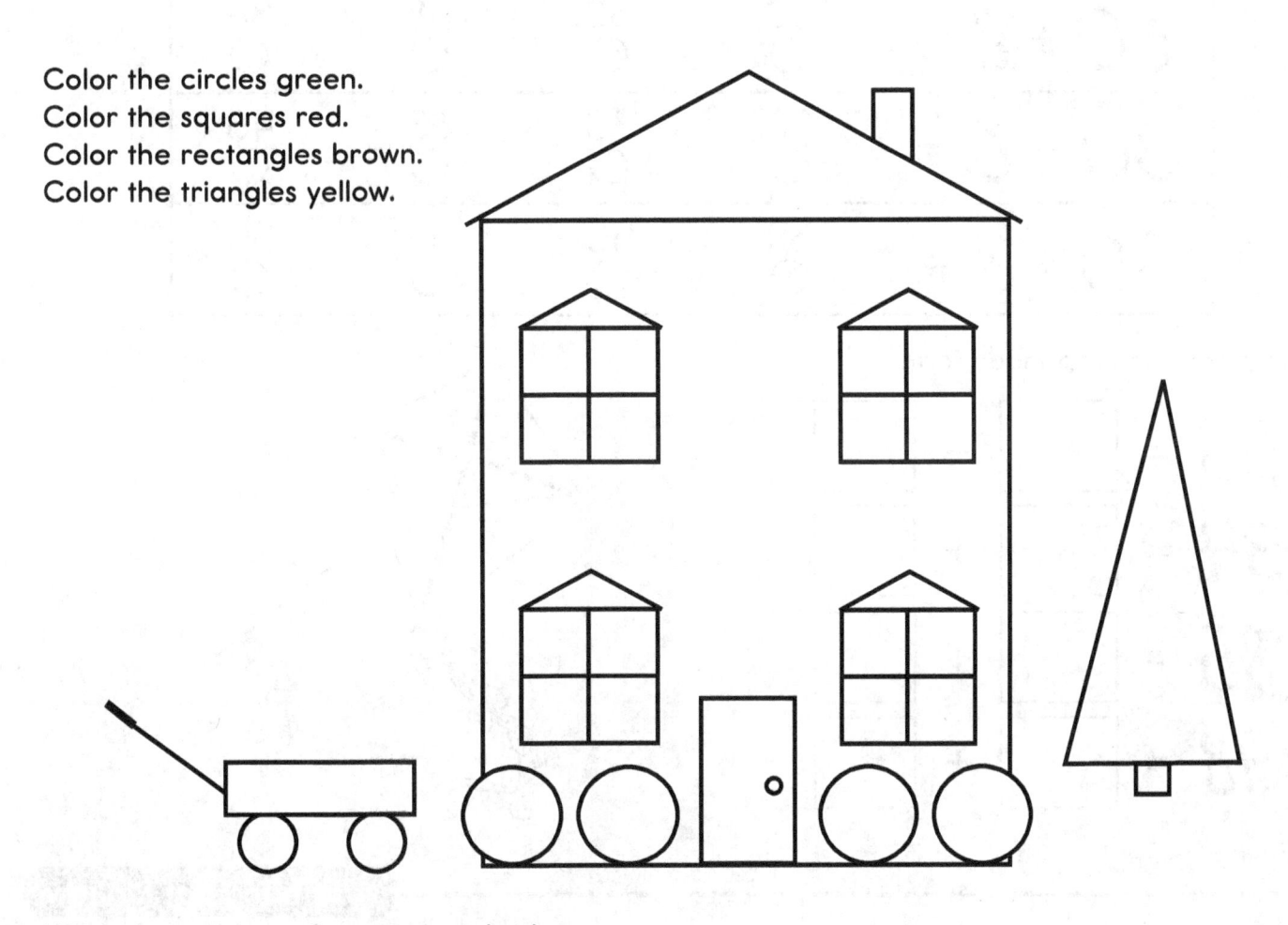

Color the circles green.
Color the squares red.
Color the rectangles brown.
Color the triangles yellow.

Write the number of corners and sides.

Shape	Corners	Sides
square		
rectangle		
triangle		

Chapter 15 • Lesson 118

Chapter 4 Review

Mark the correct answer.

60 + 8 =	63 ○	68 ○	58 ○	86 ○
30 + 6 =	36 ○	30 ○	63 ○	46 ○
90 + 3 =	39 ○	90 ○	93 ○	30 ○

Complete each expanded form.

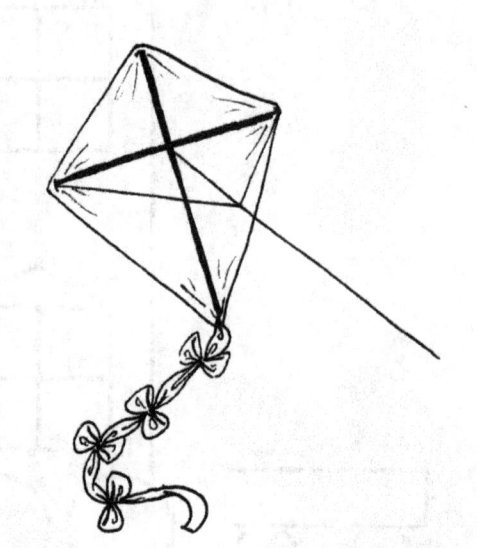

56 = ☐ + ☐

34 = ☐ + ☐

85 = ☐ + ☐

98 = ☐ + ☐

Addition Fact Review

Add.

$$\begin{array}{r}8\\+\ 2\\\hline\end{array}$$ ☐

$$\begin{array}{r}3\\+\ 7\\\hline\end{array}$$ ☐

$$\begin{array}{r}6\\+\ 6\\\hline\end{array}$$ ☐

$$\begin{array}{r}7\\+\ 2\\\hline\end{array}$$ ☐

$$\begin{array}{r}3\\+\ 6\\\hline\end{array}$$ ☐

$$\begin{array}{r}5\\+\ 3\\\hline\end{array}$$ ☐

$$\begin{array}{r}7\\+\ 4\\\hline\end{array}$$ ☐

$$\begin{array}{r}9\\+\ 3\\\hline\end{array}$$ ☐

$$\begin{array}{r}8\\+\ 0\\\hline\end{array}$$ ☐

$$\begin{array}{r}5\\+\ 6\\\hline\end{array}$$ ☐

Same Shape, Size, or Color

Color the shape that is the same shape and size as the first one.

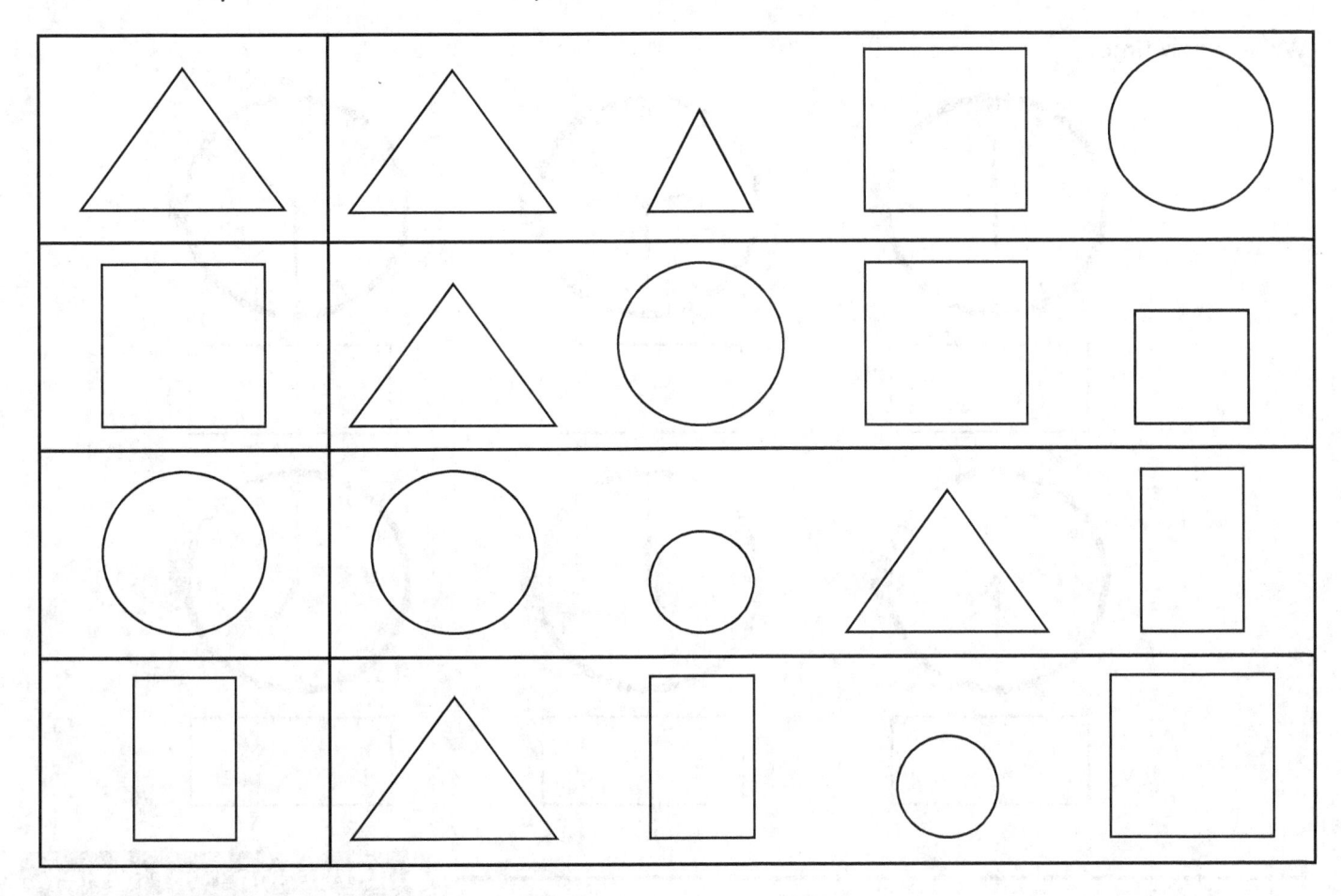

Draw a figure that is the same shape and size.

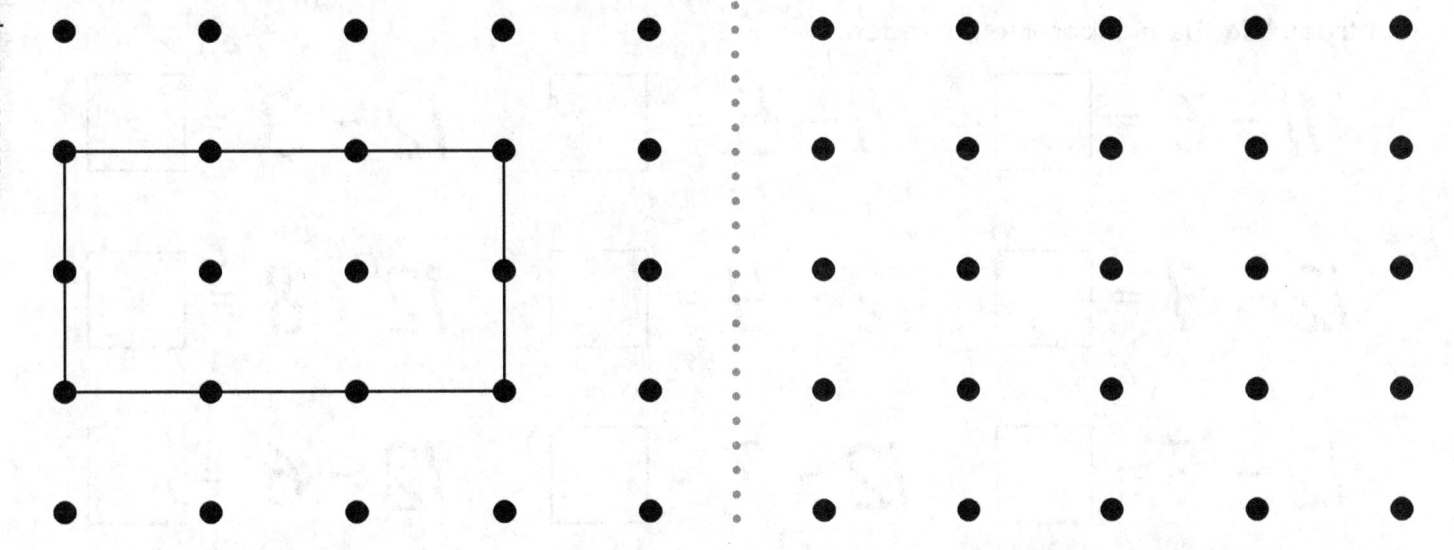

Chapter 15 • Lesson 119

Chapter 8 Review

Complete each sentence.

1 hour = ☐ minutes   half-hour = ☐ minutes

Write each time.

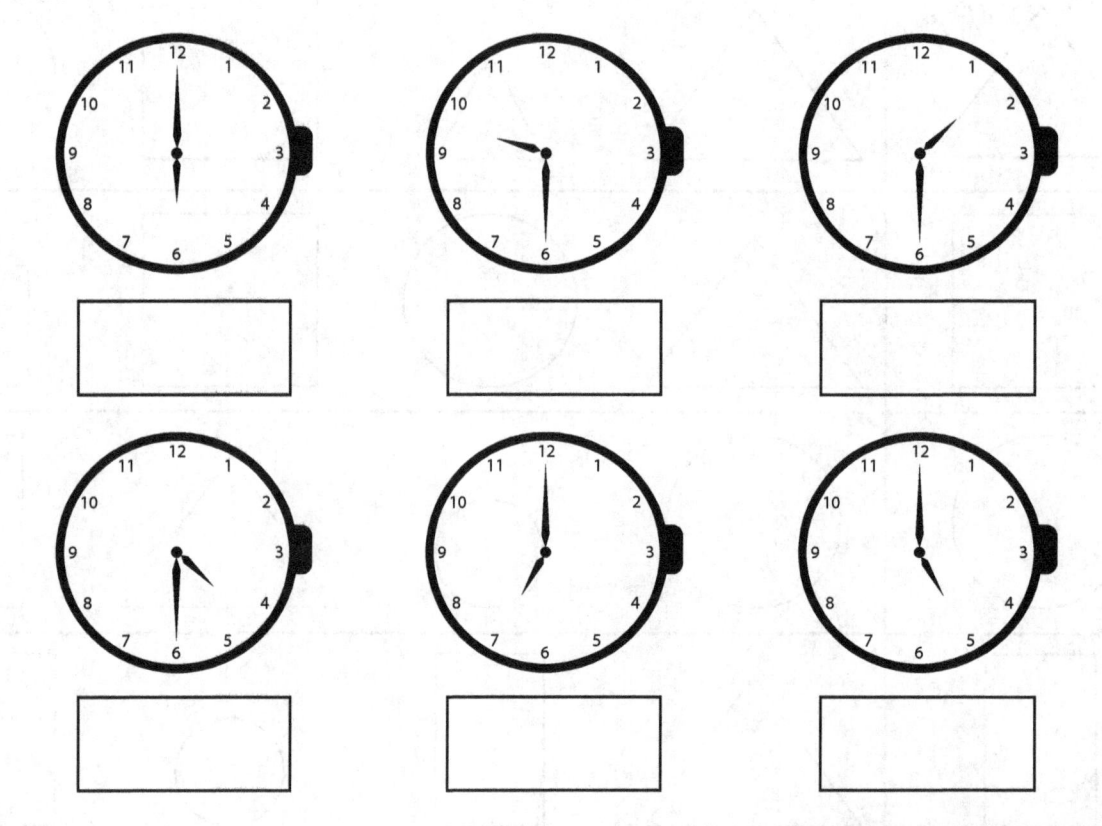

Subtraction Fact Review

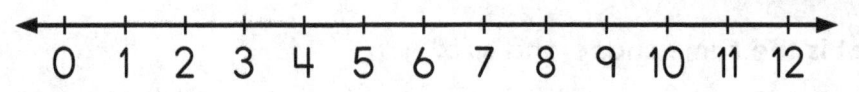

Subtract. Use the number line if needed.

11 − 6 = ☐   11 − 5 = ☐   12 − 3 = ☐

12 − 9 = ☐   12 − 4 = ☐   12 − 8 = ☐

12 − 5 = ☐   12 − 7 = ☐   12 − 6 = ☐

Symmetry

Look at the line on each figure.
Color each figure that has matching equal parts.

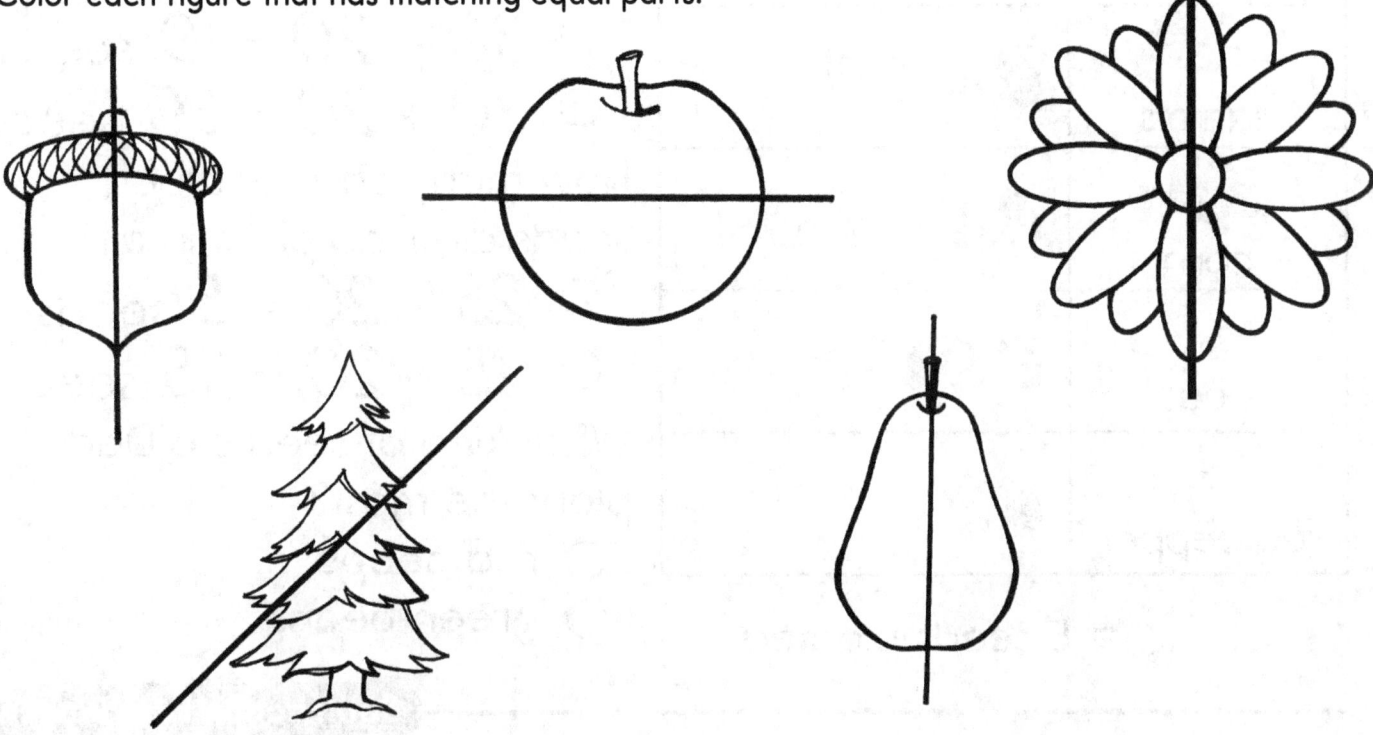

Draw a line of symmetry so each shape has matching equal parts.

Chapter 15 • Lesson 120   two hundred twenty-three **223**

Chapter 12 Review

Dad's Garden

green beans	🫘🫘🫘🫘🫘🫘
carrots	🫘🫘🫘🫘
corn	🫘🫘🫘🫘
peas	🫘🫘🫘
red pepper	🫘

🫘 = 5 seeds planted

Use the information in the graph. Mark the correct equation.

How many more green bean seeds did Dad plant than carrots?
- ○ 30 − 20 = 10 seeds
- ○ 30 + 20 = 50 seeds

How many corn and pea seeds did Dad plant in all?
- ○ 25 − 20 = 5 seeds
- ○ 25 + 20 = 45 seeds

What kind of seed did Dad plant the most?
- ○ red pepper
- ○ green beans

Subtraction Fact Review

Subtract.

8 − 6 = ☐
10 − 7 = ☐
10 − 5 = ☐
8 − 5 = ☐
11 − 9 = ☐

9 − 4 = ☐
11 − 8 = ☐
8 − 7 = ☐
10 − 9 = ☐
10 − 4 = ☐

7 − 6 = ☐
8 − 4 = ☐
9 − 9 = ☐
11 − 2 = ☐
9 − 7 = ☐

Patterns

_ _ _ _ _ _ _ _ _ _ _ _

Circle the note that comes next.

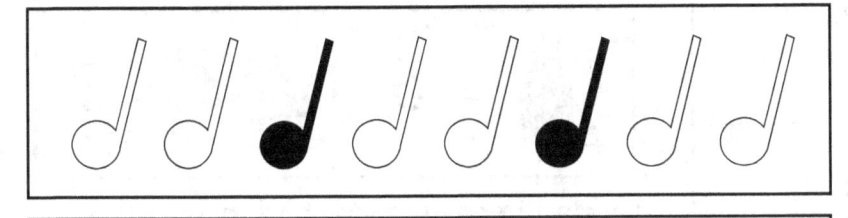

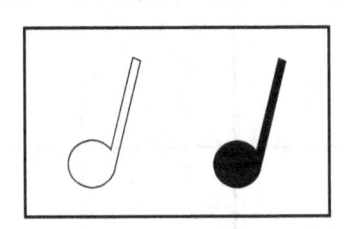

Draw the shapes to complete each pattern.

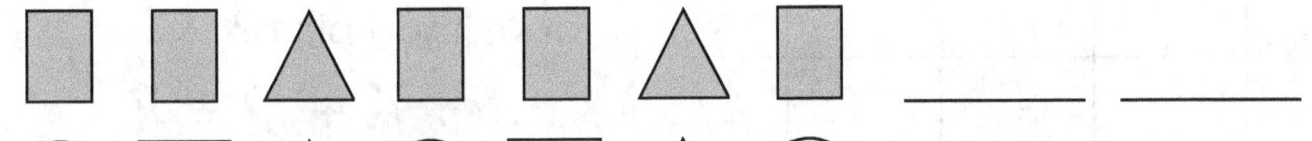

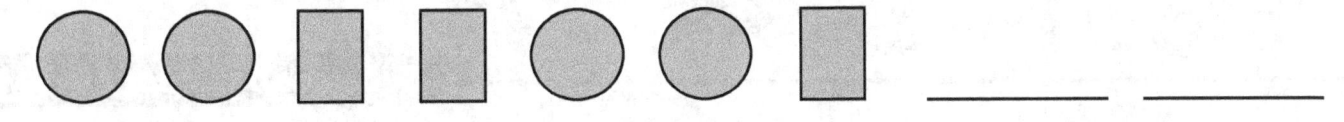

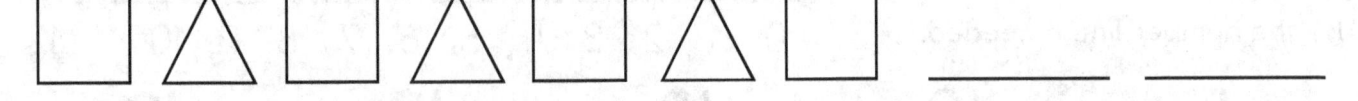

Write the letters or numbers to complete each pattern.

J C B J C B _ _ _

4 6 4 6 4 6 _ _ _

8 5 5 8 5 5 _ _ _

Chapter 15 • Lesson 121 — two hundred twenty-five 225

Chapter 12 Review

Our Pets

Use the information to make a bar graph.

= 4   = 2
= 6   = 3

Circle the pet the kids had the most.
Underline the pet the kids had the least.

cat   bird   dog   fish

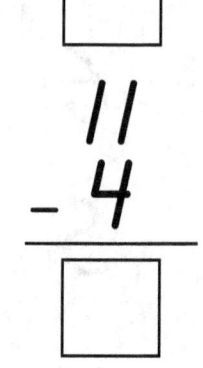

Mark the correct answer.

How many more kids had dogs than birds?

○ 6 + 2 = 8
○ 6 − 2 = 4

Subtraction Fact Review

Subtract.
Use the number line if needed.

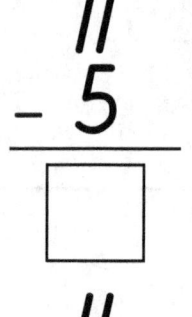

0 1 2 3 4 5 6 7 8 9 10 11 12

11 − 5 =
12 − 9 =
12 − 4 =
12 − 5 =
12 − 6 =

11 − 6 =
12 − 3 =
12 − 8 =
12 − 7 =
11 − 4 =

Chapter 15 Review

Mark the correct shape.

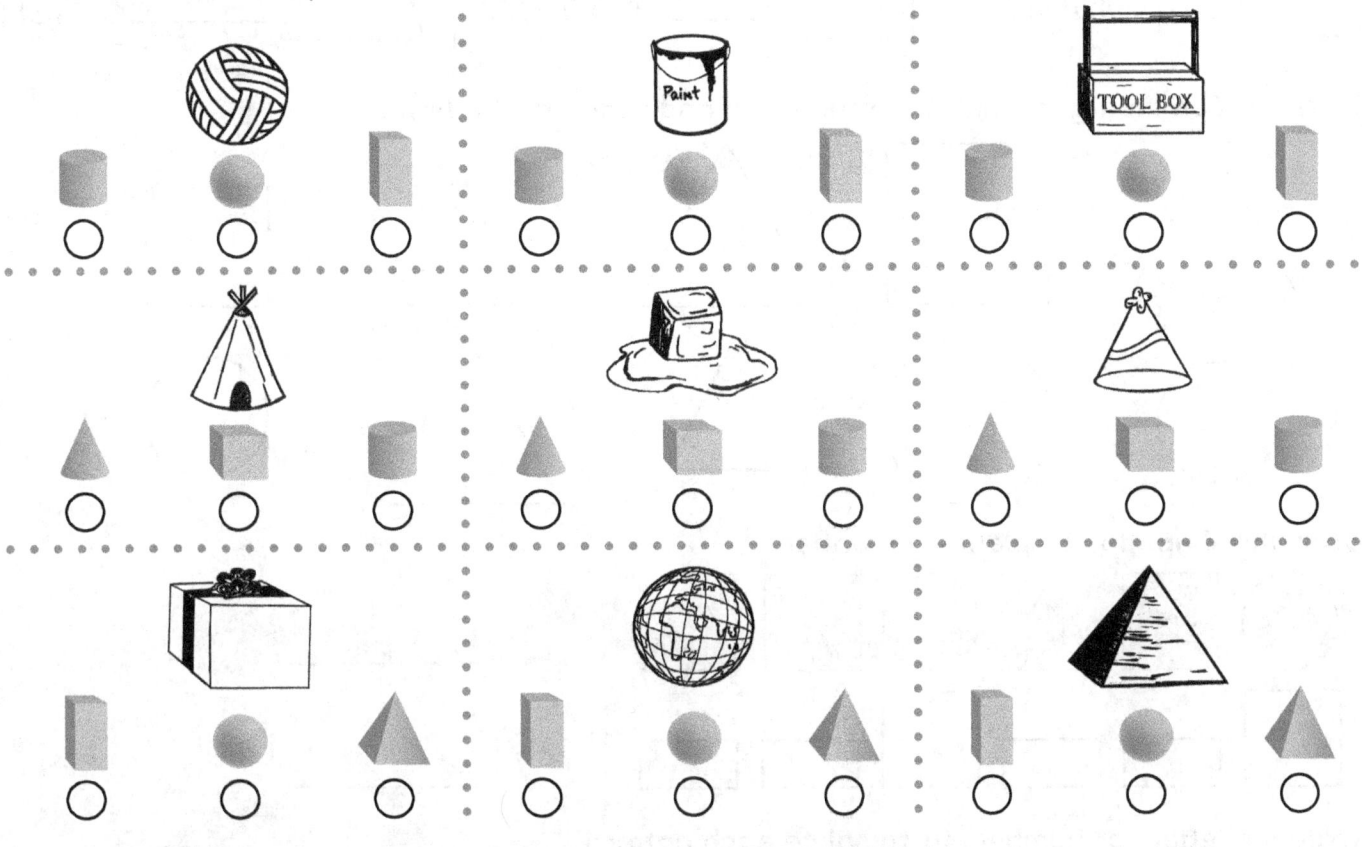

Does each solid figure have flat sides, curved sides, or both?
Mark the correct answer.

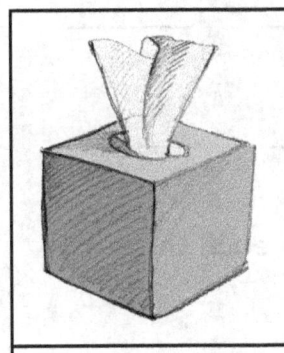

○ flat sides
○ curved sides
○ both

○ flat sides
○ curved sides
○ both

○ flat sides
○ curved sides
○ both

○ flat sides
○ curved sides
○ both

Chapter 15 Review two hundred twenty-seven 227

Write the number of sides and corners.

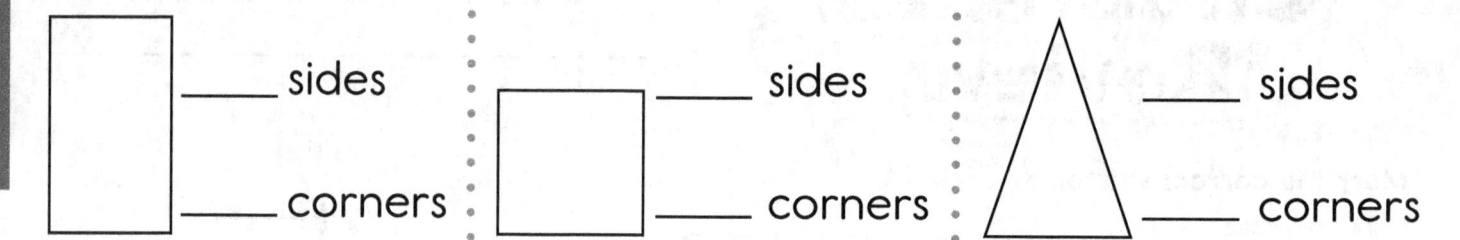

____ sides
____ corners

____ sides
____ corners

____ sides
____ corners

Mark an *X* on the figure that is the same size and shape as the first one.

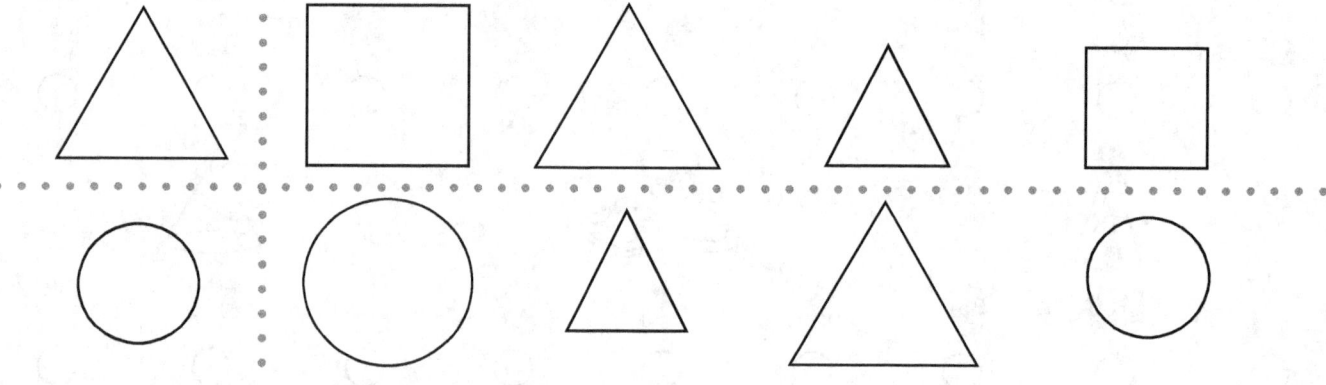

Draw the shape to complete each pattern.

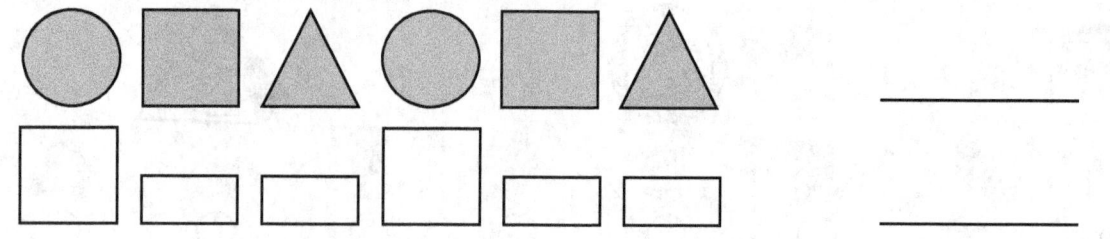

Write the letters or numbers to complete each pattern.

2 4 2 4 2 4 2 ___ ___ ___

X Y Z X Y Z X Y ___ ___ ___

Mark the correct answer.

Which shape are you *less likely* to pick?
○ triangle
○ rectangle

Draw a line of symmetry so each shape has matching parts.

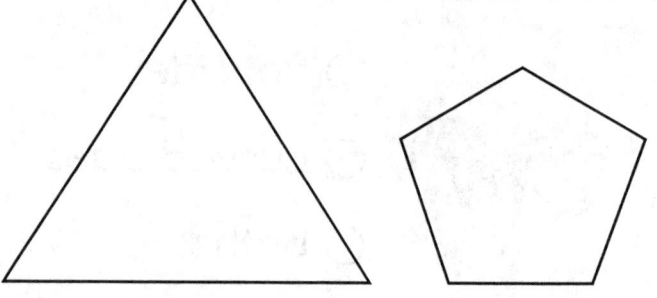

228 two hundred twenty-eight

Math 1 Reviews

© 2024 BJU Press. Reproduction prohibited.

Cumulative Review

Add.

$$\begin{array}{r}43\\+33\\\hline\end{array}$$

$$\begin{array}{r}61\\+24\\\hline\end{array}$$

$$\begin{array}{r}33\\+16\\\hline\end{array}$$

$$\begin{array}{r}74\\+25\\\hline\end{array}$$

Subtract.

$$\begin{array}{r}58\\-34\\\hline\end{array}$$

$$\begin{array}{r}75\\-43\\\hline\end{array}$$

$$\begin{array}{r}94\\-41\\\hline\end{array}$$

$$\begin{array}{r}86\\-23\\\hline\end{array}$$

Write an equation for each word problem. Solve.

Sid has a box with 24 crayons and a box with 12 crayons. How many crayons does Sid have in all?

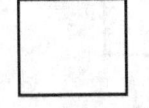

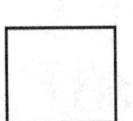

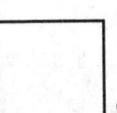

☐ ◯ ☐ ◯ ☐ crayons

Jill has a box with 48 crayons. Sue has a box with 24 crayons. How many more crayons does Jill have than Sue?

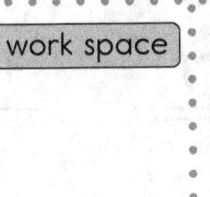

☐ ◯ ☐ ◯ ☐ crayons

Karla used 13 crayons to color her picture. Cory used 11 crayons to color his picture. How many crayons did they use in all?

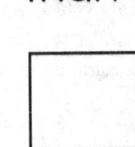

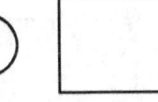

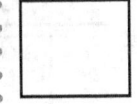

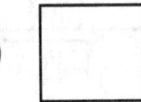

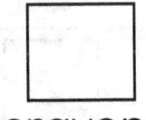

☐ ◯ ☐ ◯ ☐ crayons

Chapter 15 Cumulative Review

Addition Fact Review

Add.

7	3	6	5	7
+1	+4	+6	+4	+5
☐	☐	☐	☐	☐

4	5	6	2	3
+2	+7	+1	+8	+5
☐	☐	☐	☐	☐

9	4	5	9	2
+3	+4	+2	+1	+2
☐	☐	☐	☐	☐

6	8	4	0	6
+5	+2	+3	+0	+3
☐	☐	☐	☐	☐

5	7	6	5	4
+1	+3	+4	+5	+5
☐	☐	☐	☐	☐

3	8	5	4	0
+3	+4	+3	+6	+8
☐	☐	☐	☐	☐

Hundreds, Tens, Ones

Count by 100s to 1,000.

100 ▭ ▭ 400 ▭
600 ▭ 800 ▭ 1,000

Write each number.

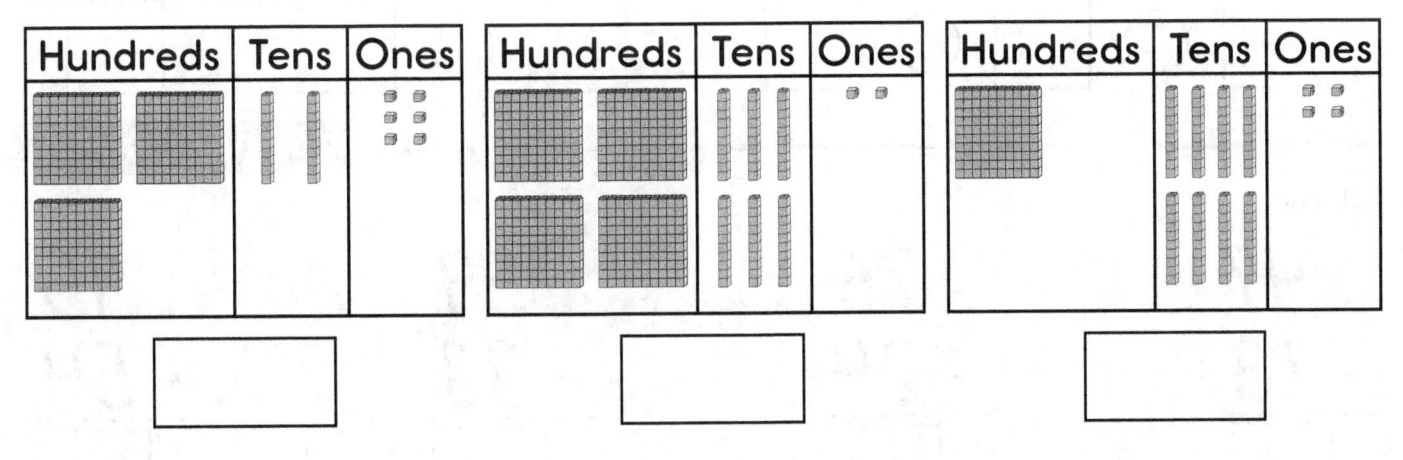

Hundreds	Tens	Ones

Hundreds	Tens	Ones

Hundreds	Tens	Ones

Write the number of hundreds, tens, and ones. Write the number.

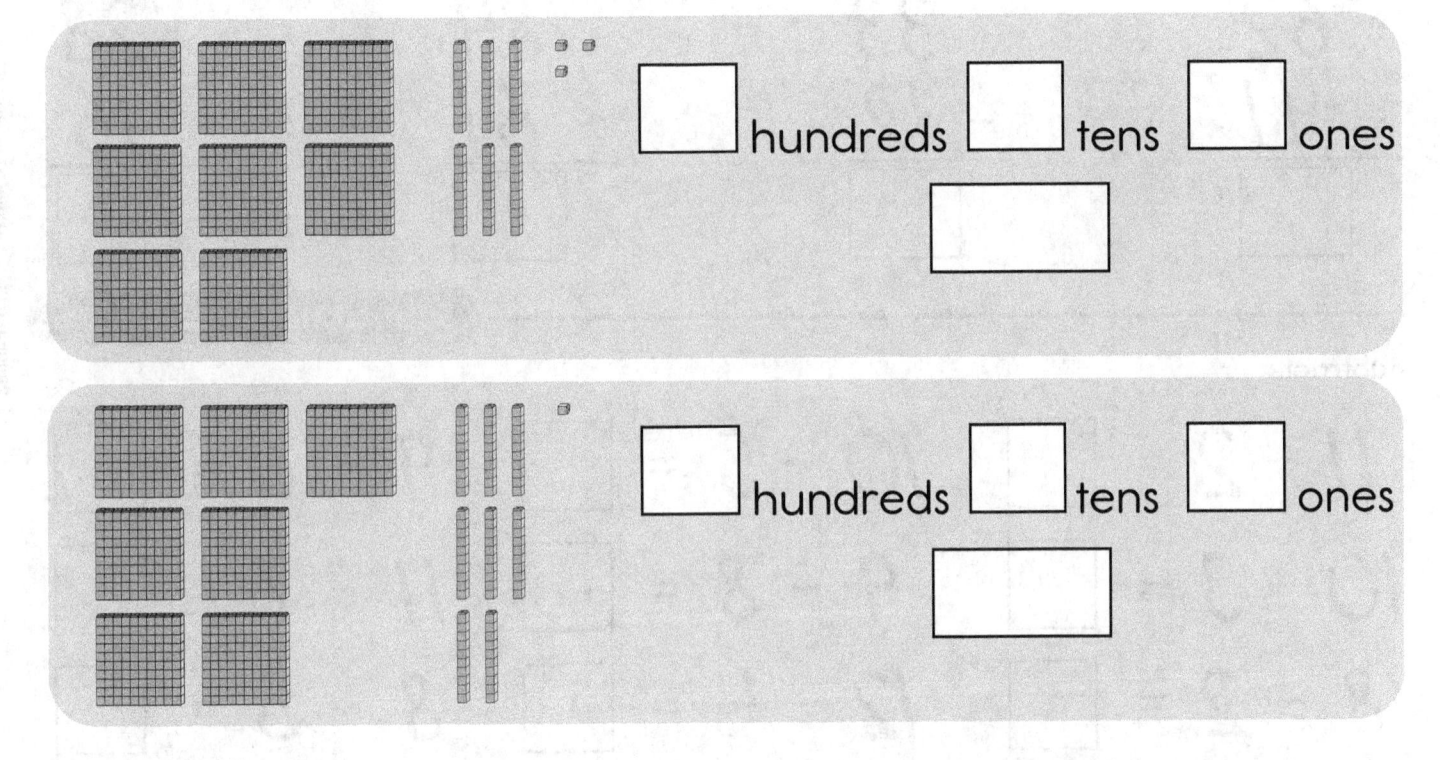

▭ hundreds ▭ tens ▭ ones

▭ hundreds ▭ tens ▭ ones

Chapter 16 • Lesson 124

Chapter 10 Review

Write each fact family.

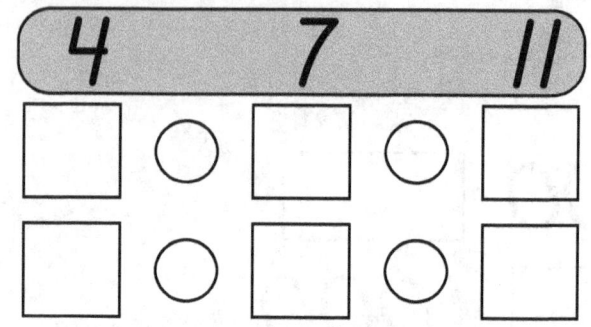

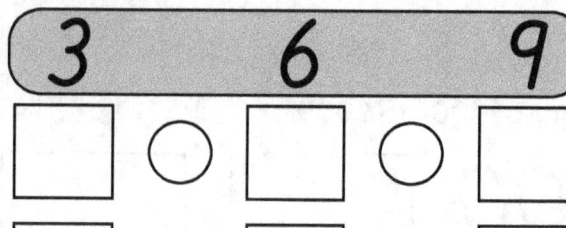

| 4 | 7 | 11 |

☐ ○ ☐ ○ ☐
☐ ○ ☐ ○ ☐
☐ ○ ☐ ○ ☐
☐ ○ ☐ ○ ☐

| 3 | 6 | 9 |

☐ ○ ☐ ○ ☐
☐ ○ ☐ ○ ☐
☐ ○ ☐ ○ ☐
☐ ○ ☐ ○ ☐

Chapter 13 Review

Subtract.

```
  47        76        59        95
- 13      - 24      - 38      - 54
----      ----      ----      ----
```

```
  62        33        84        28
- 41      - 12      - 61      - 16
----      ----      ----      ----
```

Subtraction Fact Review

Subtract.

11 − 2 = ☐   10 − 5 = ☐   9 − 6 = ☐

10 − 3 = ☐   9 − 8 = ☐   11 − 7 = ☐

8 − 2 = ☐   12 − 4 = ☐   8 − 5 = ☐

Representing 3-Digit Numbers

_____

Write each missing number.

345   623   461   901

347   625   463   903

___   ___   ___   ___

Write the number of hundreds, tens, and ones.
Write the number.

☐ hundreds
☐ tens
☐ ones

☐ hundreds
☐ tens
☐ ones

☐ hundreds
☐ tens
☐ ones

Chapter 16 • Lesson 125

Chapter 8 Review

Complete each sentence.

1 hour = ☐ minutes   a half-hour = ☐ minutes

Write each time.

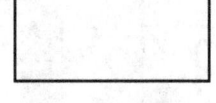

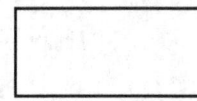

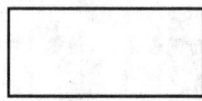

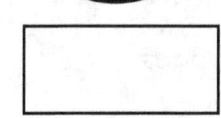

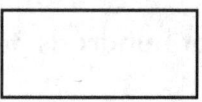

Number the days of the week in order.

☐ Thursday   ☐ Tuesday   ☐ Saturday   ☐ Sunday

☐ Monday   ☐ Friday   ☐ Wednesday

Subtraction Fact Review

Subtract.

12 − 5 = ☐   11 − 6 = ☐   7 − 4 = ☐   10 − 8 = ☐   12 − 6 = ☐

9 − 4 = ☐   8 − 5 = ☐   11 − 8 = ☐   12 − 3 = ☐   8 − 7 = ☐

Writing Numbers to 1,000

Mark the mat that shows 1 *less* than 621.

Hundreds	Tens	Ones

○

Hundreds	Tens	Ones

○

Mark the mat that shows 1 *more* than 963.

Hundreds	Tens	Ones

○

Hundreds	Tens	Ones

○

Write the number that comes just *before*.

☐ 467

Write the number that comes *between*.

684 ☐ 686

322 ☐ 324

Write the number that comes just *after*.

924 ☐

Chapter 16 • Lesson 126

Chapter 9 Review

Circle the fair share for each child.
Write the number each child gets.

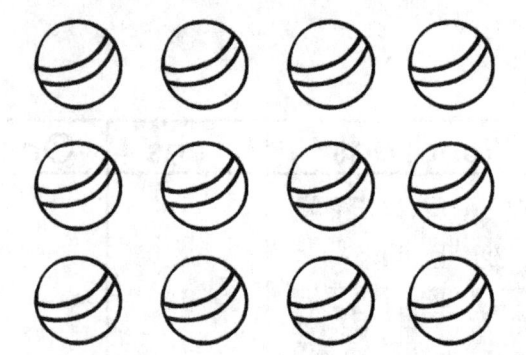

3 children get ☐ each

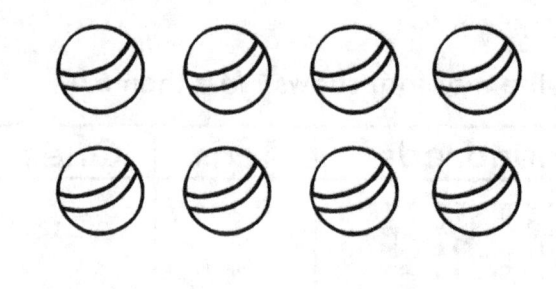

2 children get ☐ each

Color to show each fraction.

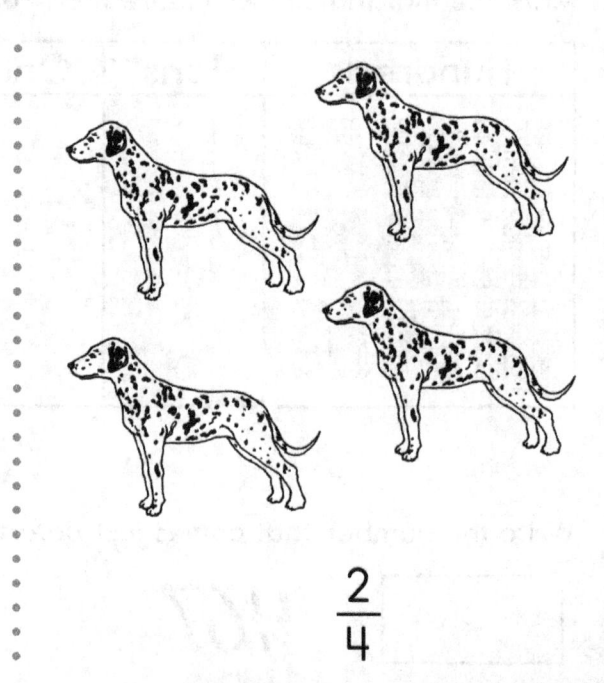

$\frac{1}{2}$ $\frac{1}{3}$ $\frac{2}{4}$

Subtraction Fact Review

Subtract.

11 − 6 = ☐   9 − 7 = ☐   8 − 6 = ☐

12 − 3 = ☐   7 − 3 = ☐   9 − 6 = ☐

10 − 3 = ☐   6 − 4 = ☐   11 − 4 = ☐

Place Value in 3-Digit Numbers

_ _ _ _ _ _ _ _ _ _ _ _

Mark the value of each underlined digit.

77<u>5</u>   ○ 500   ○ 50   ○ 5

<u>2</u>34   ○ 200   ○ 20   ○ 2

1<u>8</u>6   ○ 800   ○ 80   ○ 8

<u>9</u>23   ○ 900   ○ 90   ○ 9

Write the value of each underlined digit.

<u>3</u>75 ☐   10<u>7</u> ☐

2<u>8</u>1 ☐   <u>4</u>56 ☐

76<u>2</u> ☐   8<u>4</u>1 ☐

Read the clues and write the number.

3 hundreds, 7 tens, 2 ones   ☐

8 hundreds, 6 ones, 4 tens   ☐

5 tens, 9 hundreds, 1 one   ☐

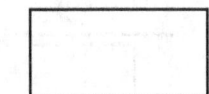

Chapter 16 • Lesson 127

Chapter 10 Review

Write each fact family.

| 5 | 6 | 11 |

| 5 | 7 | 12 |

Mark the circle next to the fact that does *not* belong in each family.

| 3 | 9 | 12 |

○ $9 + 3 = 12$
○ $3 + 9 = 12$
○ $12 - 3 = 9$
○ $8 - 4 = 4$

| 4 | 6 | 10 |

○ $6 + 4 = 10$
○ $4 + 6 = 10$
○ $10 - 5 = 5$
○ $10 - 4 = 6$

| 3 | 6 | 9 |

○ $3 + 6 = 9$
○ $6 + 2 = 8$
○ $9 - 3 = 6$
○ $9 - 6 = 3$

Addition Fact Review

Add.

$8 + 4 =$ ☐
$9 + 3 =$ ☐
$7 + 6 =$ ☐
$8 + 2 =$ ☐
$4 + 5 =$ ☐

$7 + 3 =$ ☐
$6 + 6 =$ ☐
$5 + 7 =$ ☐
$7 + 7 =$ ☐
$8 + 3 =$ ☐

Counting More & Less

Write the number that is 1 *more*.

322   687   919

Write the number that is 1 *less*.

233   579   830

Write the number that is 10 *more*.

320   275   623

Write the number that is 10 *less*.

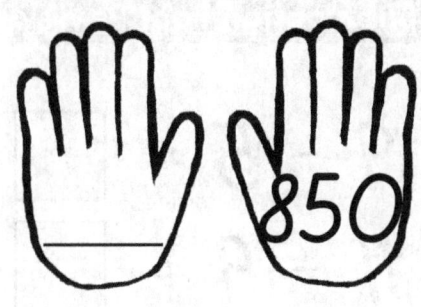

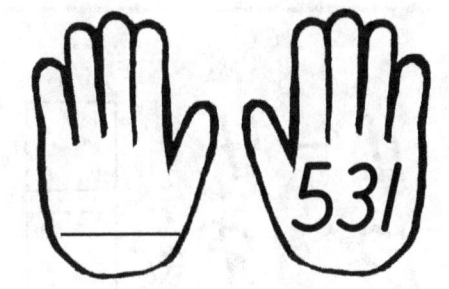

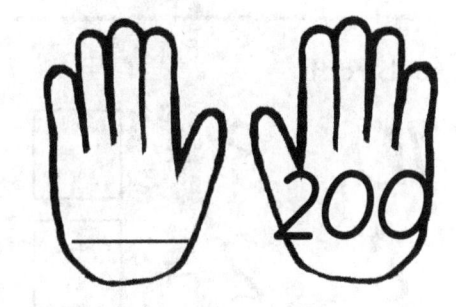

850   531   200

Chapter 16 • Lesson 128

Chapter 13 Review

Subtract.

65	74	87	64	38
−32	−21	−22	−30	−16

96	85	68	54	81
−24	−14	−52	−43	−20

Chapter 11 Review

Guess each length. Use an inch ruler to measure.

My Guess	My Measure
☐ inches	☐ inches
☐ inches	☐ inches
☐ inches	☐ inches

Use an inch ruler to draw the length. Start at the star.

5 inches ★

Subtraction Fact Review

Subtract.

10 − 3 = ☐   11 − 4 = ☐   9 − 5 = ☐

8 − 2 = ☐   12 − 9 = ☐   11 − 9 = ☐

7 − 4 = ☐   11 − 6 = ☐   9 − 7 = ☐

Comparing 3-Digit Numbers

Mark the correct answer.

345 ? 223
○ is less than <
○ is greater than >

473 ? 178
○ is less than <
○ is greater than >

149 ? 396
○ is less than <
○ is greater than >

721 ? 819
○ is less than <
○ is greater than >

421 ? 423
○ is less than <
○ is greater than >

596 ? 581
○ is less than <
○ is greater than >

Draw a dot next to the smaller number.
Draw the correct sign.

300 > 200	626 • 738
120 140	396 400
238 220	840 830

Chapter 16 • Lesson 129 — two hundred forty-one 241

Chapter 12 Review

Add.

```
  42      64      57      80      71
+ 23    + 31    + 42    + 16    + 26
____    ____    ____    ____    ____


  60      72      81      30      55
+ 26    + 27    + 16    + 59    + 43
____    ____    ____    ____    ____
```

Chapter 11 Review

Circle the correct temperature.

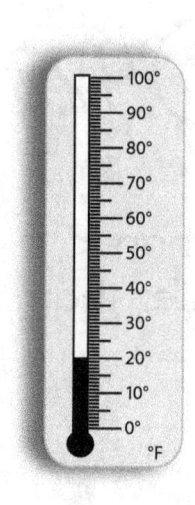

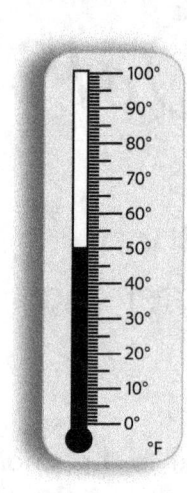

80°F   20°F   30°F
60°F   40°F   50°F
40°F   60°F   70°F

Mark the correct measuring tool.

What does it weigh?

How much milk do I add?

Adding Large Numbers

_____

Add the Ones first.
Add the Tens next.
Add the Hundreds last.

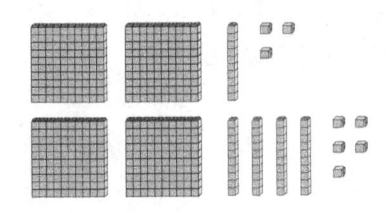

Hundreds	Tens	Ones
2	1	3
+ 2	4	5

Add.

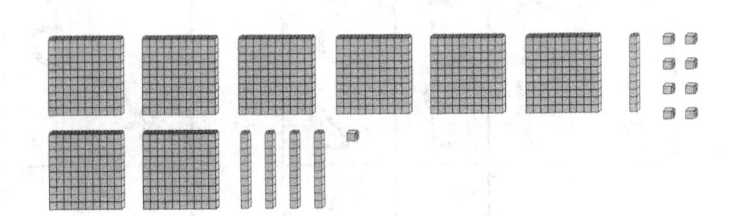

Hundreds	Tens	Ones
6	1	8
+ 2	4	1

Add.

Hundreds	Tens	Ones
1	3	4
+ 4	2	3

Hundreds	Tens	Ones
5	3	2
+ 4	1	5

Hundreds	Tens	Ones
1	4	2
+ 2	4	5

Write an equation for the word problem.
Solve by using the Hundreds/Tens/Ones frame.

Noah lived for 600 years before the Flood. He lived for 350 more years after the Flood. How many years did Noah live in all?

Hundreds	Tens	Ones

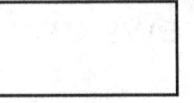

☐ ○ ☐ ○ ☐ years

Chapter 16 • Lessons 130–31   two hundred forty-three **243**

Add.

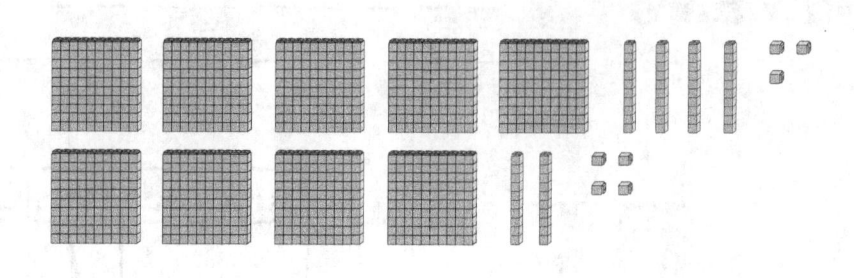

Hundreds	Tens	Ones
5	4	3
+ 4	2	4

Add.

Hundreds	Tens	Ones
6	8	4
+ 2	1	1

Hundreds	Tens	Ones
2	3	5
+ 4	3	2

Hundreds	Tens	Ones
5	7	4
+ 1	0	5

Hundreds	Tens	Ones
4	3	5
+ 1	4	2

Hundreds	Tens	Ones
7	3	5
+ 2	4	2

Hundreds	Tens	Ones
3	9	3
+ 2	0	4

Write an equation for the word problem.
Solve by using the Hundreds/Tens/Ones frame.

Dad has 150 nails and 125 screws in his tool kit. How many screws and nails does Dad have in all?

☐ ○ ☐ ○ ☐ screws and nails

Hundreds	Tens	Ones
+		

Practicing Adding Large Numbers

Add.

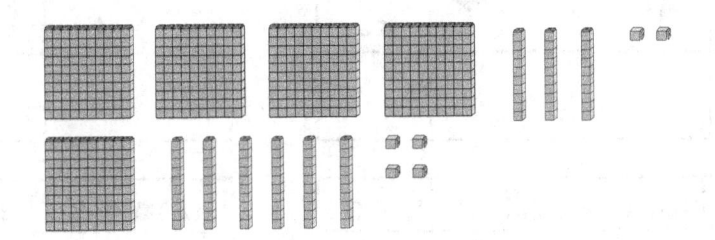

Hundreds	Tens	Ones
4	3	2
+1	6	4

Add.

Hundreds	Tens	Ones
2	6	4
+1	2	5

Hundreds	Tens	Ones
6	7	5
+3	2	1

Hundreds	Tens	Ones
3	4	2
+4	1	6

Hundreds	Tens	Ones
7	5	4
+2	4	3

Write an equation for the word problem.
Solve by using the Hundreds/Tens/Ones frame.

On Friday 618 people came to the circus.
On Saturday 371 people came to the circus.
How many people came to see the circus?

Hundreds	Tens	Ones
+		

☐ ○ ☐ ○ ☐ people

Chapter 16 • Lesson 132 two hundred forty-five **245**

Add.

Hundreds	Tens	Ones
6	3	5
+ 3	4	1

Hundreds	Tens	Ones
4	5	3
+ 2	4	2

Hundreds	Tens	Ones
8	2	3
+ 1	6	3

Hundreds	Tens	Ones
7	0	5
+ 1	9	3

Hundreds	Tens	Ones
2	5	4
+ 3	2	1

Hundreds	Tens	Ones
3	8	2
+ 4	0	7

Write an equation for the word problem.
Solve by using the Hundreds/Tens/Ones frame.

The students sold 156 bags of cotton candy and 342 bags of popcorn. How many bags of food did the students sell in all?

☐ ○ ☐ ○ ☐ bags

Hundreds	Tens	Ones
+		

Chapter 16 Review

Count by 100s to 1,000.

100 _____ _____ 400 _____

600 _____ 800 _____ 1,000

Write each missing number.

455 _____ 457 _____ 459 _____ 461

739 _____ 741 _____ 743 _____ 745

Write each number.

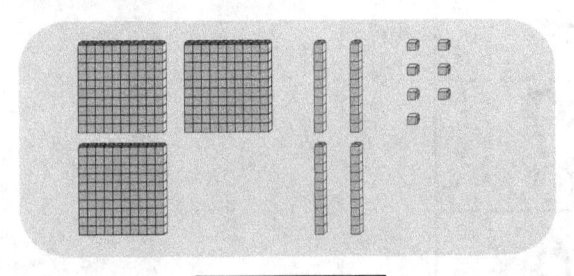

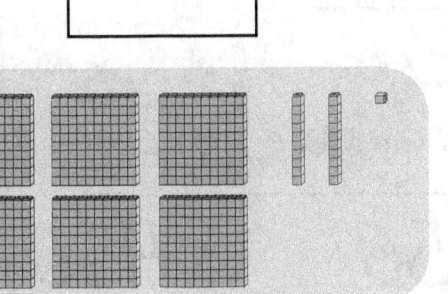

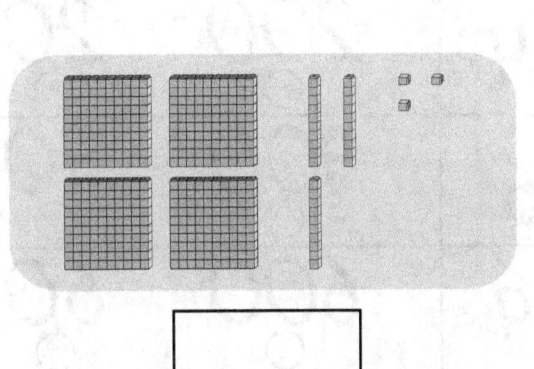

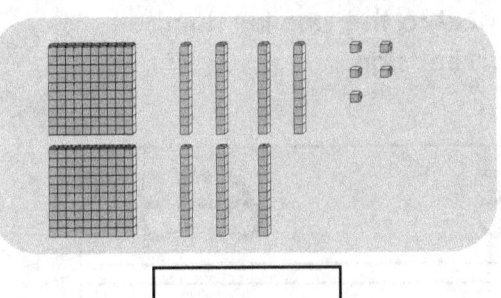

Chapter 16 Review two hundred forty-seven 247

Write the number that is 1 more, 1 less, 10 more, or 10 less.

1 more		1 less		10 more		10 less	
245			725	480			828
300			560	326			240
861			193	655			945

Draw a circle around the correct number.

8 in the Ones place   5 in the Hundreds place   2 in the Tens place

483   608   563   452   623   392

Mark the value of each underlined digit.

75<u>3</u>	300 ○	30 ○	3 ○
3<u>2</u>4	200 ○	20 ○	2 ○
<u>9</u>16	900 ○	90 ○	9 ○
56<u>9</u>	600 ○	60 ○	6 ○

Draw a dot next to the smaller number.
Draw the correct sign.

> is greater than   < is less than

195	643
700	500

356	456
903	890

248   two hundred forty-eight   Math 1 Reviews

Cumulative Review

- - - - - - - - - - - - - - - -

Add.

$$\begin{array}{r}2\\4\\+2\\\hline\square\end{array}$$ $$\begin{array}{r}7\\1\\+4\\\hline\square\end{array}$$ $$\begin{array}{r}6\\3\\+2\\\hline\square\end{array}$$ $$\begin{array}{r}5\\2\\+3\\\hline\square\end{array}$$

Write an equation for each word problem. Solve.

Mother used 42 apples to make applesauce. She used 23 apples to make pies. How many apples did Mother use in all?

work space

□ ○ □ ○ □ apples

Paul has 27 apples in his basket. He sold 16 apples. How many apples does Paul have left?

work space

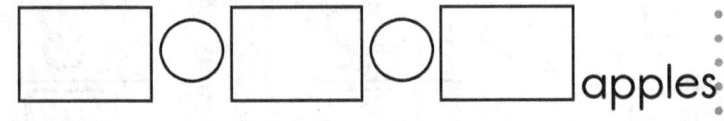

□ ○ □ ○ □ apples

Write the total value.

_____ ¢

_____ ¢

_____ ¢

Chapter 16 Cumulative Review

Subtraction Fact Review

Subtract.

5 − 1 = ☐	9 − 9 = ☐	6 − 2 = ☐	10 − 8 = ☐	11 − 2 = ☐
7 − 5 = ☐	10 − 6 = ☐	11 − 3 = ☐	9 − 7 = ☐	6 − 4 = ☐
9 − 4 = ☐	9 − 5 = ☐	7 − 4 = ☐	5 − 2 = ☐	8 − 5 = ☐
5 − 3 = ☐	7 − 6 = ☐	5 − 4 = ☐	10 − 2 = ☐	8 − 3 = ☐
4 − 4 = ☐	11 − 9 = ☐	8 − 0 = ☐	6 − 1 = ☐	11 − 7 = ☐
12 − 7 = ☐	6 − 5 = ☐	6 − 3 = ☐	8 − 4 = ☐	12 − 8 = ☐

Measure in Centimeters

Mark an *X* on the shorter object.

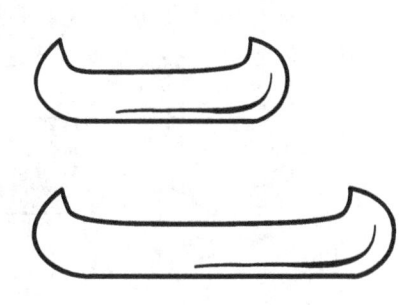

Write the number of centimeters.

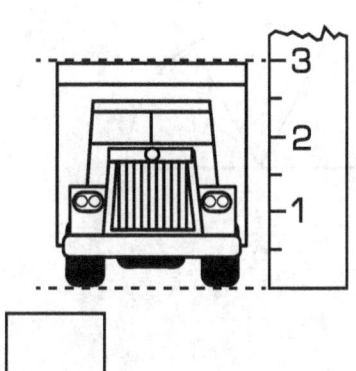

☐ centimeters

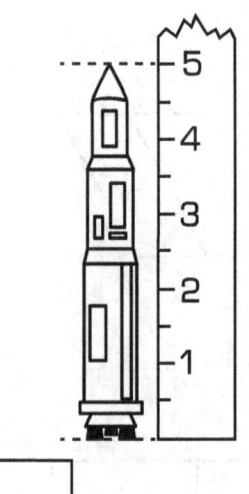

☐ centimeters

☐ centimeters

Use a centimeter ruler to measure.

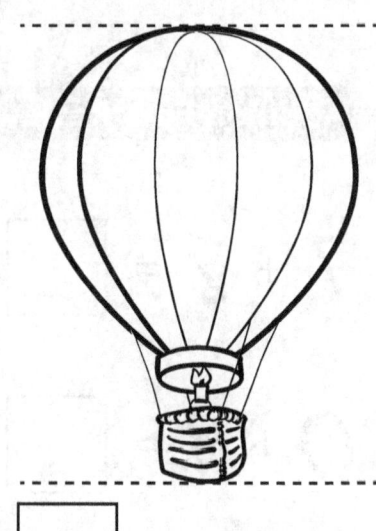

☐ centimeters

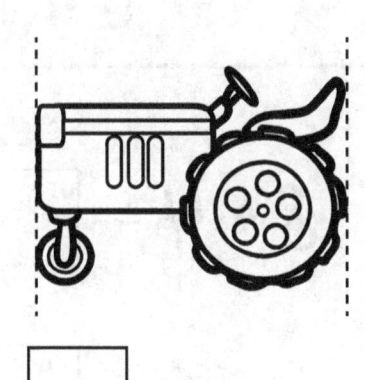

☐ centimeters

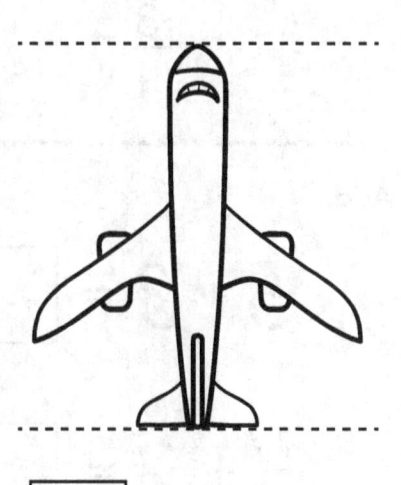

☐ centimeters

Chapter 17 • Lesson 136

Chapter 9 Review

Match the shape to the correct fraction.

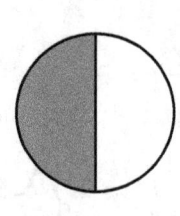

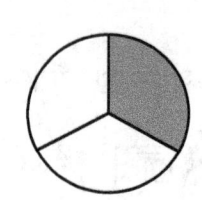

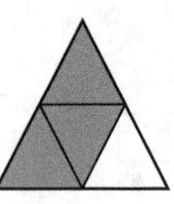

$$\frac{2}{3}$$
$$\frac{3}{4}$$
$$\frac{1}{2}$$
$$\frac{1}{3}$$

Color the shape to match the fraction.

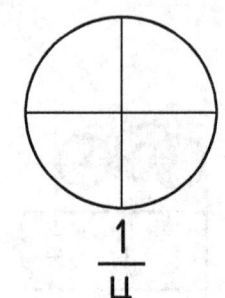

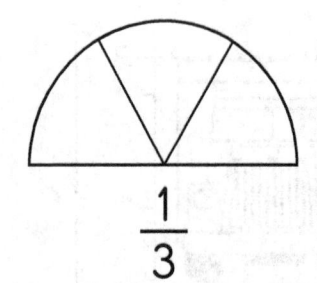

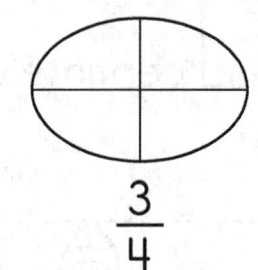

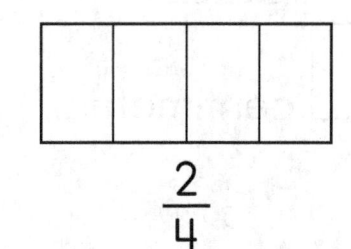

$$\frac{1}{4}$$

$$\frac{2}{3}$$

$$\frac{1}{3}$$

$$\frac{3}{4}$$

$$\frac{2}{4}$$

Addition Fact Review

Add.

$5 + 3 = \square$ $8 + 4 = \square$ $7 + 2 = \square$

$4 + 2 = \square$ $3 + 6 = \square$ $0 + 9 = \square$

Estimate & Measure in Centimeters

Guess each length. Use a centimeter ruler to measure.

? cm

? cm

? cm

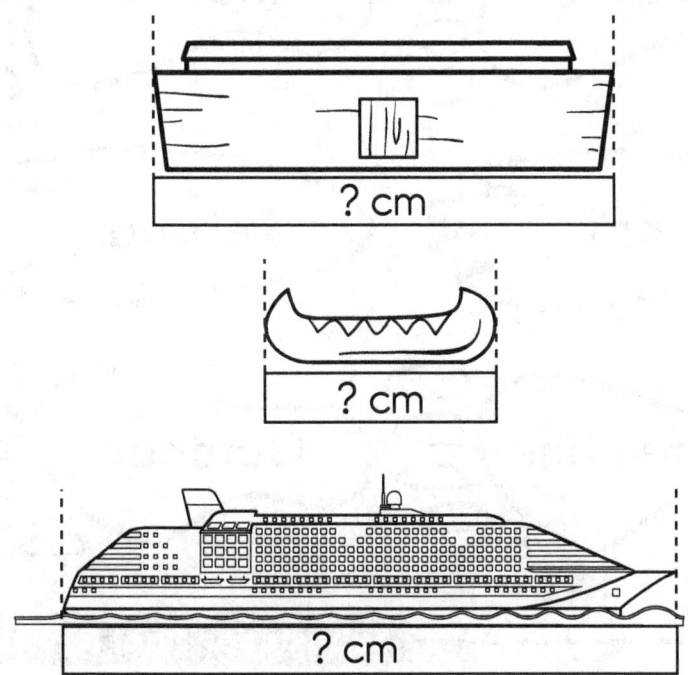

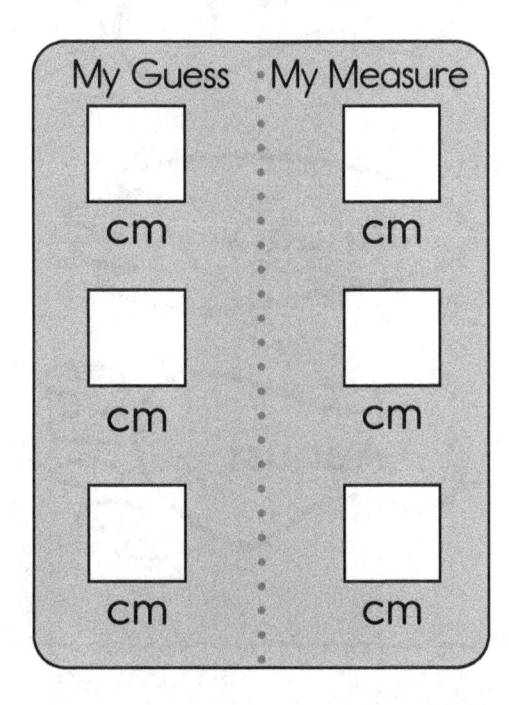

My Guess	My Measure
☐ cm	☐ cm
☐ cm	☐ cm
☐ cm	☐ cm

Use a centimeter ruler to measure.

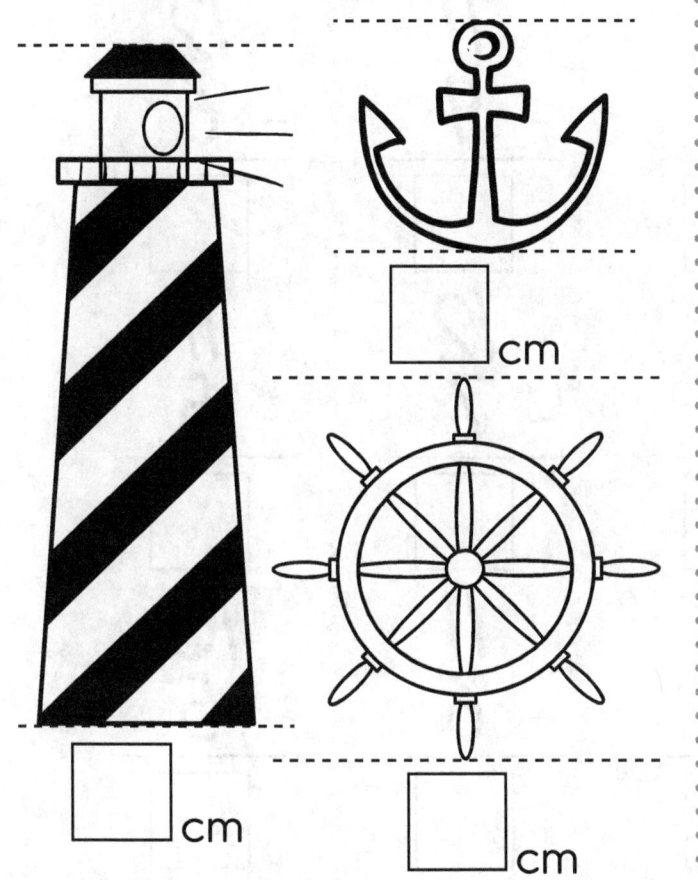

☐ cm

☐ cm

☐ cm

Write an equation for the word problem. Solve.

John's boat is 33 centimeters long. Sam's boat is 21 centimeters long. How much longer is John's boat than Sam's boat?

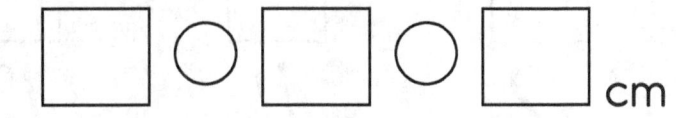

work space

☐ ◯ ☐ ◯ ☐ cm

Chapter 17 • Lesson 137

Chapter 8 Review

Color the paw print with the month that comes next.

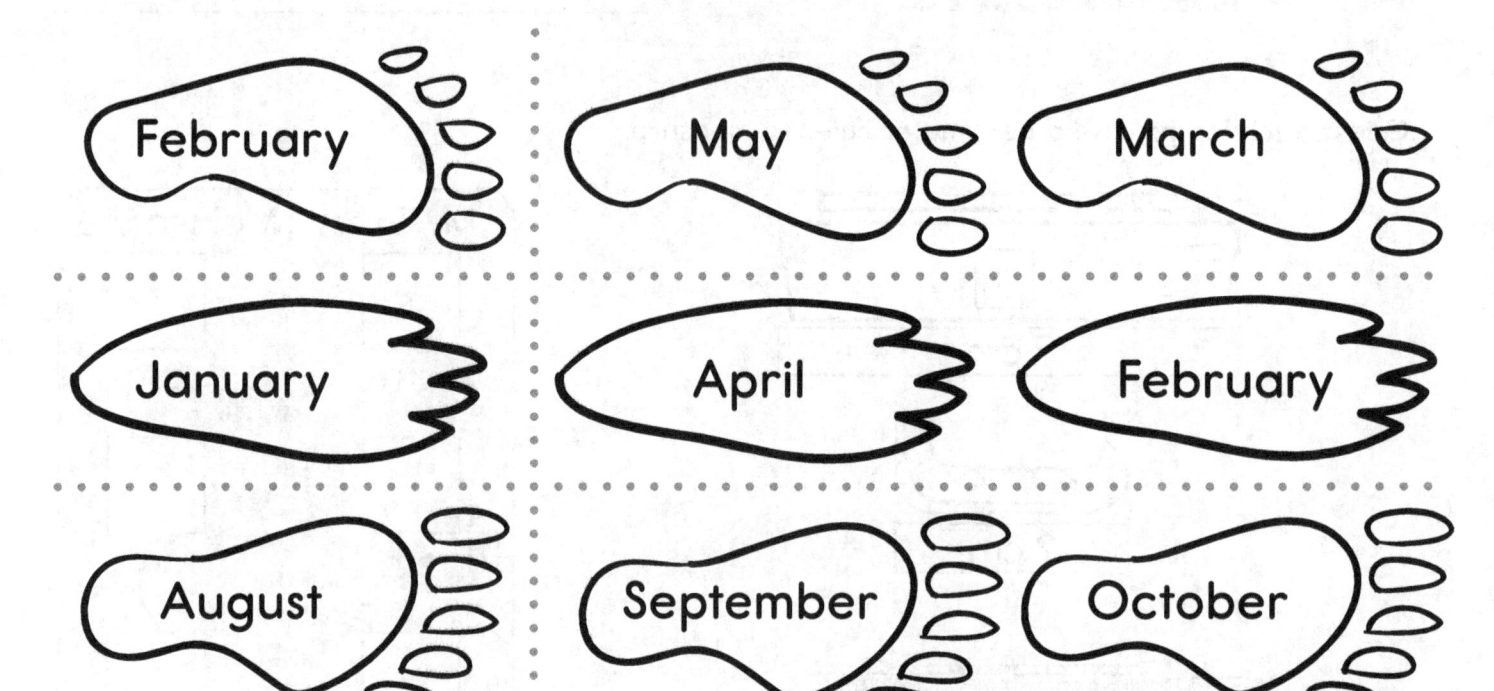

February   May   March

January   April   February

August   September   October

Subtraction Fact Review

Subtract.

12 − 3 = ☐	11 − 4 = ☐	12 − 8 = ☐	11 − 9 = ☐	12 − 6 = ☐
11 − 7 = ☐	12 − 4 = ☐	11 − 8 = ☐	12 − 7 = ☐	11 − 5 = ☐
12 − 9 = ☐	11 − 3 = ☐	12 − 5 = ☐	11 − 6 = ☐	10 − 5 = ☐

two hundred fifty-four

Math 1 Reviews

Using Liters

_____
- - - - - - - - - - -
_____

Mark an X on each container that holds *less than* 1 liter.
Circle each container that holds *more than* 1 liter.

Write an equation for each word problem. Solve.

Uncle Kyle has 9 liters of blue paint and 6 liters of green paint. How many more liters of blue paint does Uncle Kyle have than green paint?

☐ ◯ ☐ ◯ ☐ liters

Zoe is making punch for a party. She needs 5 liters of orange drink and 3 liters of ginger ale. How many more liters of orange drink than ginger ale does she need?

☐ ◯ ☐ ◯ ☐ liters

Chapter 17 • Lesson 138

Chapter 10 Review

Write each fact family.

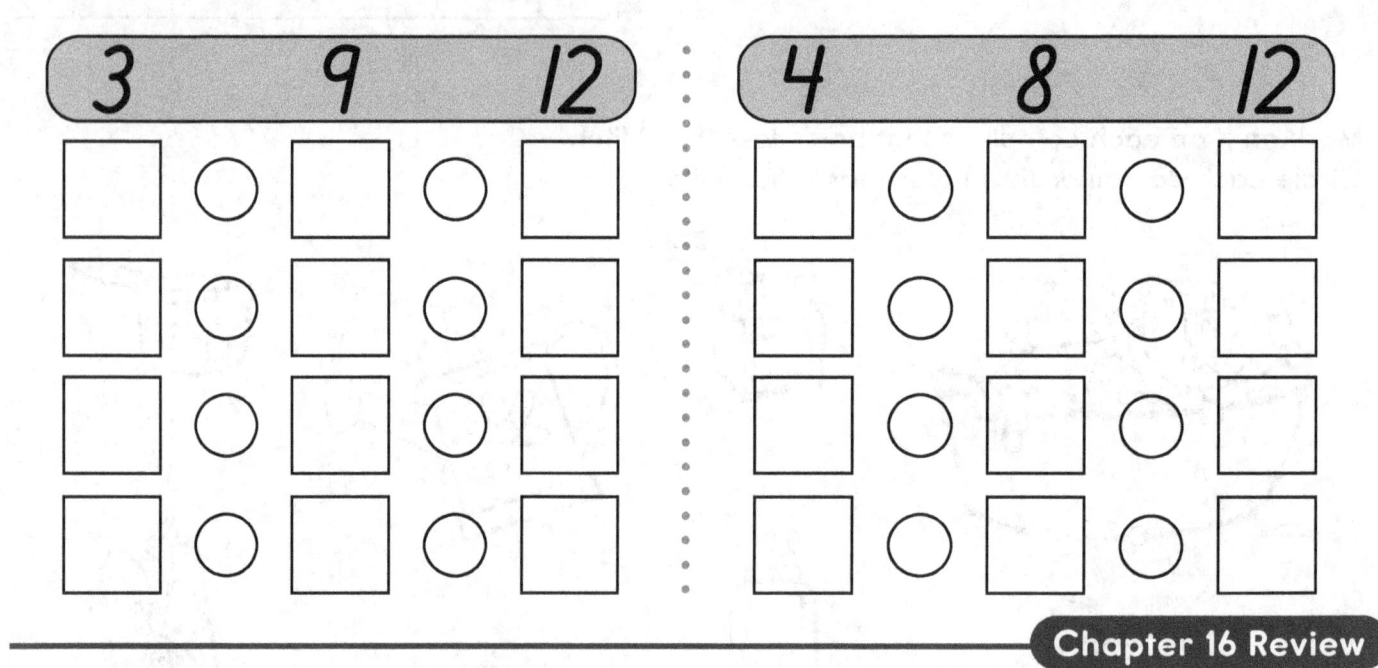

| 3 | 9 | 12 |

☐ ○ ☐ ○ ☐
☐ ○ ☐ ○ ☐
☐ ○ ☐ ○ ☐
☐ ○ ☐ ○ ☐

| 4 | 8 | 12 |

☐ ○ ☐ ○ ☐
☐ ○ ☐ ○ ☐
☐ ○ ☐ ○ ☐
☐ ○ ☐ ○ ☐

Chapter 16 Review

Mark the value of each underlined digit.

6<u>2</u>3	200 ○	20 ○	2 ○
<u>5</u>40	500 ○	50 ○	5 ○
97<u>2</u>	200 ○	20 ○	2 ○
<u>7</u>56	700 ○	70 ○	7 ○
2<u>3</u>5	300 ○	30 ○	3 ○
19<u>9</u>	900 ○	90 ○	9 ○
<u>5</u>63	500 ○	50 ○	5 ○
27<u>1</u>	100 ○	10 ○	1 ○

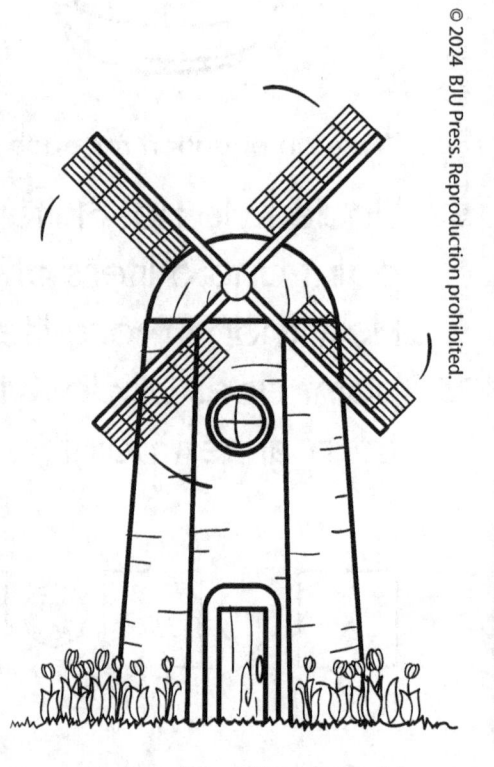

Math 1 Reviews

Using Kilograms

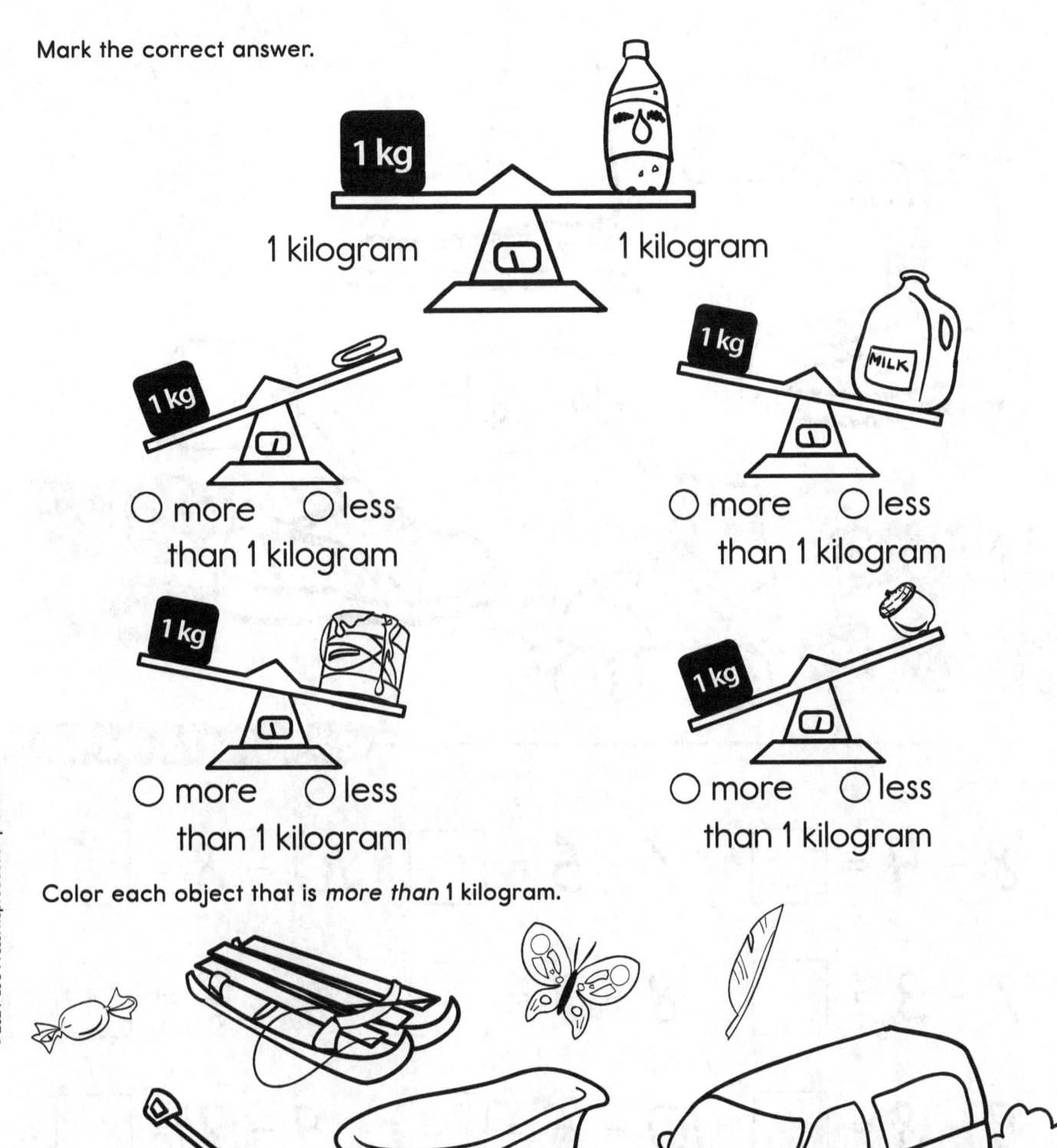

‾‾‾‾‾‾‾‾‾‾‾‾‾‾‾‾
- - - - - - - - - -
‾‾‾‾‾‾‾‾‾‾‾‾‾‾‾‾

Mark the correct answer.

1 kilogram 1 kilogram

○ more ○ less
than 1 kilogram

○ more ○ less
than 1 kilogram

○ more ○ less
than 1 kilogram

○ more ○ less
than 1 kilogram

Color each object that is *more than* 1 kilogram.

Chapter 17 • Lesson 139   two hundred fifty-seven   **257**

Chapter 14 Review

Write the total value.

Subtraction Fact Review

Subtract.

8 − 4 = ☐   7 − 5 = ☐   10 − 8 = ☐

7 − 3 = ☐   8 − 3 = ☐   11 − 7 = ☐

9 − 8 = ☐   10 − 6 = ☐   9 − 7 = ☐

10 − 2 = ☐   9 − 6 = ☐   10 − 4 = ☐

Read a Celsius Thermometer

Write each temperature.

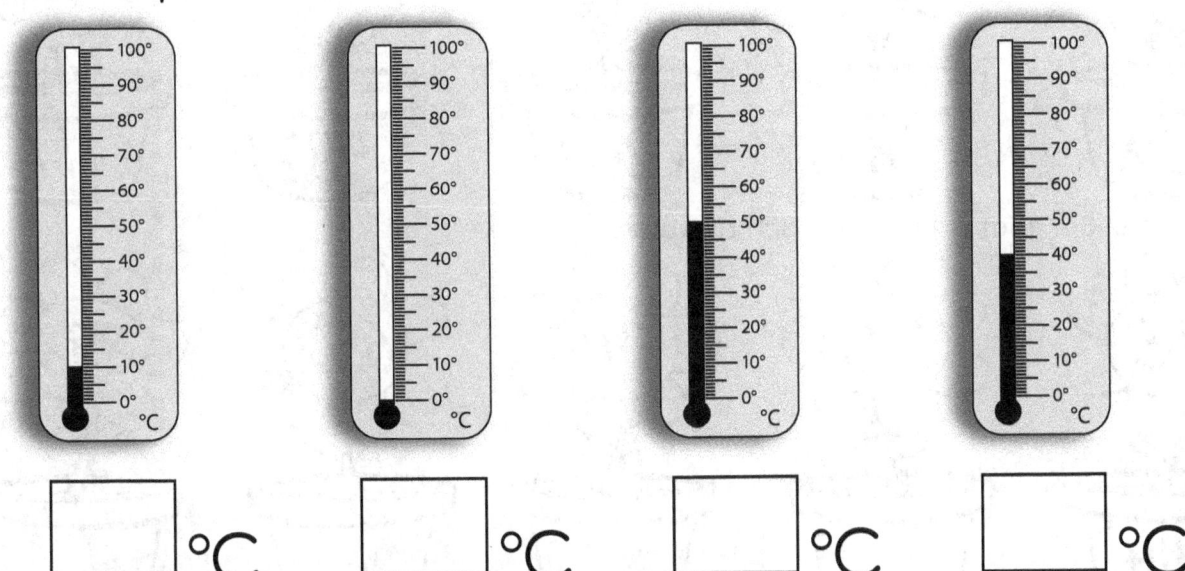

☐ °C   ☐ °C   ☐ °C   ☐ °C

Draw a line from each thermometer to the correct picture.

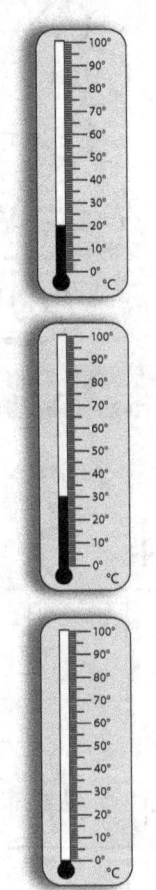

Chapter 17 • Lesson 140

Chapter 16 Review

Write the number that is *1 more* on each flower.

- 329 / 328
- ___ / 683
- ___ / 269
- ___ / 432
- ___ / 117

Write the number that is *1 less* on each flowerpot.

- 415 / 414
- 273 / ___
- 446 / ___
- 652 / ___
- 898 / ___

Write the number that is *10 more* on each flower.

- 340 / 330
- ___ / 626
- ___ / 431
- ___ / 748
- ___ / 550

Addition Fact Review

Add.

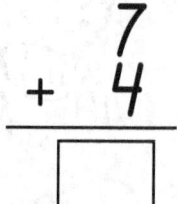

$\begin{array}{r}8\\+\;2\\\hline\square\end{array}$ $\begin{array}{r}7\\+\;4\\\hline\square\end{array}$ $\begin{array}{r}3\\+\;3\\\hline\square\end{array}$ $\begin{array}{r}5\\+\;2\\\hline\square\end{array}$ $\begin{array}{r}9\\+\;0\\\hline\square\end{array}$

$\begin{array}{r}3\\+\;6\\\hline\square\end{array}$ $\begin{array}{r}9\\+\;2\\\hline\square\end{array}$ $\begin{array}{r}6\\+\;1\\\hline\square\end{array}$ $\begin{array}{r}4\\+\;3\\\hline\square\end{array}$ $\begin{array}{r}8\\+\;4\\\hline\square\end{array}$

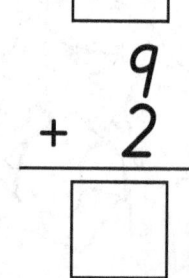

Measure Perimeter

Use a centimeter ruler to find each perimeter.
Mark an *X* on each side after you measure it.

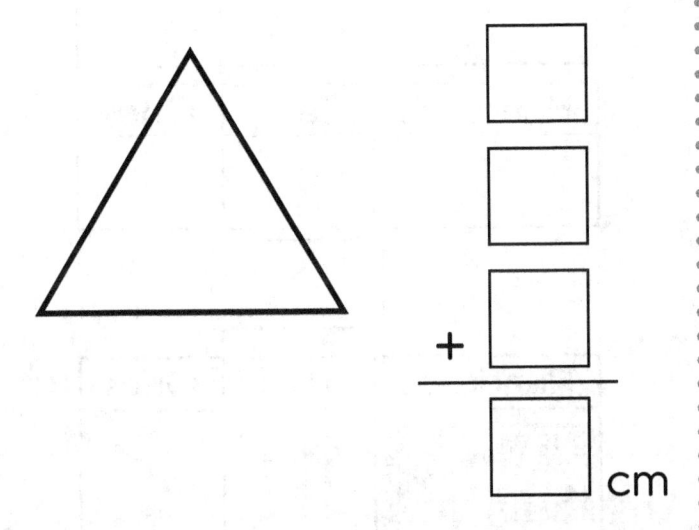

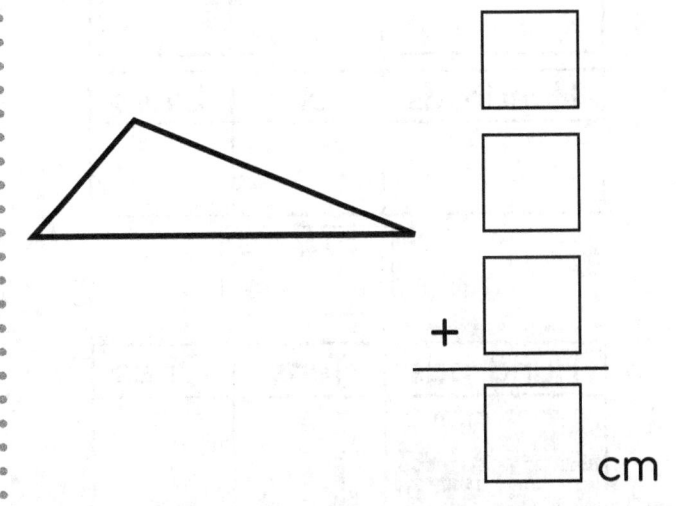

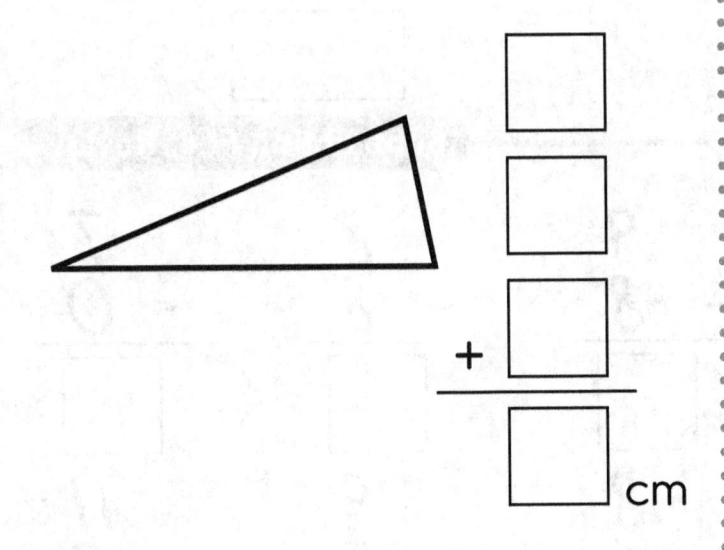

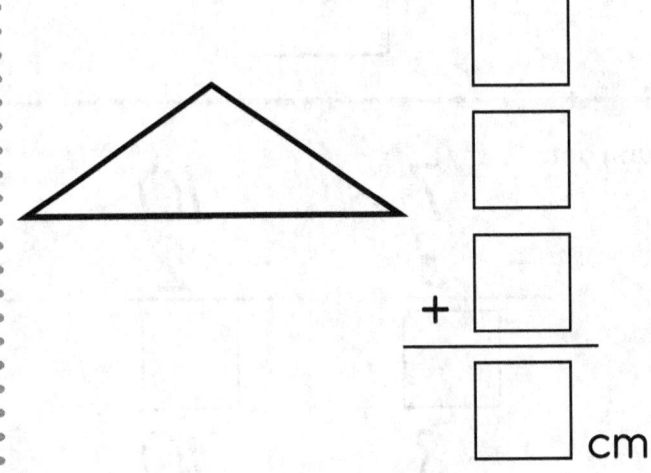

Chapter 17 • Lesson 141

Chapter 16 Review

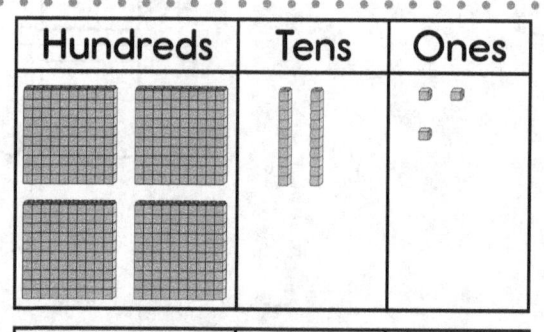

Write the hundreds, tens, and ones.
Write the number.

Hundreds	Tens	Ones

Hundreds	Tens	Ones

Hundreds	Tens	Ones

Hundreds	Tens	Ones

Hundreds	Tens	Ones

Hundreds	Tens	Ones

Subtraction Fact Review

Subtract.

$7 - 3 =$ $10 - 2 =$ $9 - 8 =$ $8 - 8 =$ $7 - 0 =$

$8 - 5 =$ $10 - 9 =$ $8 - 6 =$ $9 - 5 =$ $11 - 9 =$

262 two hundred sixty-two

Math 1 Reviews

Chapter 17 Review

Mark an *X* on the shorter object.
Circle the longer object.

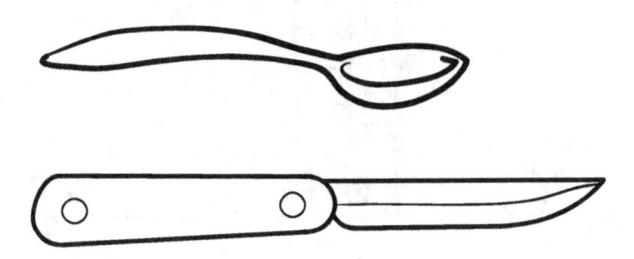

Mark an *X* on the shorter object.
Circle the taller object.

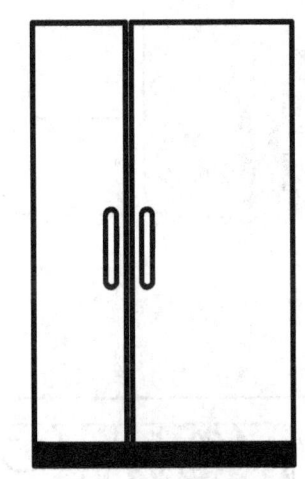

Use a centimeter ruler to measure.

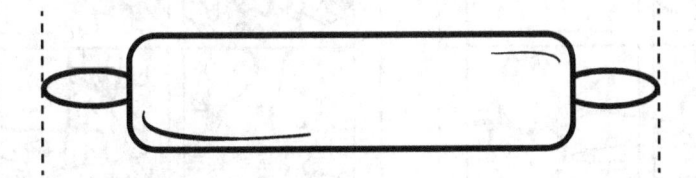

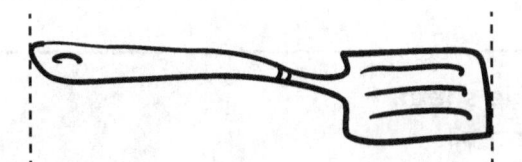

☐ cm

☐ cm

Circle the container that holds *less than* 1 liter.

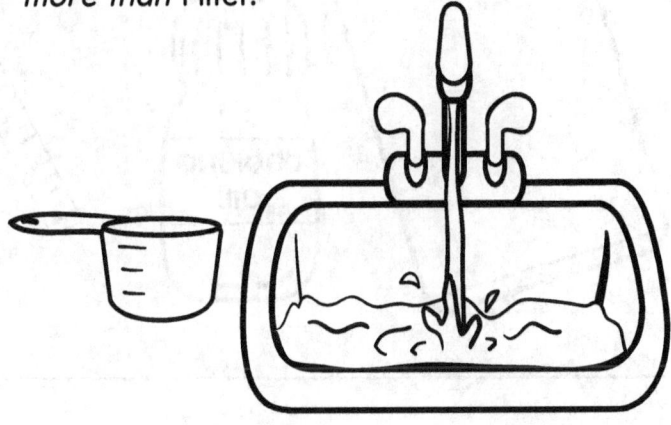

Circle the container that holds *more than* 1 liter.

Chapter 17 Review   two hundred sixty-three   **263**

Write each temperature.
Draw a line from each thermometer to the correct picture.

Circle the tool used to find the length of the cookie sheet.
Mark an *X* on the tool used to measure the cooking oil.

264 two hundred sixty-four

Math 1 Reviews

Cumulative Review

Mark the correct shape.

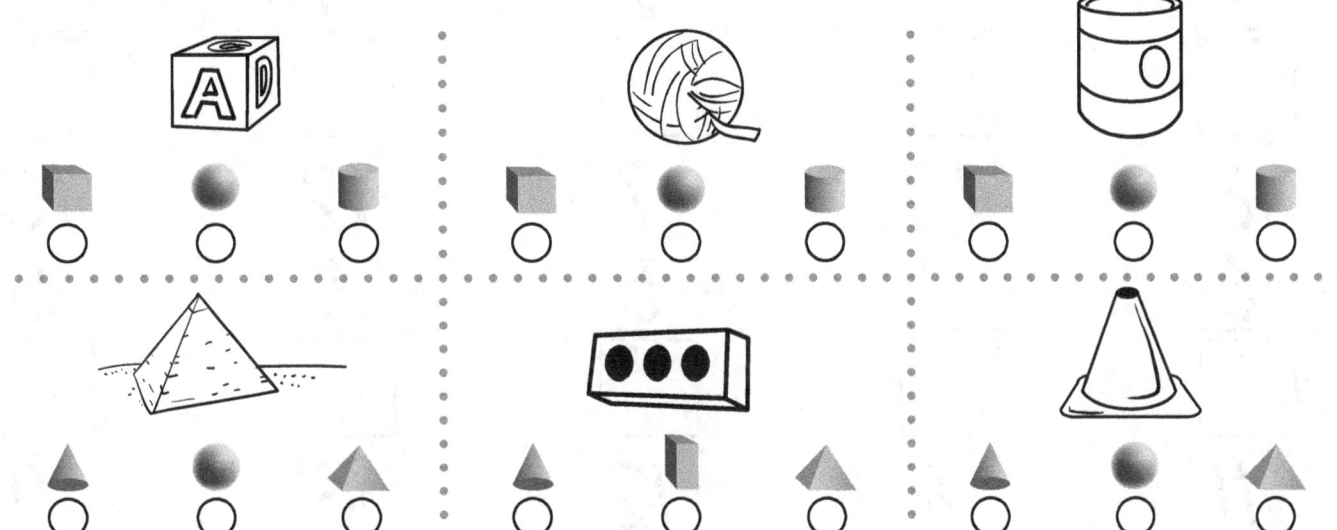

Does each solid figure have flat sides, curved sides, or both? Mark the correct answer.

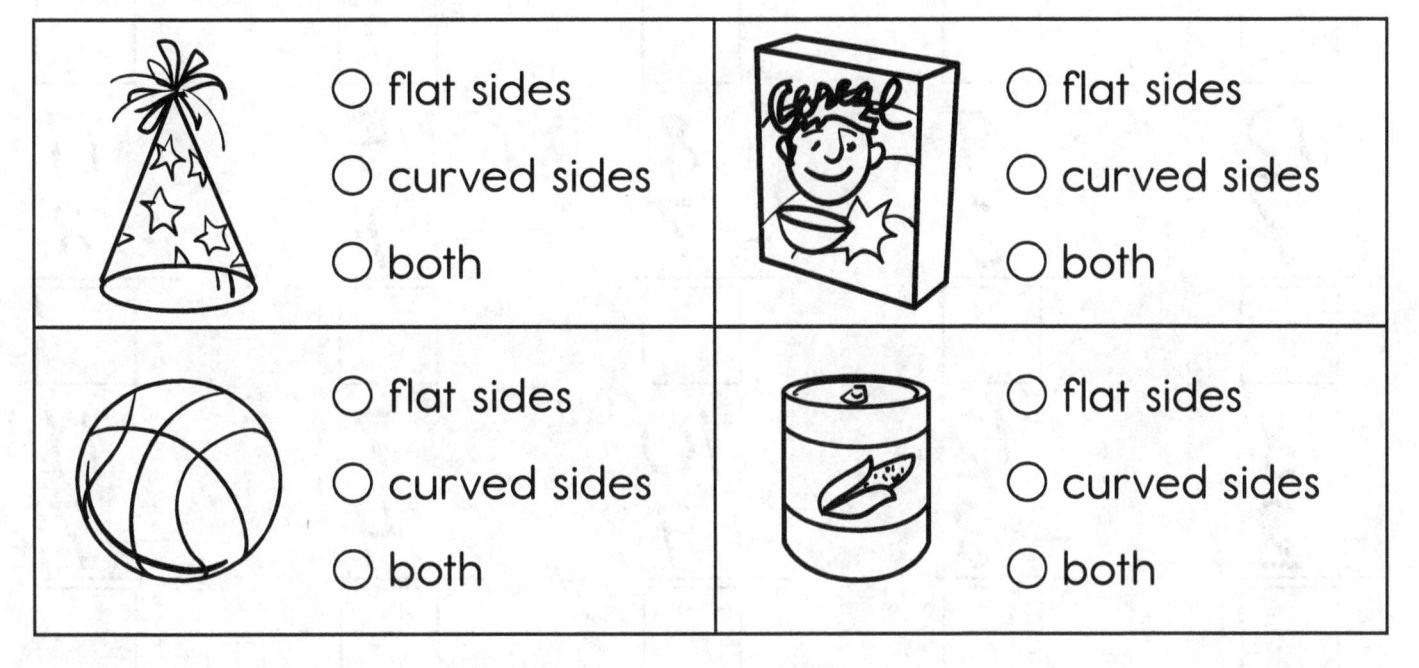

○ flat sides
○ curved sides
○ both

○ flat sides
○ curved sides
○ both

○ flat sides
○ curved sides
○ both

○ flat sides
○ curved sides
○ both

Write the number of sides and corners.

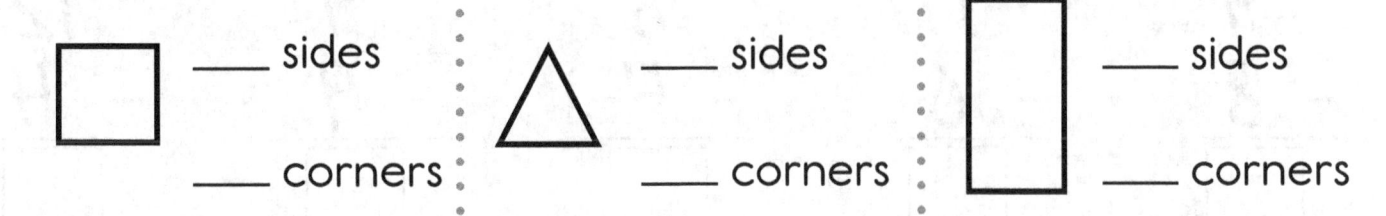

___ sides

___ corners

___ sides

___ corners

___ sides

___ corners

Chapter 17 Cumulative Review   two hundred sixty-five **265**

Subtraction Fact Review

Subtract.

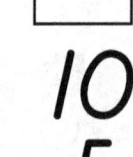

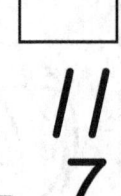

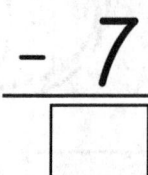

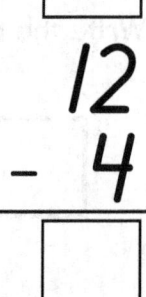

7 − 7 = ☐	12 − 9 = ☐	10 − 2 = ☐	6 − 4 = ☐	7 − 2 = ☐
8 − 6 = ☐	10 − 1 = ☐	7 − 5 = ☐	12 − 8 = ☐	10 − 5 = ☐
11 − 4 = ☐	8 − 3 = ☐	12 − 6 = ☐	6 − 5 = ☐	9 − 7 = ☐
12 − 7 = ☐	9 − 3 = ☐	8 − 7 = ☐	5 − 1 = ☐	11 − 8 = ☐
8 − 2 = ☐	12 − 3 = ☐	10 − 9 = ☐	7 − 4 = ☐	11 − 7 = ☐
9 − 8 = ☐	11 − 6 = ☐	8 − 4 = ☐	9 − 9 = ☐	12 − 4 = ☐

Add Doubles

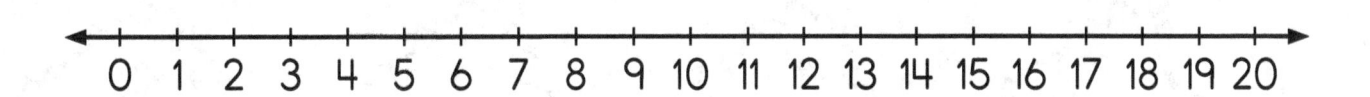

0 1 2 3 4 5 6 7 8 9 10 11 12 13 14 15 16 17 18 19 20

Add. Use the number line if needed.

8 + 8 = ☐
9 + 9 = ☐
7 + 7 = ☐

```
   9        10         8         7
+  9      + 10       + 8       + 7
 ___       ___        ___       ___
 ☐         ☐          ☐         ☐
```

Complete each missing addend equation.

9 + ☐ = 18   ☐ + 7 = 14

10 + ☐ = 20   ☐ + 8 = 16

Chapter 18 • Lesson 144

Chapter 9 Review

Read the fraction.
Color the number of objects to show the fraction.

$\frac{3}{4}$

$\frac{1}{3}$

$\frac{2}{4}$

Addition Fact Review

Add. Mark the correct circle.

7 + 5 = 10 11 12

3 + 7 = 10 11 12

8 + 2 = 10 11 12

4 + 7 = 10 11 12

7 + 7 = 12 13 14

3 + 8 = 10 11 12

5 + 6 = 10 11 12

9 + 3 = 10 11 12

6 + 6 = 10 11 12

8 + 4 = 10 11 12

Add Near Doubles

Think $3 + 3 = 6$.
$6 + 1$ ... $3 + 4 = 7$

Near Doubles Strategy
Think of the double fact for the smaller addend. Add 1 to the double fact sum.

Add.

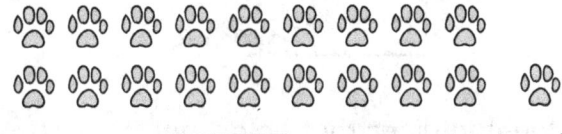

$8 + 8 = \square$ $8 + 9 = \square$ $9 + 9 = \square$ $9 + 10 = \square$

$\begin{array}{r} 7 \\ + 7 \\ \hline \square \end{array}$ $\begin{array}{r} 7 \\ + 8 \\ \hline \square \end{array}$ $\begin{array}{r} 5 \\ + 5 \\ \hline \square \end{array}$ $\begin{array}{r} 5 \\ + 6 \\ \hline \square \end{array}$

Write an equation for each word problem. Draw pictures to show each equation. Solve.

Mason counted 8 black dogs at the pet store. He counted 9 white dogs. How many dogs did he count in all?

There are 5 dogs sleeping in the pen. There are 6 dogs eating. How many dogs are there in all?

☐ ○ ☐ ○ ☐ dogs   ☐ ○ ☐ ○ ☐ dogs

Chapter 18 • Lesson 145

Chapter 11 Review

Use an inch ruler to measure.

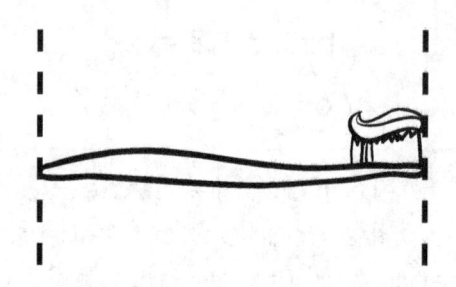

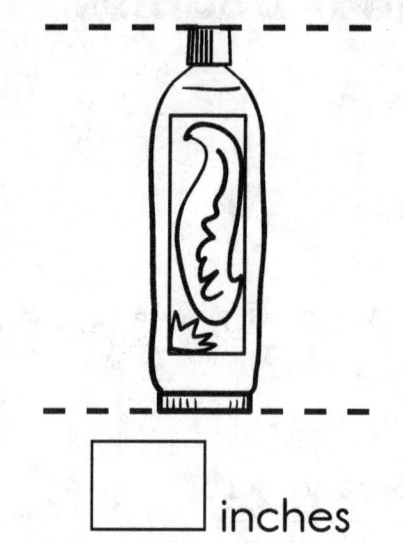

☐ inches   ☐ inches

Circle the container that holds *more*.

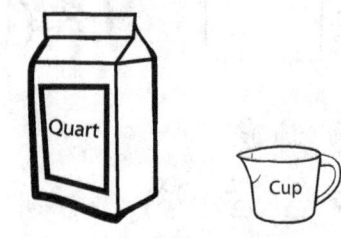

Circle the correct measuring tool.

How long is my floss?

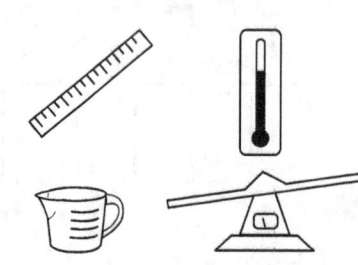

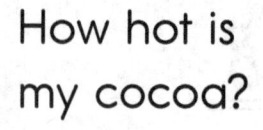

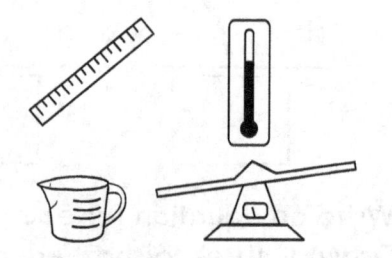

How hot is my cocoa?

Subtraction Fact Review

Subtract. Mark the correct circle.

9 − 4 =   3 4 5
           ○ ○ ○

11 − 6 = 3 4 5
           ○ ○ ○

3 − 1 =   1 2 3
           ○ ○ ○

11 − 8 = 3 4 5
           ○ ○ ○

6 − 3 =   3 4 5
           ○ ○ ○

12 − 6 = 6 7 8
           ○ ○ ○

Add 9 or 10

Color each addend. Add.
Write the answer.

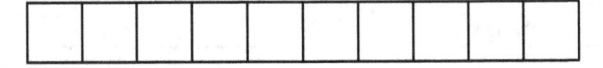

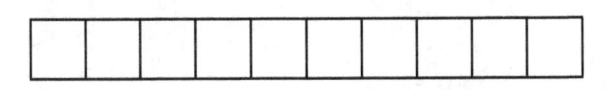

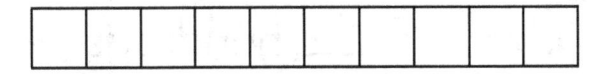

10 + 5 = ☐

10 + 9 = ☐

10 + 3 = ☐

10 + 4 = ☐

Color each addend. Complete the Ten Bar with the second addend. Add.

9 + 6 = ☐

9 + 8 = ☐

9 + 4 = ☐

9 + 7 = ☐

Add.

10   9   10   10   10   9
+ 7   + 2   + 8   + 2   + 6   + 3
☐   ☐   ☐   ☐   ☐   ☐

Chapter 16 Review

Write the number of hundreds, tens, and ones.

278 ____ hundreds   ____ tens   ____ ones

461 ____ hundreds   ____ tens   ____ ones

589 ____ hundreds   ____ tens   ____ ones

Match the clues to the correct number.

7 ones, 4 tens, 3 hundreds   864

6 tens, 8 hundreds, 4 ones   728

7 hundreds, 8 ones, 2 tens   347

Addition Fact Review

Add.

$5 + 7 =$ ☐

$3 + 8 =$ ☐

$6 + 5 =$ ☐

$6 + 4 =$ ☐

$3 + 2 =$ ☐

$2 + 7 =$ ☐

$9 + 1 =$ ☐

$5 + 4 =$ ☐

$3 + 9 =$ ☐

$8 + 4 =$ ☐

Add 6, 7, 8

Color each addend. Complete the Ten Bar with the second addend. Add.

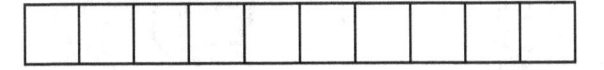

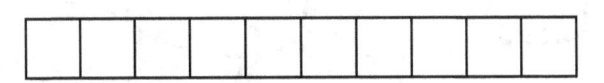

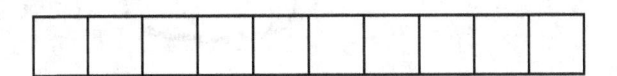

8 + 3 = ☐

7 + 5 = ☐

6 + 8 = ☐

8 + 4 = ☐

7 + 4 = ☐

8 + 5 = ☐

Write an equation for the word problem. Solve.

Ava caught 8 bugs.
She put them in a jar.
She caught 6 more bugs.
She put them in a different jar.
How many bugs did Ava catch?

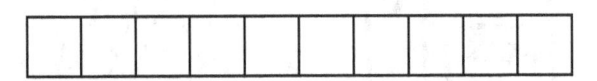

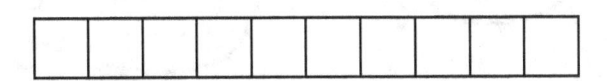

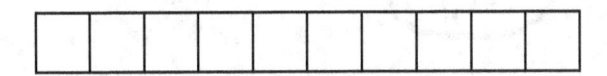

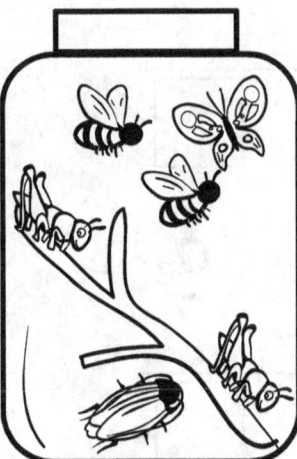

☐ ◯ ☐ ◯ ☐ bugs

Chapter 17 Review

Circle each container that holds *more than* 1 liter.
Mark an *X* on each container that holds *less than* 1 liter.

Subtraction Fact Review

Subtract.

$$\begin{array}{r} 9 \\ -3 \\ \hline \end{array} \qquad \begin{array}{r} 12 \\ -6 \\ \hline \end{array} \qquad \begin{array}{r} 11 \\ -4 \\ \hline \end{array} \qquad \begin{array}{r} 8 \\ -6 \\ \hline \end{array} \qquad \begin{array}{r} 10 \\ -9 \\ \hline \end{array}$$

$$\begin{array}{r} 8 \\ -7 \\ \hline \end{array} \qquad \begin{array}{r} 9 \\ -4 \\ \hline \end{array} \qquad \begin{array}{r} 7 \\ -6 \\ \hline \end{array} \qquad \begin{array}{r} 12 \\ -8 \\ \hline \end{array} \qquad \begin{array}{r} 9 \\ -7 \\ \hline \end{array}$$

274   two hundred seventy-four   Math 1 Reviews

3 Addends

There are 4 children praying.

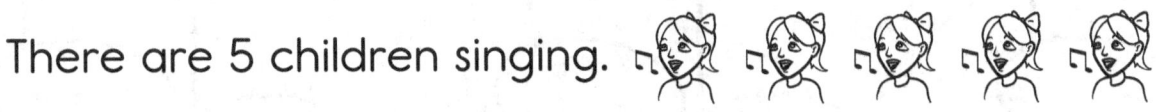

There are 3 children reading.

There are 5 children singing.

4 + 3 + 5 = ☐ children

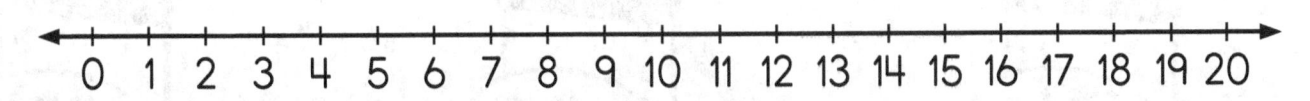

Use the number line to add.
Write the sum.
Use the code to find the answer.

10	11	12	13	14	15	16	17
d	f	g	i	l	o	r	y

We should

```
  5      4      9      4      4      5      9
  4      8      3      4      2      4      1
+ 3    + 2    + 3    + 8    + 7    + 2    + 7
```

God.
1 Corinthians 6:20

Chapter 18 • Lesson 148

Chapter 16 Review

Write the number that comes *after*.

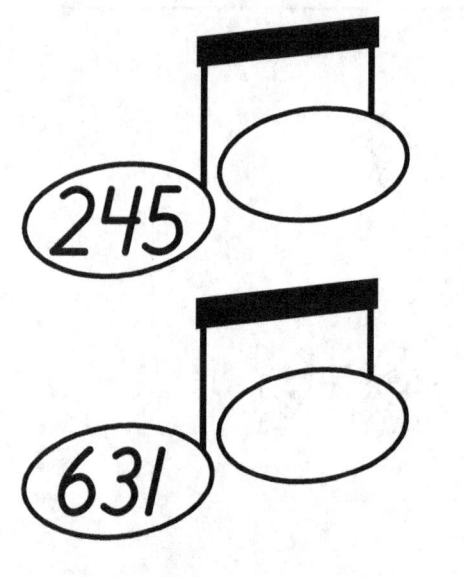

245 473 752
631 302 899

Write the number that comes *before*.

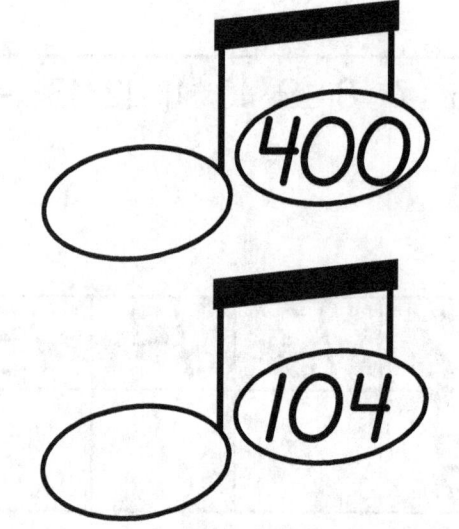

346 400 571
988 104 655

Addition Fact Review

Mark the correct circle.

8 + 1 =   9 10 11
7 + 4 =   9 10 11
9 + 1 =   9 10 11
6 + 5 =   9 10 11

5 + 5 =   8 9 10
8 + 3 =   9 10 11
7 + 3 =   9 10 11
6 + 2 =   8 9 10

276   two hundred seventy-six

Math 1 Reviews

Double Fact Families

Write an addition and subtraction equation for each dot pattern.

☐ + ☐ = 20
20 − ☐ = ☐

☐ + ☐ = 18
18 − ☐ = ☐

☐ + ☐ = ☐
☐ − ☐ = ☐

☐ + ☐ = ☐
☐ − ☐ = ☐

Add or subtract.
Write the numbers for each fact family.

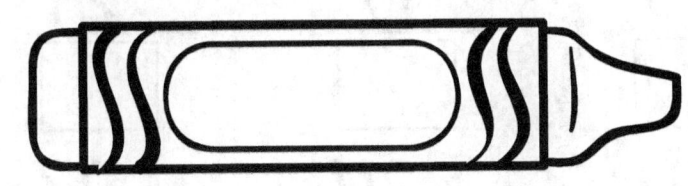

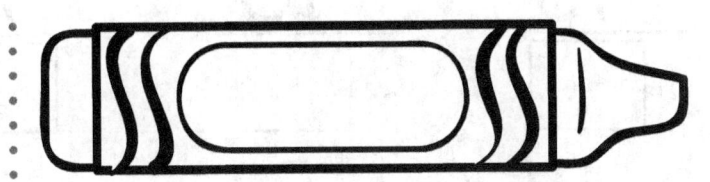

7 + 7 = ☐
14 − 7 = ☐

8 + 8 = ☐
16 − 8 = ☐

Write an addition and subtraction equation for each double fact family.

10 10 20

☐ + ☐ = ☐
☐ − ☐ = ☐

9 9 18

☐ + ☐ = ☐
☐ − ☐ = ☐

Chapter 18 • Lesson 149

Chapter 17 Review

Draw a line to match each thermometer with the correct picture.

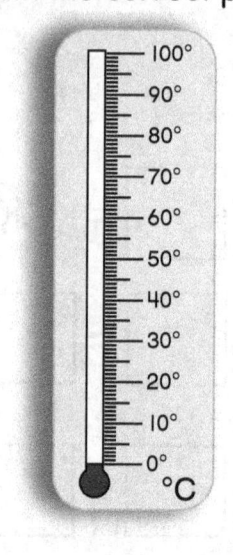

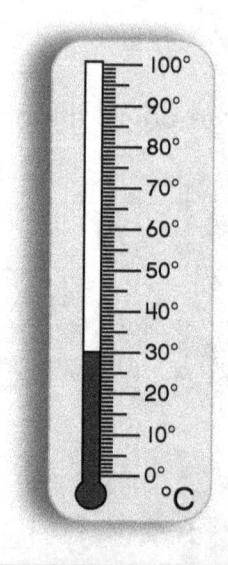

Chapter 12 Review

Add.

```
  32        25        43        81        74
+ 15      + 10      + 16      + 17      + 21
____      ____      ____      ____      ____


  64        53        21        34        27
+ 33      + 40      + 43      + 14      + 62
____      ____      ____      ____      ____
```

Subtraction Fact Review

Mark the correct circle.

10 − 1 = ○ 9 ○ 10 ○ 11

8 − 6 = ○ 1 ○ 2 ○ 3

11 − 4 = ○ 7 ○ 8 ○ 9

12 − 6 = ○ 4 ○ 5 ○ 6

9 − 7 = ○ 1 ○ 2 ○ 3

11 − 8 = ○ 1 ○ 2 ○ 3

278 two hundred seventy-eight

Math 1 Reviews

Fact Families for 13, 14; Missing Addend

Add or subtract.
Write the number for each fact family.

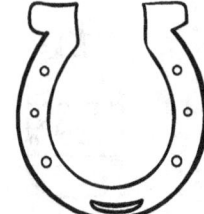

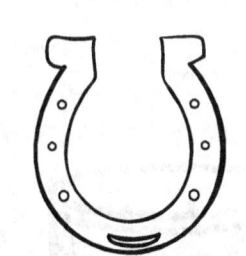

6 + 8 =
8 + 6 =
14 − 6 =
14 − 8 =

5 + 9 =
9 + 5 =
14 − 5 =
14 − 9 =

Write each fact family.

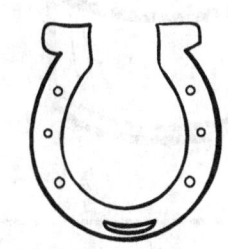

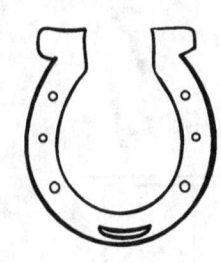

| 6 | 7 | 13 |

☐ + ☐ = ☐
☐ + ☐ = ☐
☐ − ☐ = ☐
☐ − ☐ = ☐

| 5 | 8 | 13 |

☐ + ☐ = ☐
☐ + ☐ = ☐
☐ − ☐ = ☐
☐ − ☐ = ☐

Complete the equation for the word problem.

The cowboy needs to catch 14 horses. He has caught 7 horses. How many more horses does the cowboy need to catch?

7 + ☐ = 14
horses

Chapter 18 • Lesson 150

Chapter 7 Review

Circle each container that holds *more than* 1 liter.

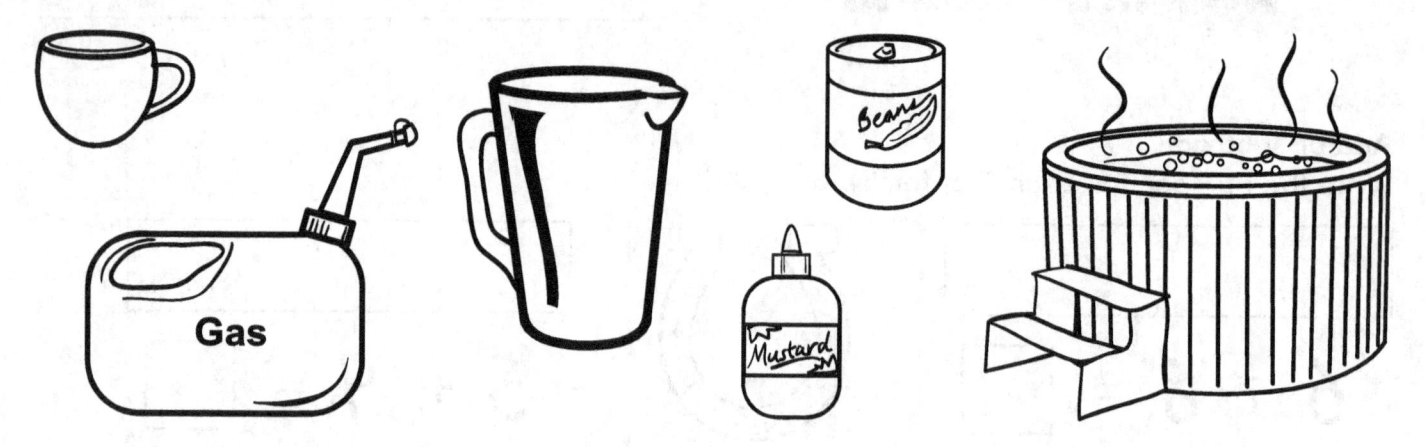

Mark an *X* on each container that holds *less than* 1 liter.

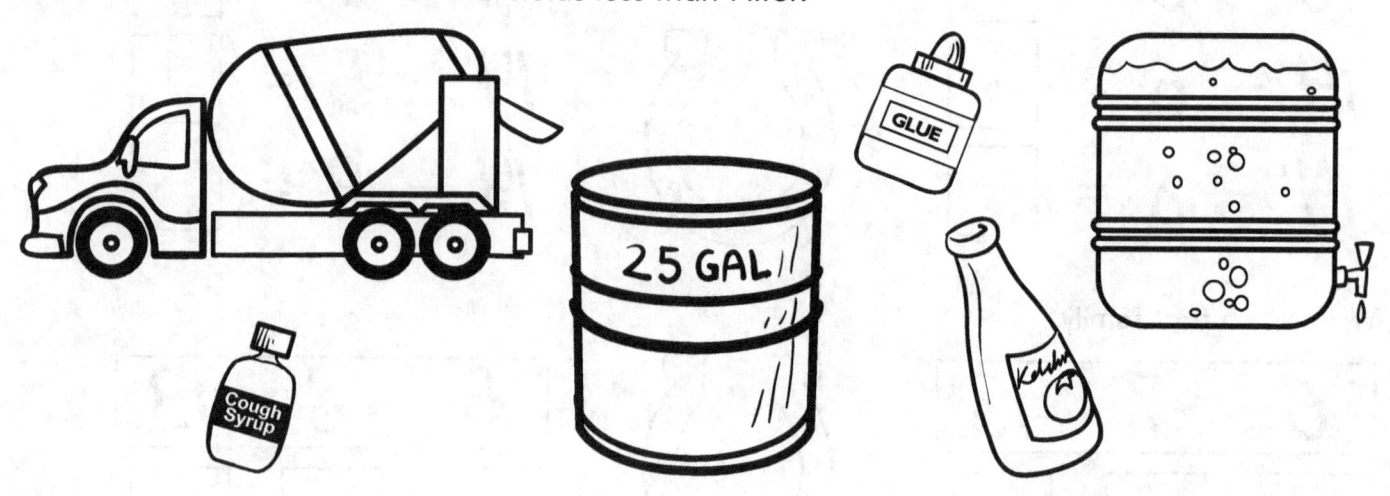

Addition Fact Review

Add.

8	9	7	6	4
+ 3	+ 0	+ 4	+ 6	+ 5
☐	☐	☐	☐	☐

7	5	0	3	2
+ 3	+ 2	+ 6	+ 6	+ 8
☐	☐	☐	☐	☐

Fact Families for 15, 16

_____
- - - - - - - - - - -
_____

Add or subtract.
Write the numbers for the fact family.

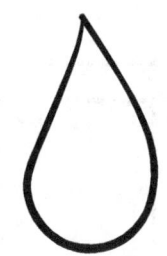

8 + 8 = ☐
16 − 8 = ☐

Write the fact family.

6   9   15

☐ + ☐ = ☐
☐ + ☐ = ☐
☐ − ☐ = ☐
☐ − ☐ = ☐

Add or subtract. Use the key to color.

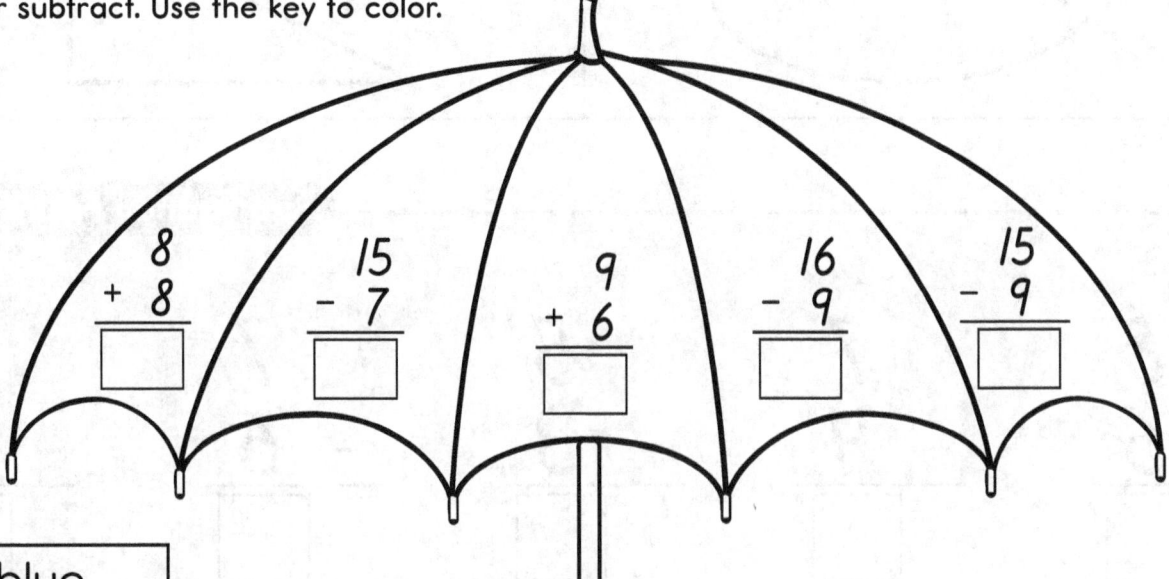

  8   15   9   16   15
+ 8   − 7   + 6   − 9   − 9
____   ____   ____ ____   ____

6	blue
7	red
8	yellow
15	orange
16	purple

Chapter 18 • Lesson 151   two hundred eighty-one   **281**

Color each shape.

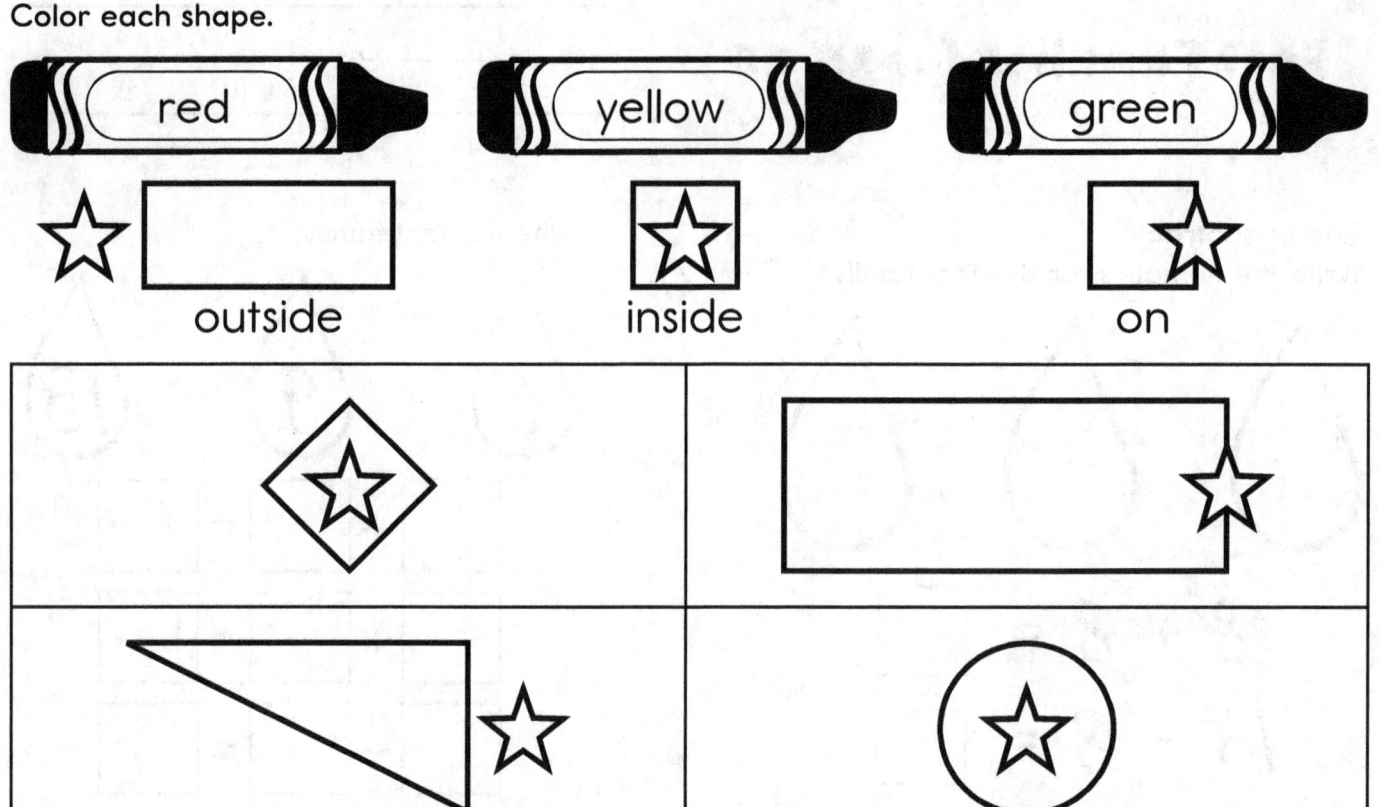

red — ☆ ▭ outside

yellow — ★ inside

green — ☆ on

Subtraction Fact Review

Subtract.

8	7	5	7	6
− 6	− 4	− 3	− 6	− 3
☐	☐	☐	☐	☐

9	10	7	12	8
− 7	− 8	− 5	− 3	− 5
☐	☐	☐	☐	☐

282 two hundred eighty-two · Math 1 Reviews

Fact Families for 17, 18; Multistep Problems

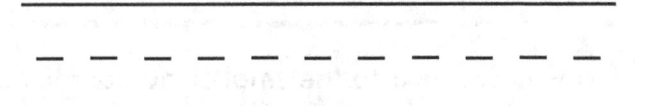

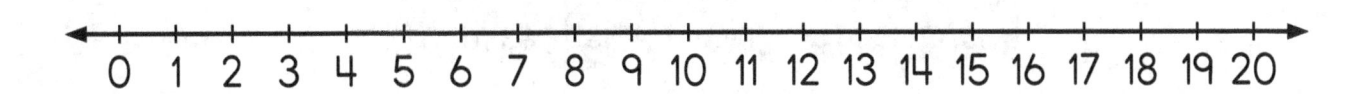

```
←——————————————————————————→
 0  1  2  3  4  5  6  7  8  9  10 11 12 13 14 15 16 17 18 19 20
```

Add or subtract. Use the number line if needed.

8 + 9 = ☐   9 + 9 = ☐   17 − 8 = ☐

18 − 9 = ☐   17 − 9 = ☐   9 + 8 = ☐

Solve the multistep problem.

Shelby stopped to feed 1 duck.
Then 3 more ducks came to be fed.
After eating, 2 ducks waddled off.
How many ducks stayed with Shelby?

Add.

☐ duck
○ ☐ ducks came
_____
☐ ducks in all

Subtract.

☐ ducks in all
○ ☐ ducks waddled off
_____
☐ ducks stayed

Chapter 18 • Lesson 152

Chapter 16 Review

Draw a dot next to the smaller number. Draw the correct sign.

> is greater than   < is less than

254	361
849	633
300	200

596	782
415	420
721	785

Subtraction Fact Review

Mark the correct circle.

8 + 4 = ○ 10 ○ 11 ○ 12

7 + 3 = ○ 10 ○ 11 ○ 12

6 + 2 = ○ 8 ○ 9 ○ 10

9 + 1 = ○ 8 ○ 9 ○ 10

5 + 5 = ○ 10 ○ 11 ○ 12

6 + 5 = ○ 9 ○ 10 ○ 11

7 + 2 = ○ 9 ○ 10 ○ 11

9 + 3 = ○ 10 ○ 11 ○ 12

6 + 3 = ○ 9 ○ 10 ○ 11

5 + 7 = ○ 10 ○ 11 ○ 12

Chapter 18 Review

Color each addend. Add. Write the answer.

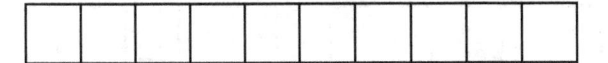

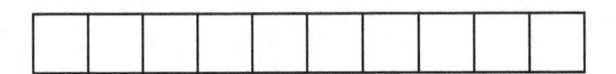

10 + 6 = ☐

7 + 4 = ☐

9 + 1 = ☐

9 + 7 = ☐

Add.

8 + 8 = ☐

7 + 7 = ☐

9 + 8 = ☐

7 + 8 = ☐

9 + 9 = ☐

6 + 6 = ☐

9 + 10 = ☐

6 + 7 = ☐

Write each fact family.

8	8	16

☐ + ☐ = ☐
☐ − ☐ = ☐

5	8	13

☐ + ☐ = ☐
☐ + ☐ = ☐
☐ − ☐ = ☐
☐ − ☐ = ☐

7	8	15

☐ + ☐ = ☐
☐ + ☐ = ☐
☐ − ☐ = ☐
☐ − ☐ = ☐

Solve the multistep problem.

The team had 8 footballs.
They lost 4 footballs.
The coach found 2 footballs in a bag.
How many footballs does the team have now?

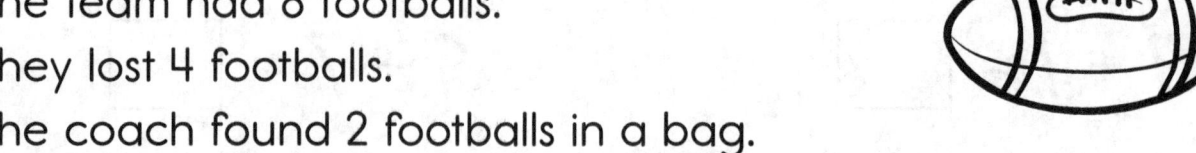

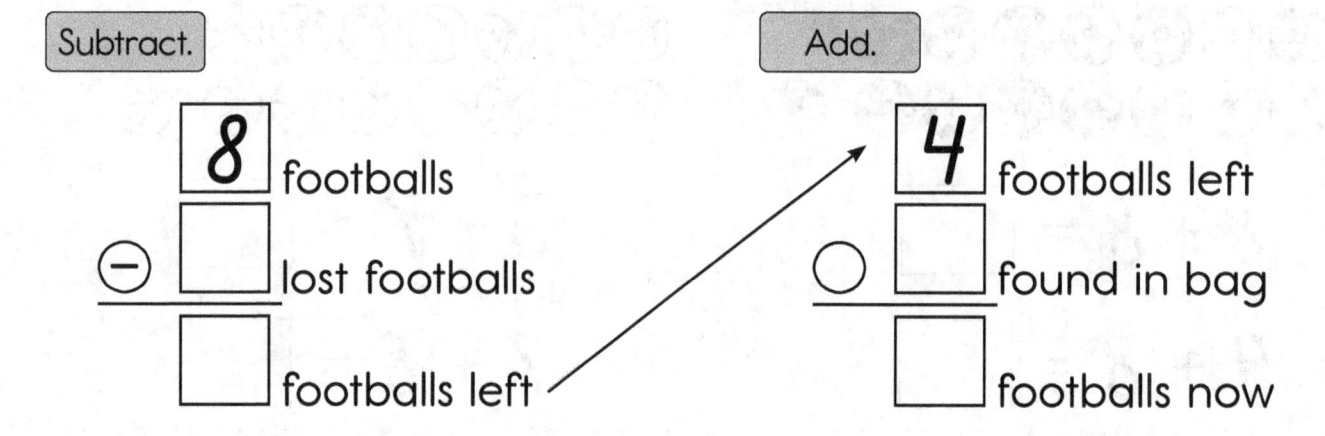

Subtract.

8 footballs
− ☐ lost footballs
☐ footballs left

Add.

4 footballs left
◯ ☐ found in bag
☐ footballs now

Add.

6
3
+ 3
☐

8
2
+ 3
☐

9
1
+ 1
☐

2
7
+ 3
☐

286 two hundred eighty-six

Math 1 Reviews

Cumulative Review

Write the total value.

_____ ¢

_____ ¢

_____ ¢

Write the value of each underlined digit.

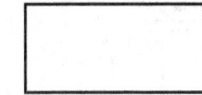

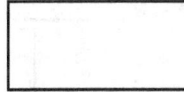

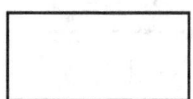

4<u>1</u>5 ☐   <u>7</u>38 ☐   19<u>0</u> ☐

Color the correct clock.

4:00

12:00

5:30

1:30

Chapter 16 Cumulative Review

Addition Fact Review

Add.

2 + 4	8 + 3	7 + 2	9 + 1	6 + 6
4 + 5	5 + 6	4 + 4	7 + 5	8 + 4
7 + 3	4 + 6	6 + 5	3 + 2	9 + 3
8 + 2	3 + 6	3 + 5	3 + 4	4 + 2
8 + 1	5 + 5	6 + 2	7 + 4	9 + 0

Counting Pennies with a Quarter

Write the value of each coin. Mark its name.

____ ¢   ○ penny
         ○ nickel

____ ¢   ○ nickel
         ○ quarter

Write the value as you *count on*. Write the total.

____ ¢ ____ ¢ ____ ¢ ____ ¢ ____ ¢

____ ¢ ____ ¢ ____ ¢ ____ ¢

____ ¢ ____ ¢ ____ ¢ ____ ¢ ____ ¢ ____ ¢

Write the total value.
Can you buy the item? Mark the correct answer.

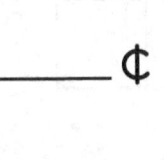

____ ¢   35¢   ○ yes
               ○ no

____ ¢   15¢   ○ yes
               ○ no

Chapter 19 • Lesson 155

Chapter 2 Review

Read the word. Circle the correct picture.

tenth

sixth

second

Addition Fact Review

Add.

5 + 4 = ☐   9 + 3 = ☐   6 + 6 = ☐

9 + 2 = ☐   4 + 6 = ☐   7 + 4 = ☐

4 + 4 = ☐   7 + 5 = ☐   2 + 8 = ☐

6 + 3 = ☐   0 + 9 = ☐   3 + 5 = ☐

Counting Nickels with a Quarter

Write the value as you *count on*.
Can you buy the item? Mark the correct answer.

<u>25</u> ¢   <u>30</u> ¢   _____ ¢   _____ ¢   _____ ¢

_____ ¢   _____ ¢   _____ ¢

_____ ¢   _____ ¢   _____ ¢   _____ ¢

- party hats 50¢ — ○ yes ○ no
- balloons 30¢ — ○ yes ○ no
- straws 40¢ — ○ yes ○ no

work space

Write an equation for each word problem. Solve.

Mark spent 49¢ on plates.
He spent 30¢ on cups.
How much money did he spend in all?

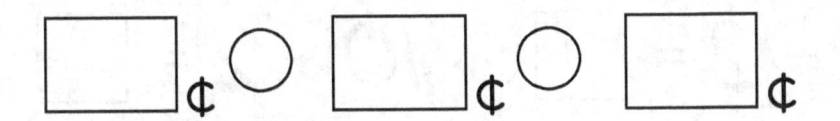

☐ ¢ ○ ☐ ¢ ○ ☐ ¢

Bryce had 69¢.
He spent 33¢ on a gift.
How much money does he have left?

☐ ¢ ○ ☐ ¢ ○ ☐ ¢

work space

Chapter 19 • Lesson 156

Chapter 8 Review

	October					
Sunday	Monday	Tuesday	Wednesday	Thursday	Friday	Saturday
				1	2	3
4	5	6	7	8	9	10
11	12	13	14	15	16	17
18	19	20	21	22	23	24
25	26	27	28	29	30	31

Mark the correct answer.

How many days are in October?   29 ○   30 ○   31 ○

How many Thursdays are in October?   3 ○   4 ○   5 ○

What day is the 21st?   Tuesday ○   Wednesday ○   Thursday ○

Subtraction Fact Review

Subtract.

12 − 6 = ☐   9 − 2 = ☐   10 − 2 = ☐

11 − 4 = ☐   10 − 6 = ☐   12 − 4 = ☐

8 − 5 = ☐   7 − 4 = ☐   9 − 5 = ☐

10 − 3 = ☐   11 − 5 = ☐   11 − 9 = ☐

Counting Dimes with a Quarter

Write the value as you *count on*. Write the total.
Draw a line from each plate to the correct item.

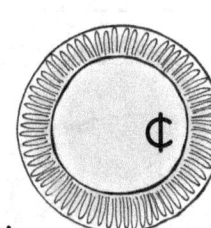

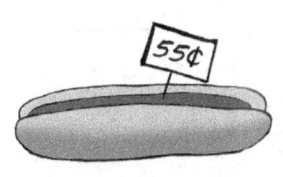

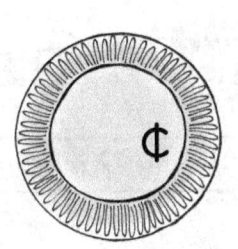

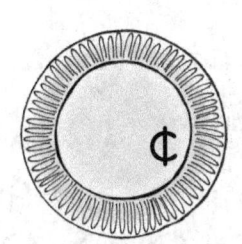

____¢ ____¢ ____¢ ____¢ ____¢

____¢ ____¢ ____¢ ____¢

____¢ ____¢ ____¢

____¢ ____¢ ____¢ ____¢

Mark an X on the coins needed to buy each item.

Chapter 19 • Lesson 157   two hundred ninety-three   293

Chapter 11 Review

Guess each length. Use an inch ruler to measure.

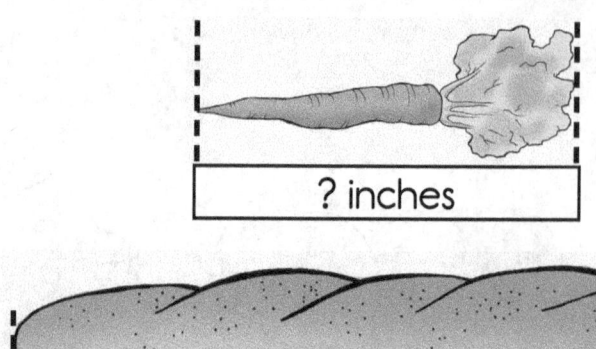

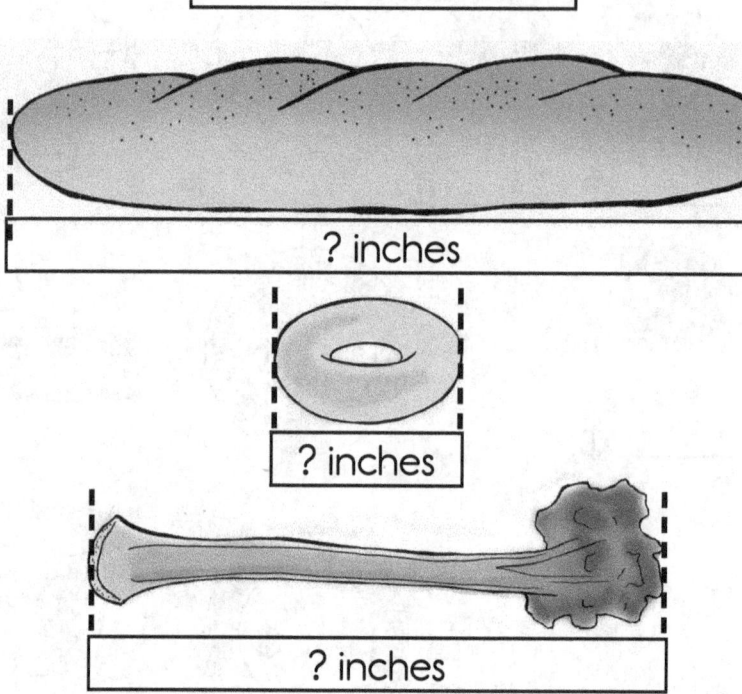

? inches

? inches

? inches

? inches

My Guess	My Measure
☐ inches	☐ inches
☐ inches	☐ inches
☐ inches	☐ inches
☐ inches	☐ inches

Addition Fact Review

Mark the correct answer.

0 + 8 = ○7 ○8 ○9

5 + 3 = ○7 ○8 ○9

3 + 6 = ○7 ○8 ○9

3 + 9 = ○10 ○11 ○12

8 + 4 = ○10 ○11 ○12

5 + 7 = ○10 ○11 ○12

2 + 8 = ○10 ○11 ○12

7 + 4 = ○10 ○11 ○12

Counting with 2 Quarters

Write the value as you *count on*.
Can you buy the item? Mark the correct answer.

____¢ ____¢ ____¢ ____¢ ____¢   car 55¢   ○ yes ○ no

____¢ ____¢ ____¢ ____¢ ____¢   bunny 70¢   ○ yes ○ no

____¢ ____¢ ____¢ ____¢   jacks 50¢   ○ yes ○ no

Mark an X on the coins needed to buy each item.

teddy bear 65¢

book 80¢

Write an equation for the word problem. Solve.

Lani had 39¢ in her purse.
Her mother gave her 3 dimes.
How much money does Lani have?

☐ ¢ ○ ☐ ¢ ○ ☐ ¢

Chapter 19 • Lesson 158   two hundred ninety-five   **295**

Chapter 10 Review

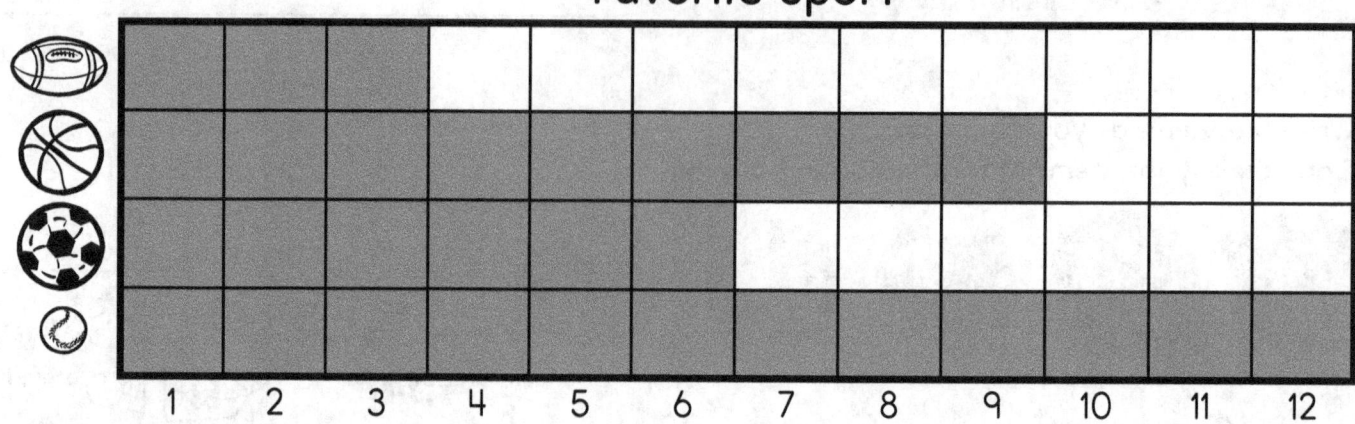

Favorite Sport

Use the graph to answer each question.

Some first graders were asked to pick the sport they liked the most.

How many children picked baseball? ☐

How many children picked football? ☐

How many more children picked baseball than football?

☐ ◯ ☐ ◯ ☐ children

How many more children picked basketball than soccer?

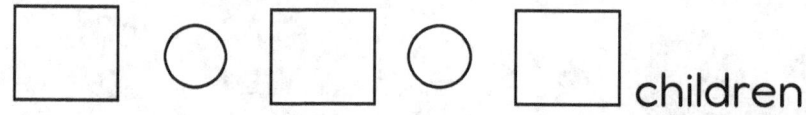

☐ ◯ ☐ ◯ ☐ children

Subtraction Fact Review

Mark the correct answer.

8 − 4 =   4 5 6
          ◯ ◯ ◯

6 − 3 =   3 4 5
          ◯ ◯ ◯

10 − 5 = 3 4 5
          ◯ ◯ ◯

12 − 6 = 4 5 6
          ◯ ◯ ◯

6 − 5 =   0 1 2
          ◯ ◯ ◯

10 − 4 = 4 5 6
          ◯ ◯ ◯

4 − 2 =   2 3 4
          ◯ ◯ ◯

11 − 5 = 4 5 6
          ◯ ◯ ◯

Equal Sets

_____
- - - - - - - - - -

Write the number of pennies in each set.
Write the total number of pennies.

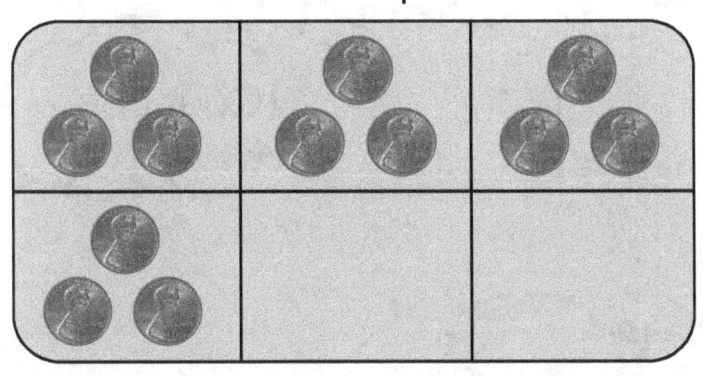

Count by 3s

4 sets of [3] pennies = [ ] ¢

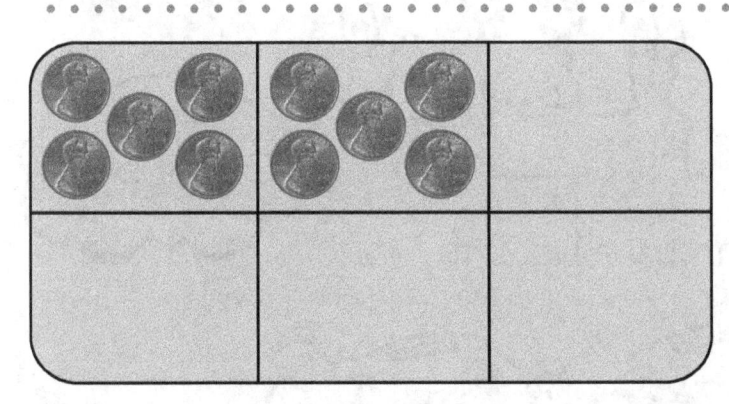

Count by 5s

2 sets of [ ] pennies = [ ] ¢

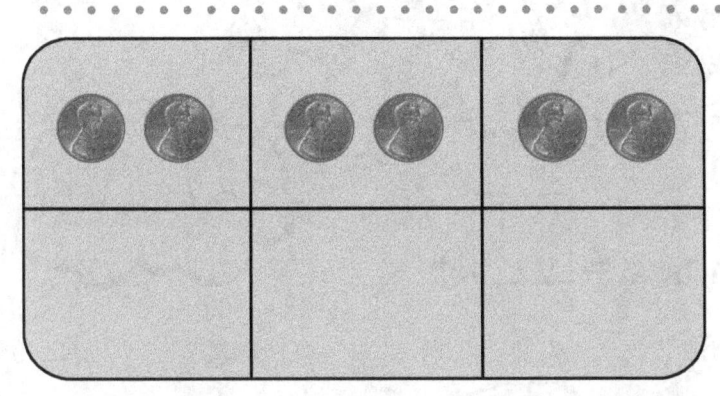

Count by 2s

3 sets of [ ] pennies = [ ] ¢

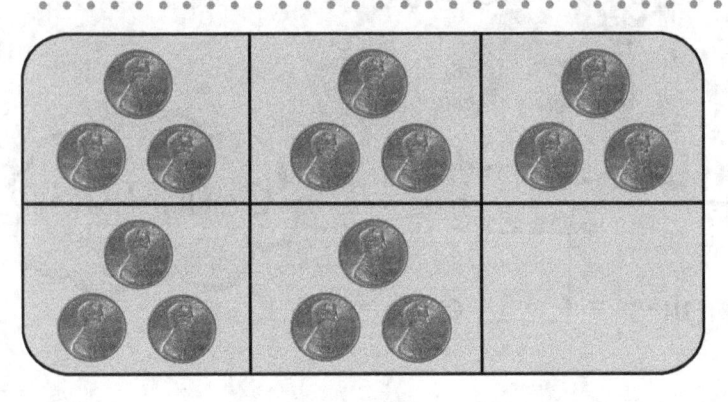

Count by 3s

5 sets of [ ] pennies = [ ] ¢

Chapter 19 • Lesson 159

Write the number of pennies in each set.
Write the total number of pennies.

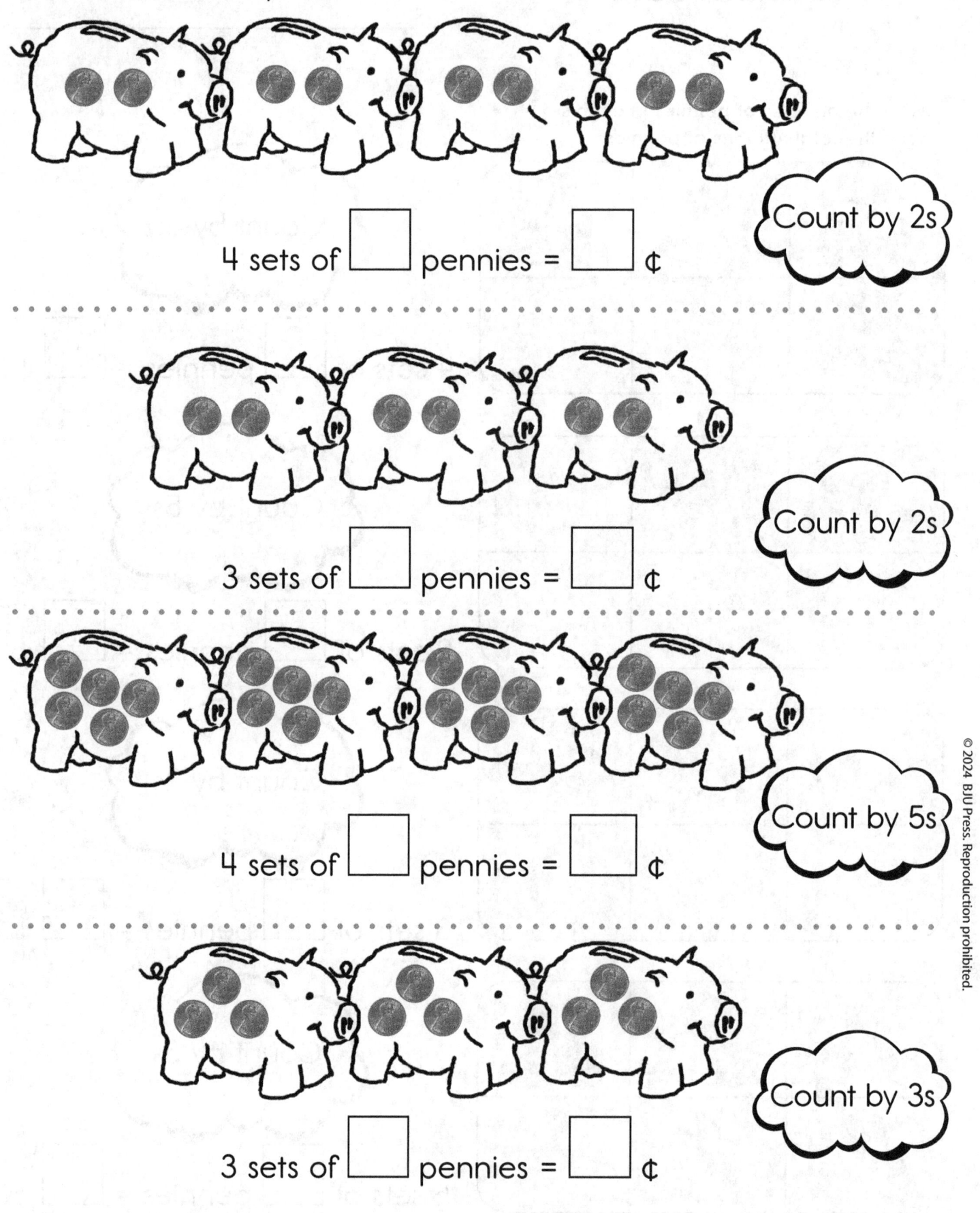

4 sets of ☐ pennies = ☐ ¢   *Count by 2s*

3 sets of ☐ pennies = ☐ ¢   *Count by 2s*

4 sets of ☐ pennies = ☐ ¢   *Count by 5s*

3 sets of ☐ pennies = ☐ ¢   *Count by 3s*

Repeated Addition

Write the number in each set. Write the total number.

How many cupcakes are there?

3 sets of [2]

2 + 2 + 2 = ☐ cupcakes

How many bags of popcorn are there?

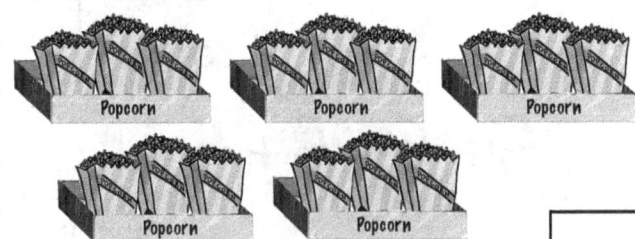

5 sets of ☐

3 + 3 + 3 + 3 + 3 = ☐
bags of popcorn

How many candy apples are there?

4 sets of ☐

3 + 3 + 3 + 3 = ☐ candy apples

How many pretzels are there?

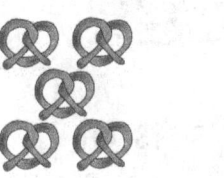

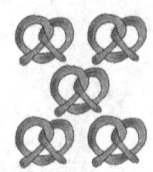

3 sets of ☐

5 + 5 + 5 = ☐ pretzels

Picture the problem. Write an equation for the word problem. Solve.

Dad bought 4 boxes of ice cream bars. There were 2 ice cream bars in each box. How many ice cream bars did he buy?

[2] + ☐ + ☐ + ☐ = ☐ ice cream bars

Chapter 19 • Lesson 160

Write the number in each set. Write the total number.

How many gumballs are there?

4 sets of ☐

5 + 5 + 5 + 5 = ☐ gumballs

How many candles are there?

3 sets of ☐

3 + 3 + 3 = ☐ candles

How many cookies are there?

5 sets of ☐

2 + 2 + 2 + 2 + 2 = ☐ cookies

How many pieces of candy are there?

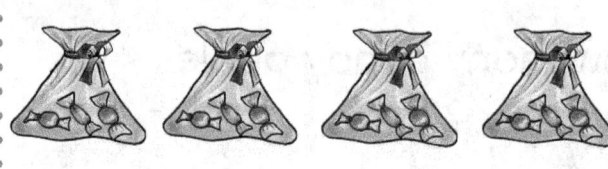

4 sets of ☐

3 + 3 + 3 + 3 = ☐ pieces of candy

Picture the problem. Write an equation for the word problem. Solve.

Uncle Blake bought 3 ice cream cones. Each cone had 2 scoops of ice cream. How many scoops did Uncle Blake buy?

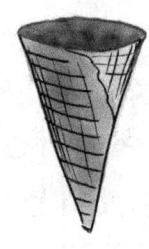

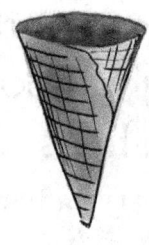

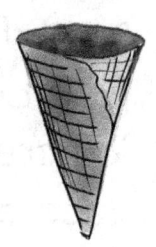

2 + ☐ + ☐ = ☐ scoops of ice cream

300 three hundred

Math 1 Reviews

Writing Repeated Addition Equations

Write the number in each set.
Complete the repeated addition equation.

How many pencils are there?

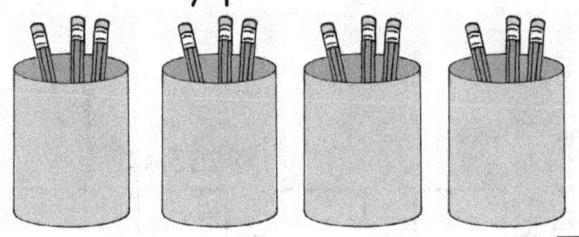

4 sets of ☐

$3 + \square + \square + \square = \square$ pencils

How many books are there?

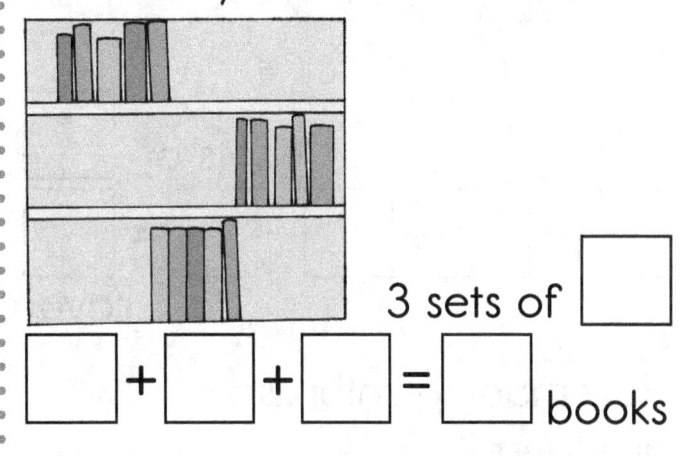

3 sets of ☐

$\square + \square + \square = \square$ books

How many paper clips are there?

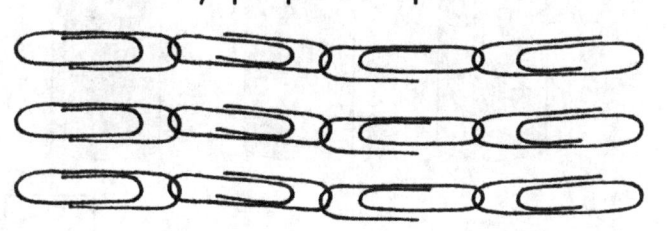

3 sets of ☐

$\square + \square + \square = \square$ paper clips

How many crayons are there?

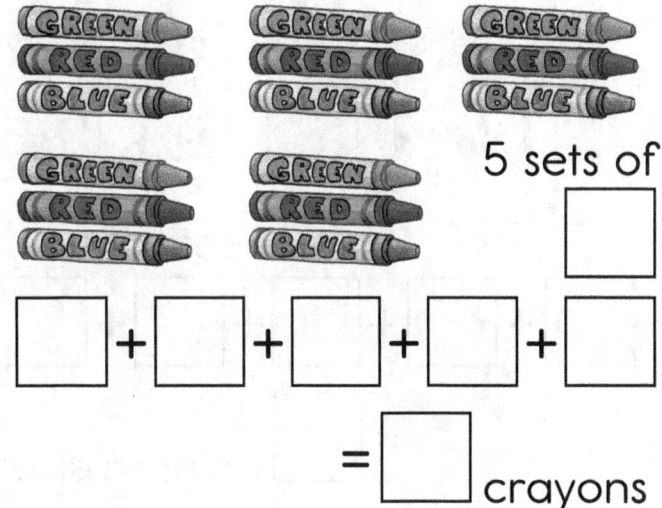

5 sets of ☐

$\square + \square + \square + \square + \square = \square$ crayons

Picture the problem. Write an equation for the word problem. Solve.

The library has 5 tables.
Each table has 2 books on it.
How many books are on the tables?

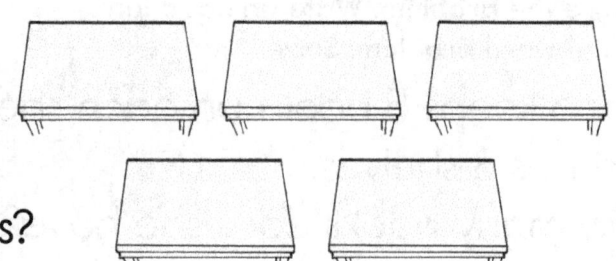

$2 + \square + \square + \square + \square = \square$ books

Chapter 19 • Lessons 161–62

Write the number in each set.
Complete the repeated addition equation.

How many crayons are there?

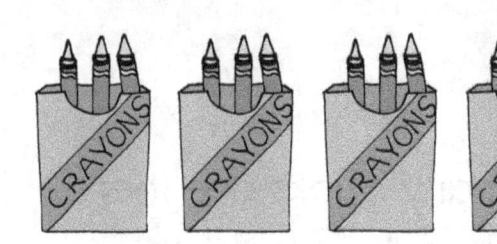

4 sets of ☐

☐ + ☐ + ☐ + ☐ = ☐ crayons

How many books are there?

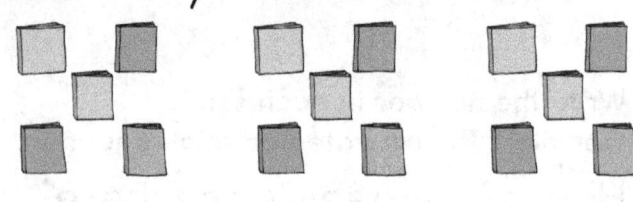

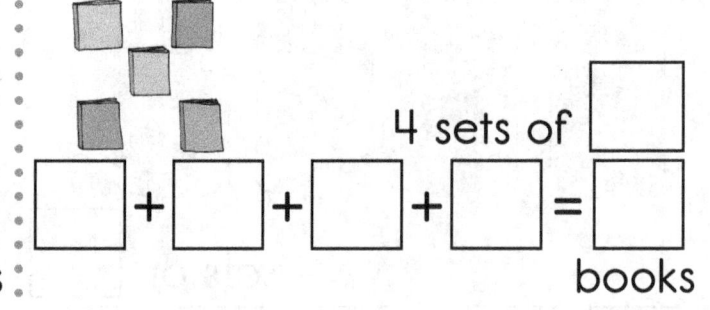

4 sets of ☐

☐ + ☐ + ☐ + ☐ = ☐ books

How many paintbrushes are there?

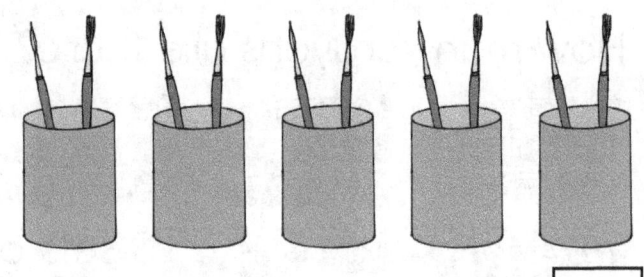

5 sets of ☐

☐ + ☐ + ☐ + ☐ + ☐ = ☐ paintbrushes

How many pens are there?

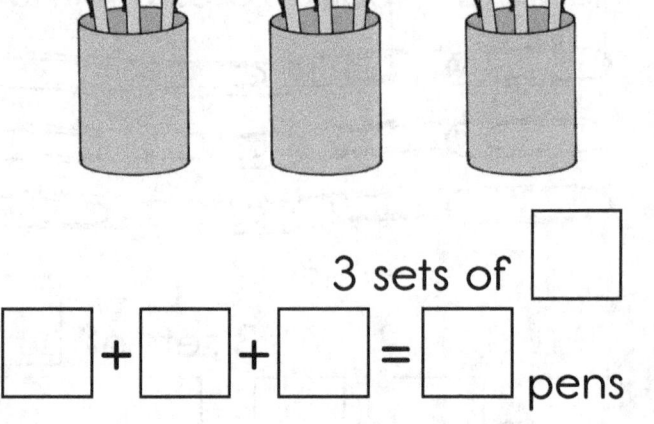

3 sets of ☐

☐ + ☐ + ☐ = ☐ pens

Picture the problem. Write an equation for the word problem. Solve.

Maria keeps 5 books on each shelf. She has 3 shelves. How many books does she have?

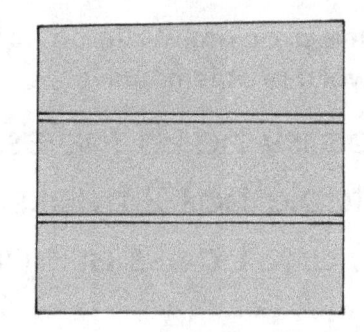

5 + ☐ + ☐ = ☐ books

Chapter 19 Review

Write the value of each coin.

_____¢   _____¢   _____¢   _____¢
quarter   nickel   penny   dime

Write the value as you *count on*.
Can you buy the item? Mark the correct answer.

_____¢ _____¢ _____¢ _____¢ _____¢   Crayons 75¢   ○ yes ○ no

_____¢ _____¢ _____¢ _____¢ _____¢   Eraser 18¢   ○ yes ○ no

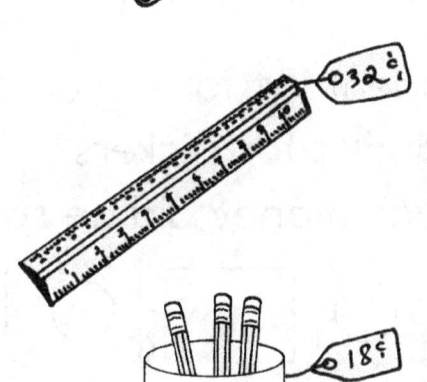

_____¢ _____¢ _____¢ _____¢   Ruler 32¢   ○ yes ○ no

_____¢ _____¢ _____¢ _____¢   Pencils 18¢ ○ yes ○ no

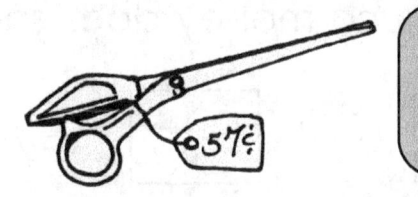

_____¢ _____¢ _____¢   Scissors 57¢ ○ yes ○ no

Chapter 19 Review   three hundred three   303

Mark an *X* on the coins needed to buy each item.

Write an equation for each word problem. Solve.

Shelby had 65¢ for a hot dog.
Her mother gave her 3 dimes.
How much money does Shelby have?

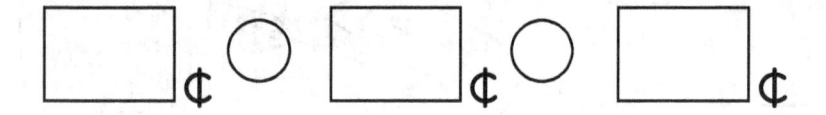

☐ ¢ ◯ ☐ ¢ ◯ ☐ ¢

work space

Daniel spent 15¢ for a glue stick.
He spent 54¢ for stickers.
How much money did he spend in all?

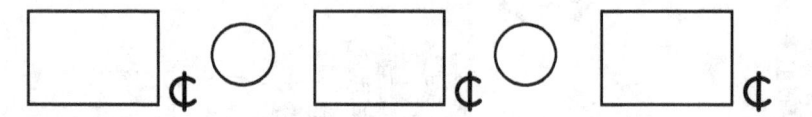

☐ ¢ ◯ ☐ ¢ ◯ ☐ ¢

work space

Tamara had 90¢ in her purse.
She gave 4 dimes to her brother.
How much money does she have left?

☐ ¢ ◯ ☐ ¢ ◯ ☐ ¢

work space

304 three hundred four   Math 1 Reviews

Cumulative Review

Write the number that comes just *before*.

☐ 348   ☐ 619

Write the number that comes just *after*.

876 ☐   293 ☐

Write the number that comes *between*.

489 ☐ 491   724 ☐ 726   560 ☐ 562

Apples Sold	
Monday	🍎🍎🍎🍎🍎
Tuesday	🍎🍎
Wednesday	🍎🍎🍎🍎
Thursday	🍎🍎🍎
Friday	🍎🍎🍎

🍎 = 5 apples

Use the pictograph to answer each question. Mark the correct answer.

How many apples did the farmer sell on Monday?

20 ○   30 ○   40 ○

How many more apples did the farmer sell on Wednesday than on Tuesday?

○ 25 − 10 = 15 apples
○ 25 + 10 = 35 apples

Chapter 19 Cumulative Review   three hundred five   **305**

Subtraction Fact Review

Subtract.

9 − 7 = ☐	5 − 2 = ☐	10 − 3 = ☐	9 − 6 = ☐	11 − 8 = ☐
7 − 3 = ☐	8 − 6 = ☐	7 − 5 = ☐	10 − 9 = ☐	8 − 5 = ☐
8 − 8 = ☐	9 − 1 = ☐	6 − 4 = ☐	7 − 0 = ☐	10 − 6 = ☐
10 − 2 = ☐	7 − 4 = ☐	9 − 8 = ☐	11 − 3 = ☐	7 − 2 = ☐
9 − 4 = ☐	8 − 2 = ☐	12 − 8 = ☐	6 − 5 = ☐	9 − 9 = ☐
8 − 4 = ☐	12 − 6 = ☐	9 − 2 = ☐	8 − 7 = ☐	12 − 5 = ☐

Time to the Hour & Half-Hour

_____

Color the hour hand red.
Color the minute hand blue.
Count the minutes by 5s.

5

There are [ ] minutes in 1 hour.

There are [ ] minutes in a half-hour.

Write each time.

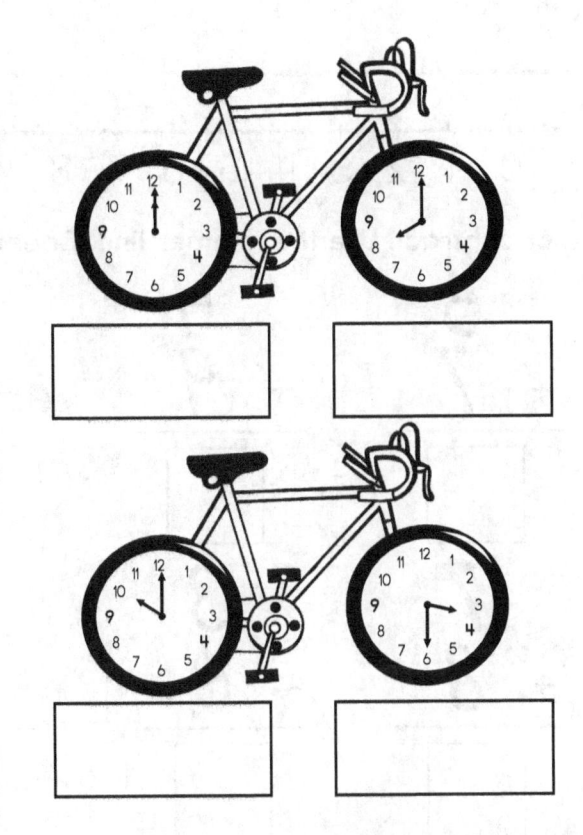

Chapter 20 • Lesson 165   three hundred seven   **307**

Chapter 17 Review

Circle each container that holds *less than* 1 liter.

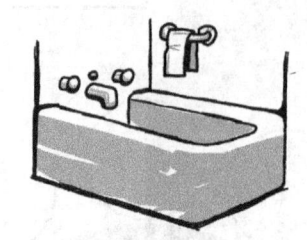

Mark an *X* on each object that is *more than* 1 kilogram.

Circle the correct temperature.

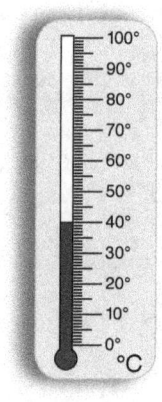

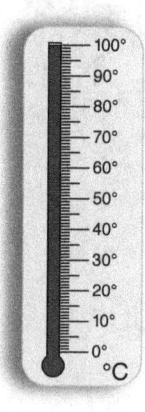

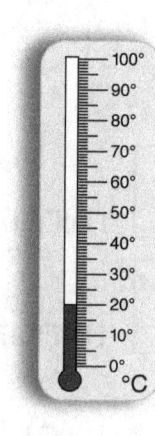

30°C　　80°C　　20°C
40°C　　90°C　　30°C
50°C　　100°C　　40°C

Fact Review

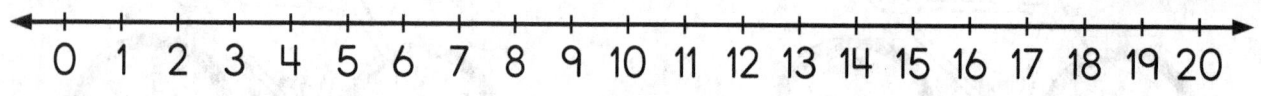

0 1 2 3 4 5 6 7 8 9 10 11 12 13 14 15 16 17 18 19 20

Add or subtract. Use the number line if needed.

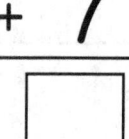

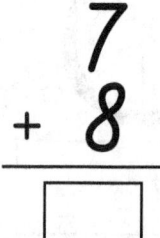

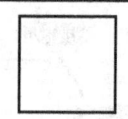

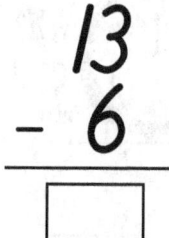

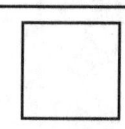

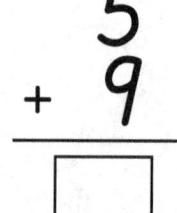

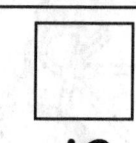

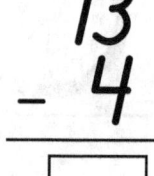

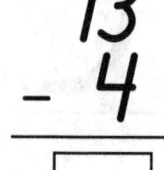

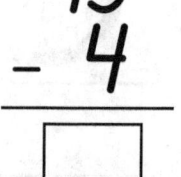

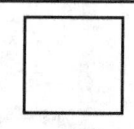

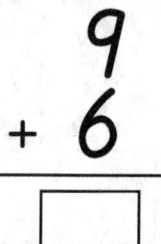

$6 + 7 = \square$

$4 + 9 = \square$

$14 - 9 = \square$

$15 - 9 = \square$

$7 + 9 = \square$

$7 + 8 = \square$

$13 - 6 = \square$

$5 + 9 = \square$

$13 - 4 = \square$

$9 + 6 = \square$

Time to 5 Minutes

2:05

Write each time.

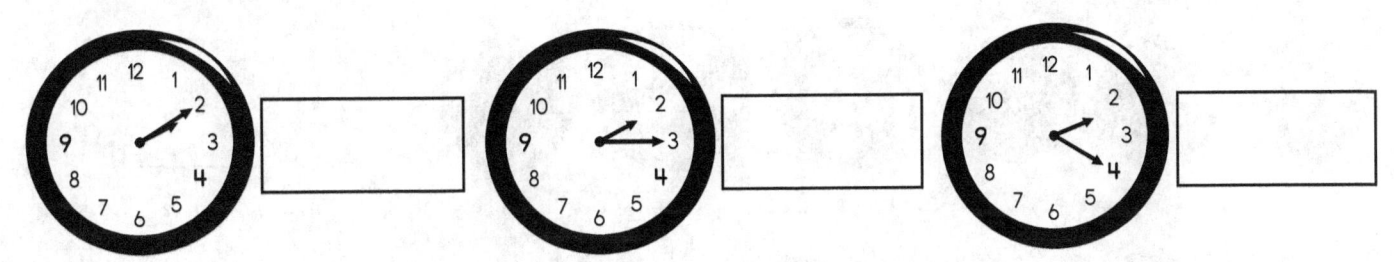

Color the correct clock yellow.

2:05

4:10

10:15

12:20

5:25

7:30

Chapter 20 • Lessons 166–67   three hundred nine   309

Chapter 17 Review

Write the total value.

_____ ¢

_____ ¢

_____ ¢

_____ ¢

_____ ¢

Fact Review

0 1 2 3 4 5 6 7 8 9 10 11 12 13 14 15 16 17 18 19 20

Add or subtract. Use the number line if needed.

7 + 8 = ☐   17 − 9 = ☐   18 − 9 = ☐

8 + 7 = ☐   17 − 8 = ☐   15 − 7 = ☐

310   three hundred ten   Math 1 Reviews

Using a Schedule

Color the ball below the correct clock orange.

Write each time.

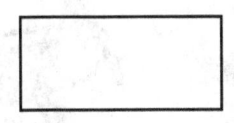

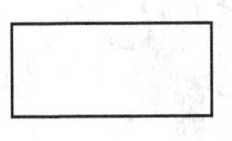

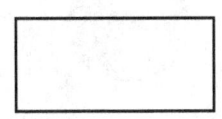

Circle the time passed.

Basketball Practice Schedule

Activity	Start	Finish	Time Passed
Dribbling	7:00	7:30	30 minutes or 1 hour
Shooting	9:00	10:00	30 minutes or 1 hour
Passing	8:00	9:00	30 minutes or 1 hour

Use the table to answer each question.

What time do they finish shooting baskets?

What time do they start passing drills?

Chapter 20 • Lessons 168–69   three hundred eleven   311

Chapter 19 Review

Mark an X on the coins needed to buy each item.

Fact Review

Add or subtract. Use the number line if needed.

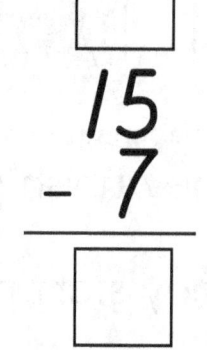

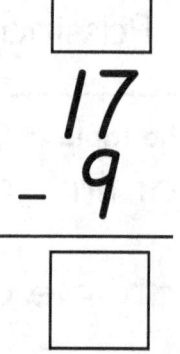

9 + 9 = ☐

7 + 8 = ☐

8 + 9 = ☐

8 + 7 = ☐

9 + 8 = ☐

15 − 8 = ☐

17 − 8 = ☐

18 − 9 = ☐

15 − 7 = ☐

17 − 9 = ☐

Using Calendars

- - - - - - - - - -

Color the sun beside the day that comes next yellow.

Monday	☀ Sunday	☀ Tuesday
Wednesday	☀ Tuesday	☀ Thursday
Friday	☀ Saturday	☀ Sunday

Circle the month that comes next.

April	March	May
July	June	August
September	October	November
November	January	December

Mark the correct answer.

What month is this?
- ○ April
- ○ July

How many days are in this month?
29 31
○   ○

What day is the 19th?
- ○ Tuesday
- ○ Wednesday

What is the date of the second Wednesday?
20 13
○   ○

| ☀ July ☀ |||||||
Sunday	Monday	Tuesday	Wednesday	Thursday	Friday	Saturday
					1	2
3	4	5	6	7	8	9
10	11	12	13	14	15	16
17	18	19	20	21	22	23
24	25	26	27	28	29	30
31						

Chapter 20 • Lesson 170

Chapter 16 Review

Write the number that is 1 *more*, 1 *less*, 10 *more*, or 10 *less*.

1 more		1 less		10 more		10 less	
764			115	570			727
800			433	243			468
598			657	924			175

Write the value of each underlined digit.

| 326 | | | 475 | | | 628 | |

Addition Fact Review

Circle the doubles.
Add.

$7 + 8 =$ ☐ $9 + 9 =$ ☐ $9 + 8 =$ ☐

$8 + 8 =$ ☐ $17 + 9 =$ ☐ $7 + 7 =$ ☐

$6 + 9 =$ ☐ $6 + 6 =$ ☐ $6 + 7 =$ ☐

Chapter 20 Review

_____
- - - - - - - - - -
_____

Mark the correct time.

○ 6:30
○ 5:30

○ 8:50
○ 7:55

○ 1:00
○ 12:00

○ 10:35
○ 10:25

Write each time.

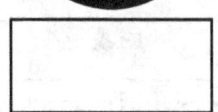

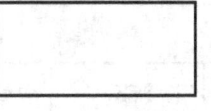

Circle the time passed.

Playground Schedule

Equipment	Start	Finish	Time Passed
Jungle Gym	2:00	3:00	30 minutes or 1 hour
Swings	1:00	1:30	30 minutes or 1 hour
Seesaw	11:00	11:30	30 minutes or 1 hour

Mark the correct time.

What time does playing on the seesaw start?   11:00 ○   1:00 ○

What time does playing on the jungle gym finish?   11:30 ○   3:00 ○

What time does playing on the swings finish?   1:00 ○   1:30 ○

Chapter 20 Review   three hundred fifteen   315

Mark the day that comes next.

Monday	○ Sunday	○ Tuesday
Wednesday	○ Tuesday	○ Thursday
Thursday	○ Friday	○ Saturday

Mark the month that comes next.

February	○ March	○ May
March	○ April	○ August
July	○ June	○ August
November	○ December	○ January

Color all the Thursdays.

June

Sunday	Monday	Tuesday	Wednesday	Thursday	Friday	Saturday
	1	2	3	4	5	6
7	8	9	10	11	12	13
14 Flag Day	15	16	17	18	19	20
21 Father's Day	22	23	24	25	26	27
28	29	30				

Mark the correct answer.

How many days are in June? 30 ○ 31 ○

What is the date of the third Sunday? 14 ○ 21 ○

What day is the 11th? Thursday ○ Friday ○

What day is the 1st? Sunday ○ Monday ○

316 three hundred sixteen Math 1 Reviews

Cumulative Review

Draw a dot next to the smaller number.
Draw the correct sign.

453   161

800   900

less than <   greater than >

594   504

399   769

Match each number to the correct number word.

1	three
2	four
3	one
4	five
5	two

6	eight
7	nine
8	seven
9	ten
10	six

11	fifteen
12	eleven
13	twelve
14	thirteen
15	fourteen

16	eighteen
17	sixteen
18	twenty
19	seventeen
20	nineteen

Subtract.

93 − 41

57 − 36

86 − 43

29 − 15

61 − 20

Addition Fact Review

Add.

6 + 7 = ☐   5 + 2 = ☐   6 + 6 = ☐

3 + 9 = ☐   4 + 6 = ☐   6 + 5 = ☐

7 + 0 = ☐   5 + 9 = ☐   4 + 8 = ☐

9 + 2 = ☐   7 + 5 = ☐   3 + 3 = ☐

6 + 3 = ☐   9 + 4 = ☐   3 + 7 = ☐

2 + 8 = ☐   9 + 3 = ☐   9 + 7 = ☐

5 + 7 = ☐   6 + 2 = ☐   6 + 9 = ☐

8 + 4 = ☐   3 + 5 = ☐   9 + 5 = ☐

4 + 2 = ☐   7 + 9 = ☐   5 + 6 = ☐

5 + 4 = ☐   9 + 6 = ☐   7 + 6 = ☐

Adding Hundreds

Add.

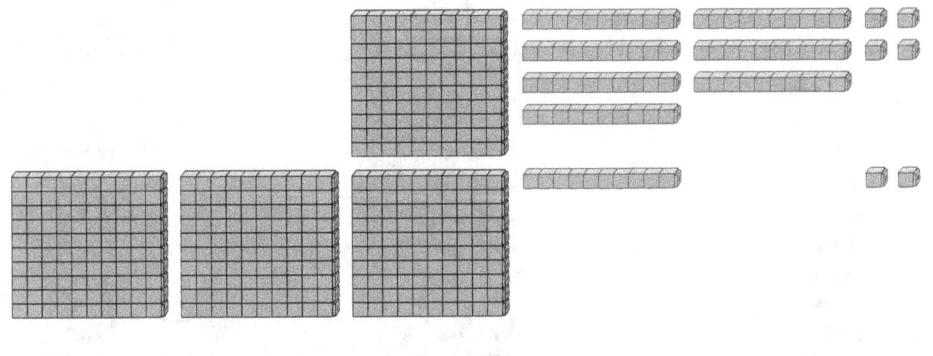

Hundreds	Tens	Ones
1	7	4
+ 3	1	2

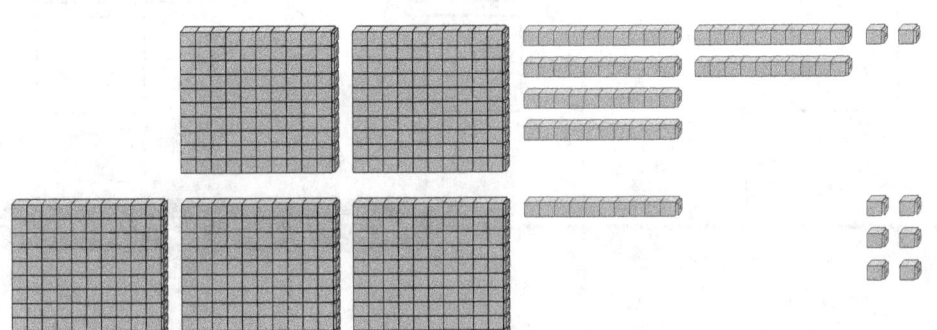

Hundreds	Tens	Ones
2	6	2
+ 3	1	6

Add.

Hundreds	Tens	Ones
4	5	3
+ 3	4	6

Hundreds	Tens	Ones
7	8	2
+ 2	0	5

Hundreds	Tens	Ones
4	2	6
+ 2	3	1

Write an equation for the word problem. Solve.

The men used 125 bricks to make some steps. They used 750 bricks to make a wall. How many bricks did they use in all?

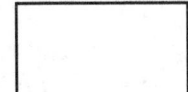

☐ ◯ ☐ ◯ ☐ bricks

Hundreds	Tens	Ones

Chapter 21 • Lesson 173

Chapter 13 Review

Subtract.

$$\begin{array}{r} 97 \\ -64 \\ \hline \square \end{array} \qquad \begin{array}{r} 38 \\ -17 \\ \hline \square \end{array} \qquad \begin{array}{r} 59 \\ -42 \\ \hline \square \end{array} \qquad \begin{array}{r} 68 \\ -21 \\ \hline \square \end{array}$$

$$\begin{array}{r} 87 \\ -46 \\ \hline \square \end{array} \qquad \begin{array}{r} 95 \\ -13 \\ \hline \square \end{array} \qquad \begin{array}{r} 86 \\ -60 \\ \hline \square \end{array} \qquad \begin{array}{r} 49 \\ -13 \\ \hline \square \end{array}$$

Subtraction Fact Review

Subtract.

$$\begin{array}{r} 14 \\ -8 \\ \hline \square \end{array} \qquad \begin{array}{r} 17 \\ -9 \\ \hline \square \end{array} \qquad \begin{array}{r} 16 \\ -8 \\ \hline \square \end{array} \qquad \begin{array}{r} 13 \\ -5 \\ \hline \square \end{array} \qquad \begin{array}{r} 13 \\ -8 \\ \hline \square \end{array}$$

$$\begin{array}{r} 15 \\ -8 \\ \hline \square \end{array} \qquad \begin{array}{r} 17 \\ -8 \\ \hline \square \end{array} \qquad \begin{array}{r} 18 \\ -9 \\ \hline \square \end{array} \qquad \begin{array}{r} 14 \\ -6 \\ \hline \square \end{array} \qquad \begin{array}{r} 15 \\ -7 \\ \hline \square \end{array}$$

Practicing Adding Hundreds

Add.

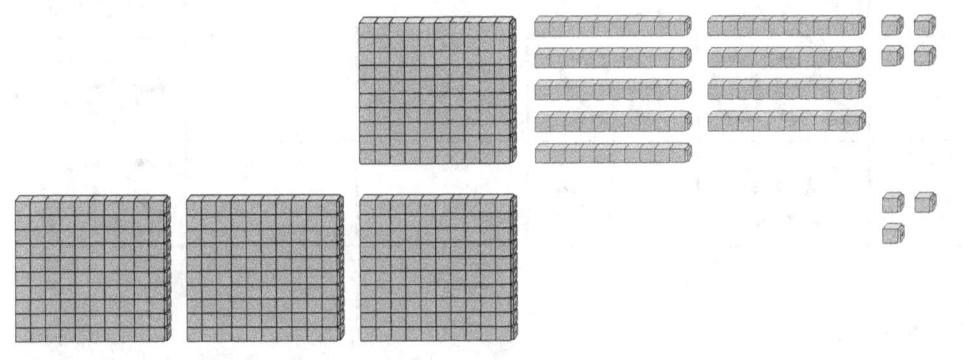

Hundreds	Tens	Ones
1	9	4
+ 3	0	3

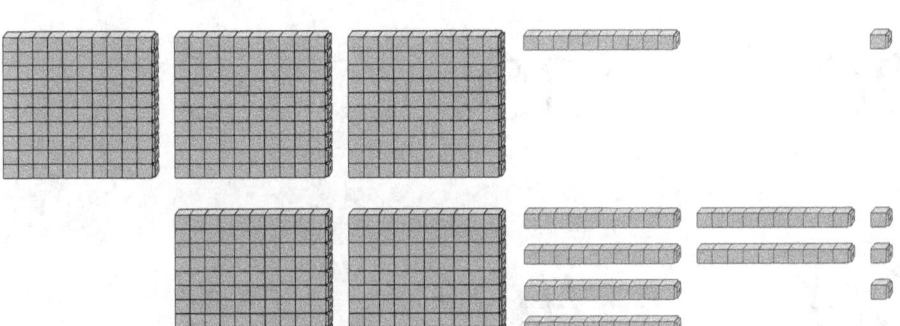

Hundreds	Tens	Ones
3	1	2
+ 2	6	5

Add.

Hundreds	Tens	Ones
7	2	5
+ 1	4	2

Hundreds	Tens	Ones
5	0	3
+ 2	9	2

Hundreds	Tens	Ones
3	4	5
+ 6	0	3

Write an equation for the word problem. Solve.

Mrs. Smith's class read 453 books.
Mr. Crow's class read 426 books.
How many books did the classes read in all?

☐ ◯ ☐ ◯ ☐ books

Hundreds	Tens	Ones

Chapter 21 • Lesson 174   three hundred twenty-one **321**

Chapter 16 Review

Write the number that is 1 *more*, 1 *less*, 10 *more*, or 10 *less*.

1 more	1 less	10 more	10 less
694		742	153			980
365		230	612			457
813		475	386			234

Mark the value of each underlined digit.

637   ○ 300   ○ 30   ○ 3

415   ○ 400   ○ 40   ○ 4

Subtraction Fact Review

Subtract.

8 − 6 = ☐   10 − 2 = ☐   9 − 4 = ☐

12 − 6 = ☐   14 − 8 = ☐   10 − 5 = ☐

11 − 6 = ☐   13 − 5 = ☐   11 − 8 = ☐

9 − 6 = ☐   15 − 7 = ☐   10 − 3 = ☐

Problem Solving

Add.

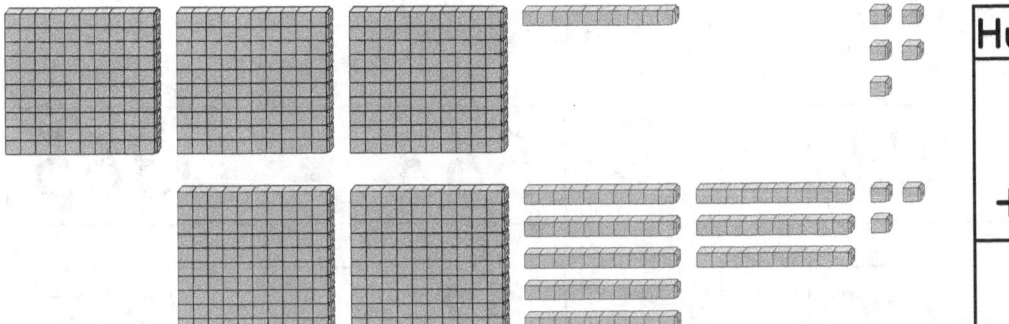

Hundreds	Tens	Ones
3	1	5
+ 2	8	3

Hundreds	Tens	Ones
6	3	4
+ 2	0	5

Hundreds	Tens	Ones
1	4	4
+ 7	2	3

Add.

503 + 472

464 + 234

231 + 538

491 + 305

work space

Write an equation for the word problem. Solve.

The hiker walked 362 miles last year.
This year he walked 435 miles.
How many miles did he walk in both years?

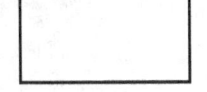

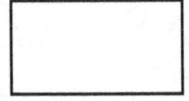

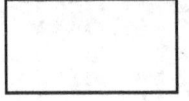

☐ ○ ☐ ○ ☐ miles

Chapter 21 • Lesson 175   three hundred twenty-three   **323**

Chapter 16 Review

Draw a dot next to the smaller number.
Draw the correct sign.

< less than   > greater than

363 413	563 365
900 500	275 375
121 122	834 814

Subtraction Fact Review

Subtract.

15 − 8 = ☐   14 − 6 = ☐   17 − 8 = ☐

17 − 9 = ☐   15 − 7 = ☐   14 − 8 = ☐

13 − 5 = ☐   9 − 9 = ☐   13 − 8 = ☐

Renaming 10 Ones

Add. Rename if needed.
Circle *yes* if you renamed.
Circle *no* if you did not rename.

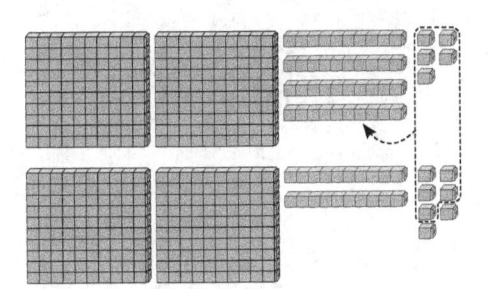

Hundreds	Tens	Ones
	[1]	
2	4	5
+2	2	7
4	7	2

Did you rename?
(yes)   no

Hundreds	Tens	Ones
	[ ]	
7	2	4
+2	4	2

Did you rename?
yes   no

Hundreds	Tens	Ones
	[ ]	
5	5	8
+3	2	3

Did you rename?
yes   no

Hundreds	Tens	Ones
	[ ]	
4	3	1
+4	1	9

Did you rename?
yes   no

Hundreds	Tens	Ones
	[ ]	
8	1	2
+1	3	4

Did you rename?
yes   no

Hundreds	Tens	Ones
	[ ]	
6	4	7
+1	2	4

Did you rename?
yes   no

Hundreds	Tens	Ones
	[ ]	
3	5	6
+6	3	6

Did you rename?
yes   no

Chapter 21 • Lesson 176   three hundred twenty-five   **325**

Add. Rename if needed.
Circle *yes* if you renamed.
Circle *no* if you did not rename.

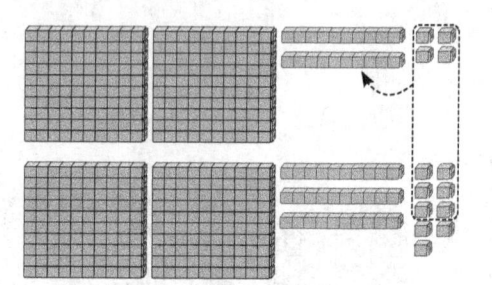

Hundreds	Tens	Ones
	1	
2	2	4
+2	3	9
4	6	3

Did you rename?
(yes)   no

Hundreds	Tens	Ones
6	3	6
+2	4	4

Did you rename?
yes   no

Hundreds	Tens	Ones
5	2	3
+3	1	1

Did you rename?
yes   no

Hundreds	Tens	Ones
4	5	7
+4	3	8

Did you rename?
yes   no

Hundreds	Tens	Ones
8	2	5
+1	6	8

Did you rename?
yes   no

Hundreds	Tens	Ones
4	3	6
+5	3	7

Did you rename?
yes   no

Hundreds	Tens	Ones
2	4	3
+7	5	2

Did you rename?
yes   no

326   three hundred twenty-six

Math 1 Reviews

Adding Hundreds with Renaming

_____

Add. Rename if needed.

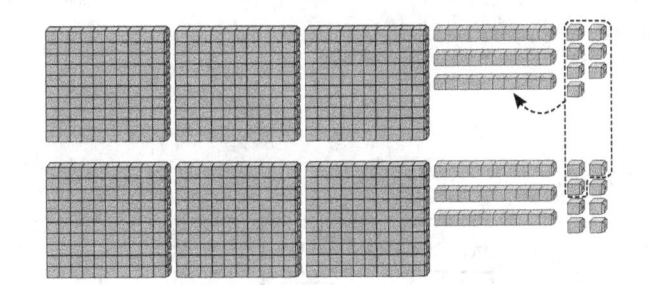

Hundreds	Tens	Ones
	☐	
3	3	7
+ 3	3	8

Hundreds	Tens	Ones
	☐	
5	3	7
+ 2	4	7

Did you rename?
yes   no

Hundreds	Tens	Ones
	☐	
2	4	5
+ 6	4	4

Did you rename?
yes   no

Hundreds	Tens	Ones
	☐	
4	2	7
+ 5	2	5

Did you rename?
yes   no

Hundreds	Tens	Ones
	☐	
3	2	8
+ 3	1	9

Did you rename?
yes   no

Hundreds	Tens	Ones
	☐	
4	3	7
+ 3	5	6

Did you rename?
yes   no

Hundreds	Tens	Ones
	☐	
5	2	9
+ 1	6	4

Did you rename?
yes   no

Write an equation for the word problem. Solve.

Dan sold 347 tickets.
Ann sold 134 tickets.
How many tickets did they sell in all?

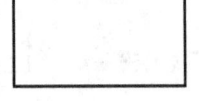

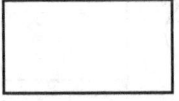

☐ ◯ ☐ ◯ ☐ tickets

Hundreds	Tens	Ones
	☐	
3	4	7
+ 1	3	4

Chapter 21 • Lessons 177–78   three hundred twenty-seven   **327**

Add. Rename if needed.

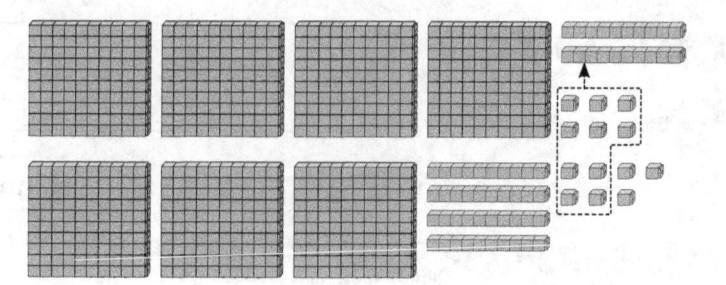

Add. Rename if needed.
Circle *yes* if you renamed.
Circle *no* if you did not rename.

Hundreds	Tens	Ones
	☐	
4	2	6
+ 3	4	7

Hundreds	Tens	Ones
	☐	
3	4	5
+ 6	2	7

Did you rename?
yes   no

Hundreds	Tens	Ones
	☐	
4	6	4
+ 2	1	5

Did you rename?
yes   no

Hundreds	Tens	Ones
	☐	
7	4	8
+ 1	4	3

Did you rename?
yes   no

Hundreds	Tens	Ones
	☐	
4	2	8
+ 3	6	1

Did you rename?
yes   no

Hundreds	Tens	Ones
	☐	
6	3	6
+ 2	4	4

Did you rename?
yes   no

Hundreds	Tens	Ones
	☐	
4	5	7
+ 4	2	6

Did you rename?
yes   no

Write an equation for the word problem. Solve.

On Friday 645 people came to the school play. On Saturday 347 people came to the school play.
How many people came in all?

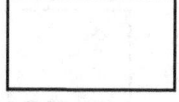

☐ ◯ ☐ ◯ ☐ people

Hundreds	Tens	Ones
	☐	
6	4	5
+ 3	4	7

328   three hundred twenty-eight

Math 1 Reviews

Chapter 21 Review

Add.

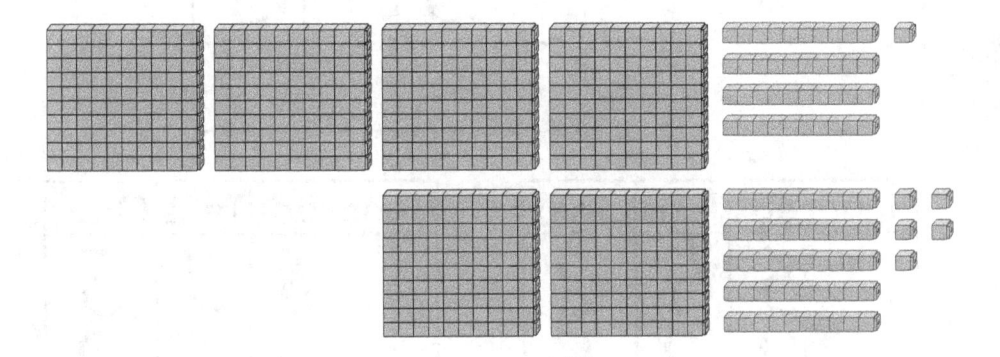

Hundreds	Tens	Ones
4	4	1
+ 2	5	5

Add.

Hundreds	Tens	Ones
7	0	7
+ 1	3	2

Hundreds	Tens	Ones
6	1	4
+ 2	5	3

Hundreds	Tens	Ones
4	3	6
+ 2	0	1

236 + 432 =

713 + 164 =

531 + 262 =

Write an equation for the word problem. Solve.

The boys set up 250 chairs this morning.
They need to set up 125 more chairs this afternoon.
How many chairs will they set up today?

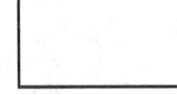

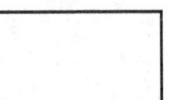

☐ ○ ☐ ○ ☐ chairs

Chapter 21 Review

Add.

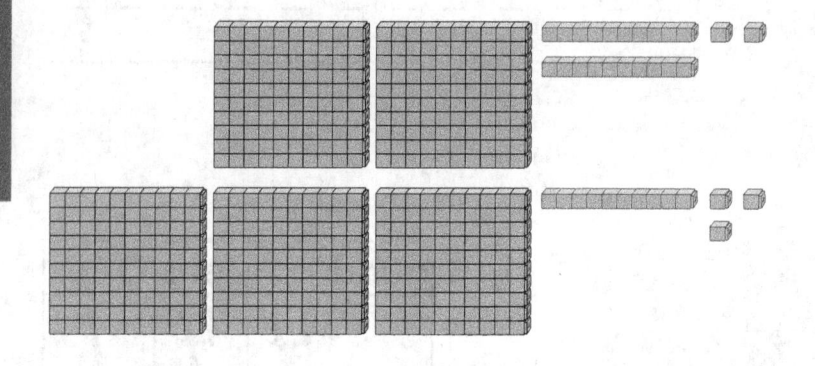

Hundreds	Tens	Ones
	2	2	2
+ 3	1	3

Hundreds	Tens	Ones
8	1	4
+ 1	6	1

Hundreds	Tens	Ones
6	4	2
+ 2	2	5

Hundreds	Tens	Ones
8	3	5
+ 1	6	2

534 + 442 =

261 + 128 =

724 + 264 =

Write an equation for the word problem. Solve.

The church had 345 song books.
They ordered 150 more.
How many song books do they have in all?

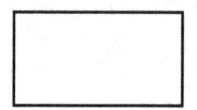

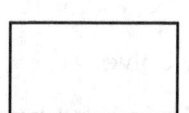

☐ ◯ ☐ ◯ ☐ song books

work space

330 three hundred thirty

Math 1 Reviews

Cumulative Review

Mark an *X* in the box if the solid figure has curves, faces, or corners.

	Curves (curved sides)	Faces (flat sides)	Corners
Sphere			
Cone			
Cylinder			
Rectangular Prism			
Pyramid			

Write the number of sides and corners.

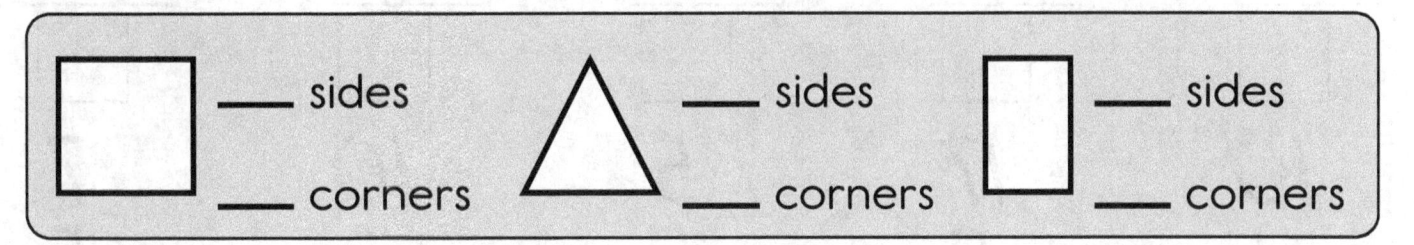

___ sides   ___ sides   ___ sides
___ corners   ___ corners   ___ corners

Look at the line on each shape.
Mark an *X* on each shape that has matching equal parts.

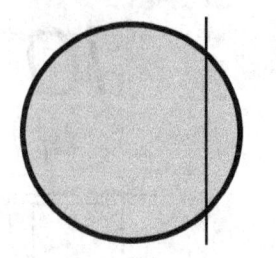

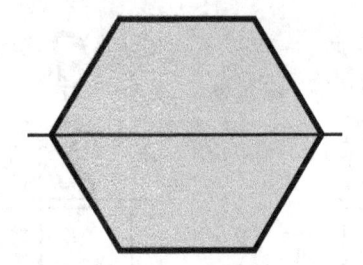

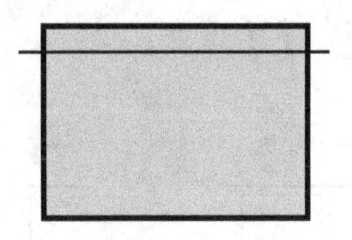

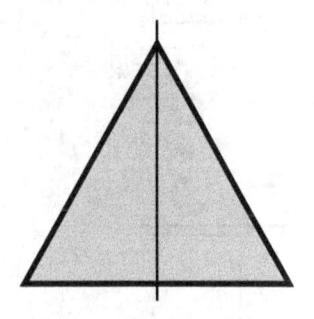

Chapter 21 Cumulative Review   three hundred thirty-one

Subtraction Fact Review

Subtract.

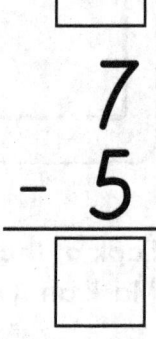

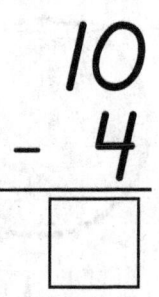

12 − 9 = ☐	7 − 2 = ☐	6 − 4 = ☐	10 − 2 = ☐	12 − 5 = ☐
11 − 6 = ☐	10 − 8 = ☐	9 − 9 = ☐	15 − 8 = ☐	12 − 6 = ☐
12 − 3 = ☐	7 − 6 = ☐	12 − 8 = ☐	7 − 0 = ☐	4 − 3 = ☐
9 − 3 = ☐	13 − 5 = ☐	10 − 5 = ☐	5 − 3 = ☐	14 − 6 = ☐
10 − 1 = ☐	17 − 9 = ☐	5 − 5 = ☐	10 − 3 = ☐	7 − 5 = ☐
8 − 3 = ☐	11 − 5 = ☐	8 − 7 = ☐	9 − 2 = ☐	10 − 4 = ☐